T0360746

Creolised Science

This rich, deeply researched study offers the first comprehensive exploration of cross-cultural plant knowledge in eighteenth-century Mauritius. Using the concept of creolisation – the process by which elements of different cultures are brought together to create entangled and evolving new entities (or not) – Brixius examines the production of knowledge on an island without long-established traditions of botany as understood by Europeans. Once foreign plants and knowledge arrived in Mauritius, they were adapted to new environmental circumstances and a new sociocultural space. Brixius explores how French colonists, settlers, mediators, labourers, and enslaved people experienced and shaped the island's botanical past, centring the contributions of subaltern actors. By foregrounding neglected non-European actors from both Africa and Asia, in a melting pot of cultivation traditions from around the world, the author presents a truly global history of botanical knowledge.

Dorit Brixius is a historian of global science and medicine interested in eighteenth-century botany and France's Indian Ocean colonies.

Science in History

Science in History is a major series of ambitious books on the history of the sciences from the mid-eighteenth century through the mid-twentieth century, highlighting work that interprets the sciences from perspectives drawn from across the discipline of history. The focus on the major epoch of global economic, industrial, and social transformations is intended to encourage the use of sophisticated historical models to make sense of the ways in which the sciences have developed and changed. The series encourages the exploration of a wide range of scientific traditions and the interrelations between them. It particularly welcomes work that takes seriously the material practices of the sciences and is broad in geographical scope.

A full list of titles in the series can be found at: www.cambridge.org/sciencehistory

Creolised Science

Knowledge in the Eighteenth-Century Indo-Pacific

Dorit Brixius

Shaftesbury Road, Cambridge CB2 8EA, United Kingdom

One Liberty Plaza, 20th Floor, New York, NY 10006, USA

477 Williamstown Road, Port Melbourne, VIC 3207, Australia

314–321, 3rd Floor, Plot 3, Splendor Forum, Jasola District Centre, New Delhi – 110025, India

103 Penang Road, #05–06/07, Visioncrest Commercial, Singapore 238467

Cambridge University Press is part of Cambridge University Press & Assessment, a department of the University of Cambridge.

We share the University's mission to contribute to society through the pursuit of education, learning and research at the highest international levels of excellence.

www.cambridge.org
Information on this title: www.cambridge.org/9781009200448

DOI: 10.1017/9781009200486

First published 2024

A catalogue record for this publication is available from the British Library

A Cataloging-in-Publication data record for this book is available from the Library of Congress

ISBN 978-1-009-20044-8 Hardback

In memory of Richard H. Grove

Contents

Figures

Acknowledgements

I am indebted to everyone who encouraged me as I wrote this book. First and foremost, my grateful thanks go to Stéphane Van Damme for his generosity, enthusiasm, and indispensable support. Likewise, I wish to thank Regina Grafe for her swift mind and essential questions – questions, moreover, which made me think more deeply about my work in general as a PhD student at the European University Institute, Florence (EUI). I owe a great deal to many people in the Centre Alexandre-Koyré, Paris (CAK), and in particular Antonella Romano and Anne Sirand; without their generosity and support, my stay as a visiting PhD student at the CAK in 2016 and 2017 would never have been as rich and fruitful as it was. I am more than grateful to them.

Without the funding of several institutions, I would never have been able to conduct this project. I am very grateful to the German Academic Exchange Service, the EUI, and the British Society for the History of Science for their generous financial support for archival missions and conference travel grants. Before I joined the EUI in 2013, the Institut historique allemand (IHA) in Paris kindly funded my dissertation project during its first steps, and in my postdoc work has likewise helped me turn my dissertation into a monograph. *Danke* and *merci* to all my former colleagues at the IHA: in particular I am indebted to Rainer Babel for believing in my dissertation project from the very start, and to Thomas Maissen, Christine Zabel, and the department of early modern history for keeping my plate clear while I was finishing the manuscript.

I would like to thank all the archivists and librarians who supported me while I was doing research in their institutions, in Paris, Aix-en-Provence, Chartres, London, Seville, Saint-Denis on La Réunion, and Coromandel in Mauritius. In particular, I wish to thank Sangeeta Mohun (National Archives of Mauritius), Pamela Dookhy (Botanic Garden in Pamplemousses, Mauritius), Kersley Pynee (Herbarium in Réduit, Mauritius), Antoine Monaque and Joëlle Garcia (Bibliothèque Centrale du Muséum National d'Histoire Naturelle Paris), Sylvie Pontillo (Archives nationales d'outre-mer Aix-en-Provence), and Ana

Hernández Callejas (Archivo General de Indias Seville) for their kind assistance.

My special thanks also go to all academics and friends in Cambridge, where I wrote parts of Chapters 5 and 6 during my stay in 2015 as a visiting PhD student at the Department of History and Philosophy of Science: Emma Spary, Sujit Sivasundaram, Simon Schaffer, Nick Jardine, Katrina Maydom, Mélanie Lamotte, Huiyi Wu, and James de Montille. In particular, I am very grateful to Šebestián Kroupa for being a friend and academic companion. His honest – some might say rather brutal! – comments helped me not only grow intellectually but also make my work more accessible to someone outside my head and my heart.

I particularly thank Raine Daston, Minakshi Menon, and Gabriela Soto Laveaga at the Max-Planck-Institut für Wissenschaftsgeschichte Berlin where I wrote part of Chapter 5 during my stay as a visiting pre-doctoral fellow in September 2015, as well as Peter Boomgaard who during the same period made possible my brief stay at the Koninklijk Instituut voor Taal-, Land- en Volkenkunde (Royal Netherlands Institute of South East Asian and Caribbean Studies) in Leiden. Like Peter Boomgaard (who sadly passed away before my dissertation was finished), Romain Bertrand helped me to navigate through the complex geographies and cultures of South East Asian islands. Without their expertise, I would have strayed too easily. I also benefited greatly from discussions with Dhruv Raina, James Delbourgo, Lissa Roberts, John Mathew, Marine Bellégo, Hélène Blais, José Beltrán, Neil Safier, Kapil Raj, Benjamin Steiner, Serge Reubi, and Kalle Kananoja, who have all shared their expertise with me. Nor should I neglect Pierre-Étienne Stockland, Kit Heintzman, Genie Yoo, and Rafaël Thiebaut for providing full references to archival material and even digital copies of manuscripts. Finally, I am indebted to Mikkel Munthe Jensen, Lenny Hodges, James White, Maxence Gévaudan, Simon Dumas Primbault, and Marius Müller for their assistance with the English language and formatting this text. Also, my language editor, Caroline Petherick – you rock! And, lastly, a big thank you to Martin Barr for his copy-editing and his assistance during the final push!

I benefited greatly, too, from discussions in seminars at the EUI, the CAK's 'atelier d'écriture' (thank you, Oury Goldman!), the IHA's 'séminaire de recherche sur les Lumières' (thank you, Pascal Firges and Antoine Liliti!) and at several doctoral workshops, including 'Plants, Politics, and Power' at the Munich Spring School (Kochel am See, March 2015), the Fourteenth Ischia Summer School, 'History of the Life Sciences' (Ischia, June 2015), the Göttingen Spirit Summer School, 'Academic Collecting and the Knowledge of Objects, 1700–1900'

(Göttingen, September 2016), and the Atelier doctorale interdisciplinaire Méditerranée 2, 'Le travail forcé' at the École française de Rome (Rome, October 2016). I owe a debt to the organisers of these events, and to Kärin Nickelsen in Kochel, Dominik Hünniger in Göttingen, and Silvia Sebastiani in Rome in particular.

I owe a huge debt to Cambridge University Press, starting with the series editors, Lissa Roberts, Simon Schaffer and Jim Secord. Two anonymous readers took their time to read my manuscript thoroughly, and provided guidance and very helpful suggestions to improve my work. Lucy Rhymer's patience and enthusiasm were likewise indispensable. Nor should I neglect Rosa Martin, Lisa Carter or Geethanjali Rangaraj.

I thank the following publishers for granting permission to reproduce published material in this book: SAGE Publications for permitting a partial reproduction in Chapter 4 of Dorit Brixius, 'From Ethnobotany to Emancipation: Slaves, Plant Knowledge, and Gardens on Eighteenth-Century Isle de France', *History of Science* 58 (1): 51–75. © [2020] (SAGE Publications), https://doi.org/10.1177/0073275319835431. Chapter 5 is based on my findings first published in Dorit Brixius, 'A Pepper Acquiring Nutmeg: Pierre Poivre, the French Spice Quest and the Role of Mediators in Southeast Asia, 1740s to 1770s', *Journal of the Western Society for French History* 43 (2015): 68–77, http://hdl.handle.net/2027/spo.0642292.0043.006. This article is licensed under CC BY-NC-ND 3.0 license, https://creativecommons.org/licenses/by-nc-nd/3.0. 2015 © Dorit Brixius. Cambridge University Press for allowing me to reproduce in Chapter 6 here my analysis of archival material originally published in Dorit Brixius. 'A Hard Nut to Crack: Nutmeg Cultivation and the Application of Natural History between the Maluku Islands and Isle de France (1750s–1780s)', *British Journal for the History of Science* 51(4): 585–606, 2018 © Cambridge University Press.

My dissertation, which this book has been based on, was written in many different places (mainly Florence, Paris, Cambridge, Berlin, and thousands of metres above the Atlantic or Indian Ocean). My family and loved ones showed a great deal of patience and understanding while I was travelling – hunting in the archives and hiding in libraries – without seeing them for months. My mum Brigit and dad Gerold kept me grounded even though I was so far away. I owe them so much, because they have taught me to be myself and have always encouraged me to do things my way. It is also important to mention my nan, Helene Klee, who sadly passed away before my work was completed. It was her stories that had sparked my interest in history; when I was little, I listened to them while eating her special 'Marmorkuchen' (marble bundt cake). Furthermore, I thank my sister Maren and my brother Lennart, and their families: they never understood my academic drivel: when I was lost in clouds of overthinking, they brought me back down to earth.

Nor should I forget my dear friends who provided moral support all along the way: Abbed Kanoor, Francesca Iurlaro, Henrik Forshamn, Jennifer Blanke, Morvarid Ayaz Naimian, Arash Naimian, Déborah Dubald, Catarina Madruga, Pablo Hernández Sau, Bruno Martinho, Jean-Baptiste Goyard, Lavinia Maddaluno, Julie Marquet, Mirjam Dageförde, Tom Tölle, Francis Poku, Rutger Birnie, Maria José Schmidt-Kessen, Frida Pritzkow, Qassem Massri, Sarah Haas, and Andreas Seifen.

Last but not least, I wish to express my sincere gratitude to you, my wonderful daughter Fiona, for reminding me what life should be about, and I apologise for dealing (intellectually, at least) with dead people so much of the time. Especially because, my love, you are not the past but the future.

Author's Note

The sources I have used were written in French, English, Spanish, Dutch, Latin, and German; in relation to plants they include eighteenth-century Malay, Chinese, and Malagasy, and there are in addition expressions stemming from other Asian and African origins. Where I have been unable to trace any words or phrases to their roots, I have indicated uncertainty. Unless otherwise indicated, all translations are my own.

Abbreviations

ADL	Archives départementales Eure et Loire
ADR	Archives départementales de la Réunion
ADS	Académie des sciences, Institut de France, Service des Archives, Paris
ADS DB	ADS-Dossiers biographiques
AGI	Archivo General de Indias, Seville
AMAE	Archives du Ministère des Affaires Étrangères, Paris
AN	Archives nationales, Paris
ANOM	Archives nationales d'outre-mer, Aix-en-Provence
BCMNHN	Bibliothèque Centrale du Muséum National d'Histoire Naturelle, Paris
BL	British Library, London
BN	Bibliothèque nationale de France, Paris
BSG	Bibliothèque Sainte-Geneviève, Paris
CIO	Compagnie française pour le commerce des Indes orientales, or French East India Company
DMB	*Dictionary of Mauritian Biography*
DSB	Gillispie, Charles Coulston, ed., *Dictionary of Scientific Biography* (New York: Scribner's, 1981)
EE BSP	Electronic Enlightenment, Project Bernardin de Saint-Pierre, www.e-enlightenment.com
FGB	Fonds Grandet-Bailly Papiers de Claude Louis Bailly (1679–1797), ADL
HARS	*Histoire de l'Académie royale des sciences*, Paris
KEW	National Archives, Kew
MA	National Archives of Mauritius, Coromandel
MARS	*Mémoires de l'Académie royale des sciences*, Paris

NAF	Manuscrits français et des Nouvelles acquisitions françaises, collection at the Département des manuscrits, BN
SE	Savants Étrangers series of the Académie royale des Sciences, Paris
VOC	Verenigde Oost-Indische Compagnie, or Dutch East India Company

Introduction

> What a sweet and delicious life I could still have if I had not been so foolish as to go to this dog of a country.[1]
>
> Madame Renouf La Coudraye, resident (by choice) of Mauritius, 1778

The best-known novel set in Isle de France, the island that we now call Mauritius, is without doubt the utopian yet tragic love story *Paul et Virginie* (1788), written by a naturalist and manager of the Parisian Botanical Garden, Jacques-Henri Bernardin de Saint-Pierre (1737–88; Bernardin for short). Inspired by his own stay in Mauritius in the late 1760s, he wrote his novel at a time when the island was vulnerable and isolated, particularly with respect to subsistence and sustainability; it was seriously short of food and medical supplies. Its colonial administrators tried to create a better life for the inhabitants by introducing plants, mostly collected from around the Indo-Pacific region, together with knowledge related to their cultivation and usage.[2] The acclimatisation of those plants and the related knowledge throughout the eighteenth century was a localised, cross-cultural, slow, and complex process – a process explored in this book.

For historians of science *Paul et Virginie* has served as a literary source in that it sheds light on the everyday life of and activities conducted by the enslaved people, the residents born on the island, and the settlers who had decided to make their lives in that remote island space.[3] The account is full of descriptions of cultivation know-how and the use of plant-based materials, describing as it does settlers emulating the everyday rituals carried out by the enslaved people, along with the production of plant-based remedies and tropical agriculture, and enduring the same plagues, famine, cyclones, violence, and the emancipation of some of the enslaved people. This monograph, based on archival research both broad and deep, is about those environmental conditions and the plant-based

[1] NA Kew, HCA 30/290, unfol., Madame Renouf La Coudraye (Mauritius) to her female friend in Aury (Brittany), 27 March 1778.

[2] See also Grove, *Green Imperialism*.

[3] See ibid. and Pimentel, *Testigos del mundo*, 299–328.

activities of the people who struggled with island life in the second half of the eighteenth century. I examine the processes involved in the collection, acclimatisation, and cultivation of plants within the French Indian Ocean colony between *c.*1750 and 1775. Although the book formally covers that relatively short twenty-five-year period, its range extends forwards and backwards through the eighteenth century in order to explain and investigate, as appropriate, the shifting dynamics of French colonial rule and the development of plant-based knowledge within, and in relation to, the island.

Mauritius is one of the Mascarene Islands, an Indian Ocean archipelago located about 700 miles due east of Madagascar, and consisting of several small islands plus three major ones: Mauritius (previously Isle de France), Réunion (previously Isle de Bourbon), and Rodrigues.[4] Unlike many other Indo-Pacific islands, Mauritius had no indigenous human population (or there is at least no evidence of any), and this makes the island a unique case of historical inquiry for knowledge production.[5] During the course of its colonisation it was an artificially created island in terms of its highly heterogeneous population characterised by social, cultural, and religious diversity, leading to creolisation.[6] The island had no native plants, either, to provide suitable food for humans and their livestock; these plants, together with related knowledge and materials, had to be imported there, and it is upon them that this book focuses.

Historians of science have claimed for a while now that knowledge was produced via cross-cultural encounters and entanglements.[7] But there is a lack of detailed connected case studies that show precisely how those interactions would have come about, and how these processes would have operated. In fact, there are very few full-length monograph studies of science *within* colonial spaces at all, let alone in the francophone Indian Ocean.[8] This absence is particularly notable in French historiography.

This is where *Creolised Science* comes in. Mauritius provides an excellent, and arguably unique, site for the examination of those processes. As explained above, both the human population, and the flora and fauna to

[4] To keep matters simple, I have used the terms 'Mauritius' and 'Réunion', rather than the eighteenth-century names, throughout this book. Only in primary source quotations have I left in place the names 'Isle de France' and 'Isle de Bourbon'.

[5] On islands and science in the Indo-Pacific, see in particular Kroupa, Mawson, and Brixius, 'Science and Islands'.

[6] Vaughan, *Creating the Creole Island*; Servan-Schreiber, *Indianité et créolité à l'île Maurice.*

[7] The most relevant are Sivasundaram, 'Sciences and the Global'; Roberts, 'Situating Science in Global History'; Bauer and Norton, 'Introduction'; Schaffer et al., *Brokered World*; Sivasundaram, *Islanded*; Safier, *Measuring the New World*; Raj, *Relocating Modern Science*; Bertrand et al., *L'Exploration du monde.*

[8] Rare examples are McClellan, *Colonialism and Science* or Regourd, 'Sciences et Colonisation sous l'Ancien Régime'.

support that population, were imported and acclimatised. The arrival of the flora and fauna was rarely coordinated with that of the humans who had the relevant knowledge of their cultivation and propagation. Thus, the processes involved in the successful acclimatisation of the plants on the island demanded a considerable degree of experimentation, drawing on the disparate skills and expertise of a wide range of people.

This book is about the practices of plant knowledge; that is, the knowing *how*, but not the knowing *that*. By looking at the interconnections between people, materials, and nature, I argue in this book that creolisation took place in the social, cultural, epistemological, and material terms that determined the application of knowledge. As a creolising process, the knowledge of plants derived from cultures all over the world was integrated into the emerging practices within the island space. The cultivators included the Europeans who had migrated to Mauritius by choice, settlers born there, labourers, and enslaved people brought in by forced migration. Unsurprisingly, the agricultural knowledge of these individuals varied widely. Consequently, cultivation turned out to be a complex process of creolising the expertise that had originated in the local populations of the plants' native habitats with the varying degrees of horticultural knowledge of the people living in Mauritius.

Once the non-native plants had arrived in Mauritius, they had to be propagated in climate and soil conditions very different from those of their original habitats. Most of the people involved had no experience in the cultivation of any plants, let alone these new and unknown ones. The few experienced horticulturalists would learn to experiment with their own traditional methods; although to begin with they had relied on their original assumptions, they also came to question them.

The aim of this book is threefold: first, to shine a light on the importance of local knowledge within global science; second, to reveal the knowledge traditions of the colonists, settlers, labourers, and enslaved people living in Mauritius; third, to explain knowledge production on that island as a creolising process. In so doing, I examine how imported materials and knowledge, including that of its enslaved people was handled and utilised; how European knowledge lost its value; and how knowledge traditions came to be adapted in the new island context.

In order to achieve these goals, I follow some of the botanical expeditions around the Indo-Pacific while basing my narrative on Mauritius as a place in which to analyse in depth the production of plant knowledge. Historian of science Sujit Sivasundaram, focusing on a unit, such as an island, under imperial strategies, suggests that 'the danger of being stranded in the sea, without a sense of perspective or without a limit to the local, is avoided'.[9] At the same time, as Mauritius is located at a crossroad of the Indian Ocean between the African east coast and China, my research looks both at the places of origin of plants (and related knowledge), where the plants and seeds were collected, and at their new island home. This study is needed in the historiography of the French colonial world, specifically because it extends outside the Atlantic basin to take into consideration the Indian Ocean (and by extension the Pacific) spheres of France's colonial expansion.[10] At the same time, I do not restrict this book to French colonial history. Rather, I tease out global, cross-imperial networks as connected histories, by using a range of European archives and the peregrinations of historical actors to steer my narrative.[11]

Global Science and Creolisation as a Concept

Sociological actor–network theories have been used to explain the processes in the history of science, traditionally centred, as it has been thus far, on Europe.[12] These models have attracted extensive criticism, particularly through their neglect of local agency, their hierarchisation of knowledge, and their telescopic view of an overall picture of empire.[13] Historians of science have acknowledged the circulatory property of knowledge, and this has led to a challenge to the institutionalisation and cultural bounds of knowledge production.[14] Indeed, this approach enables us to visualise the movements of knowledge, people, and objects. Criticisms of this approach, however, raise an awareness that a focus on the mobility of knowledge creates the risk of prioritising it overmuch.[15]

The establishment of models other than Eurocentrism has been on the agenda of historians of science for quite some time.[16] Over the last thirty

[9] Sivasundaram, *Islanded*, 26.

[10] Important exceptions on the French expansion in the Indian Ocean being Vaughan, *Creating the Creole Island*; Peabody, *Madeleine's Children*; Agmon, *Colonial Affair*; Lamotte, *Making Race*. See also the ongoing work by Lenny Hodges or Greg Mole.

[11] I am not the first to apply a cross-imperial approach. For South East Asia, see in particular Bertrand, *L'Histoire à parts égales*. On the calls for connected histories, see Subrahmanyam, *Explorations in Connected History*; Conrad, *What Is Global History?*

[12] Callon, 'Some Elements of a Sociology of Translation'; Latour, *Science in Action*; McClellan and Regourd, *Colonial Machine*.

[13] Bravo, 'Ethnographic Navigation'; Cruikshank, *Do Glaciers Listen?*, 127–53; Spary, 'Botanical Networks Revisited'; Charles and Cheney, 'Colonial Machine Dismantled'; Roberts, 'Le Centre de Toutes Choses'; Davids, 'Machines, Self-Organization, and the Global Traveling of Knowledge'.

[14] Secord, 'Knowledge in Transit'; Findlen, *Empires of Knowledge*.

[15] Easterby-Smith and Senior, 'Cultural Production of Natural Knowledge', 472.

[16] See, for instance, Raj, *Relocating Modern Science*; Schaffer et al., *Brokered World*; Sivasundaram, 'Sciences and the Global'.

years, new approaches have been taken, including: the study of different scientific disciplines; new interpretations between geographical and territorial spaces; and the historiographical treatment of the 'local'.[17] Here, the concepts of the 'middle ground'[18], 'contact zone',[19] and 'hybridity'[20] have been critically reviewed.[21] Recent claims that knowledge production was polycentric have allowed us to view a plurality of knowledge systems that had specific forms in specific times and specific places.[22] The discipline, however, is historically predetermined by European visions and traditions.[23] Sujit Sivasundaram has asserted that one cannot consider that 'science is a bundle of things or practices which comes fully formed or with great intentional force to intrude into the . . . periphery'.[24] Caution is required regarding the risks of overestimating the importance of European innovations for the practice of colonial science abroad.[25] While French interests did indeed shape certain early conditions in the intercolonial networks, they had far less impact on the form and content of the specific interactions in those networks.

Recent work on Atlantic history has suggested the term 'entanglements' – a loanword from historical anthropology in the Pacific[26] – between cultures as a way of describing global histories. The term implies the continuous and unresolved processes of cross-cultural interactions, as opposed to the 'hybridity' (of knowledge) that indicates a (smooth) fusion of two entities.[27] The fact, however, that knowledge is hybrid 'may be considerably less important than how such hybridity is enacted and regulated'.[28] Unlike 'hybridity', the term 'entanglements' does not suggest an outcome but instead the 'ongoing confrontations, shifts, and revisions: a state of mutual learning and pushback which does not dissolve into a final product'.[29] In this sense, 'entanglements' implies the highly complex and multifaceted nature of cross-cultural interactions, which has

[17] Bayly, *Birth of the Modern World*; Raj, *Relocating Modern Science*; Schaffer et al., *Brokered World*; Elshakr, 'When Science Became Western'; Safier, *Measuring the New World*; Sivasundaram, 'Sciences and the Global'; Chakrabarti, *Medicine and Empire*.

[18] White, *Middle Ground*. [19] Pratt, *Imperial Eyes*; Schiebinger, *Plants and Empire*.

[20] Bauer and Norton, 'Introduction'; Thomas, *Entangled Objects*; Knörr, 'Contemporary Creoleness'.

[21] Robert, 'Situating Science in Global History'; Raj, 'Historical Anatomy'.

[22] See for instance Ganeri, 'Well-Ordered Science'.

[23] See in particular Basalla, 'Spread of Western Science'.

[24] Sivasundaram, 'Science', 237.

[25] See on this train of thought in particular Charles and Cheney, 'Colonial Machine Dismantled'; Roberts, 'Le Centre de Toutes Choses'; Davids, 'Machines, Self-Organization, and the Global Traveling of Knowledge'.

[26] Thomas, *Entangled Objects*.

[27] Bauer and Norton, 'Introduction'; Kroupa, Mawson, and Brixius, 'Introduction'.

[28] Seth, 'Putting Knowledge in Its Place', 380. [29] Graubart, 'Shifting Landscapes'.

in turn produced multiple possible reactions, ranging from resistance to cooperation.[30]

In the context of eighteenth-century Mauritius, the concept of creolisation is the most relevant framework in which I can consider my findings and their historical significance. Studying creolisation as a rich and complex process of knowledge production, as opposed to a result or a description, allows the telling of much richer stories in the global histories of science. Here, it becomes important to analyse the asymmetrical process of the interacting Asian, African, and European traditions on the island. When we consider that creolisation means more than just a mixture, we can bring to light the asymmetries, hierarchies, and ruptures of knowledge production.[31]

The meaning of the term 'creolisation' as used in this book aligns with but also differs from current understandings of the concept. There is still a major debate around the precise nature of creolisation and whether it is a useful process to consider as a framework for an understanding of the dynamics of historical change in colonial contexts.

With this case study on Mauritius, my answer to the question as to the usefulness of this concept is clearly yes. I understand knowledge creolisation as developing, evolving, and expanding, reflecting accommodation or resistance, or a combination of the two, within knowledge production in the midst of the injustices of colonialism. At this point, I assert that hierarchies of knowledge can be rethought: instead of considering creolised knowledge as an end product in which subjugated traditions meld with the dominant form to generate a new blend, *Creolised Science* considers the asymmetries of knowledge within the creolising process in eighteenth-century Mauritius. All forms of knowledge were constantly recontextualised in this new environment, and European knowledge, long considered the dominant form in the creolising processes, lost its value – and not only once, as I will show throughout this book.

In Mauritius, the creolisation of knowledge, emerging amid asymmetrical power relations and colonial violence, needs to be considered as a generative and agile process. The term 'creole' was used by historical actors to define plants, animals, and people of different origins and born on the island, and that is how I use it in this book. I do not use it to denote a person of African or partial African descent, as in contemporary parlance in Mauritius and elsewhere. And I do *not* use it as a loaded political

[30] See also Winterbottom, *Hybrid Knowledge*; Raj, *Relocating Modern Science*; Bose, *Hundred Horizons*; Subrahmayam, *Explorations in Connected History*; Matsuda, *Pacific Worlds*; Armitage, Bashford, and Sivasundaram, *Oceanic Histories*; and on the interaction between nature and humans during Indian Ocean colonialism, Grove, *Green Imperialism*.

[31] Voicua, 'Caribbean Cultural Creolization', 998.

term, as it has become in the contemporary identity politics of Mauritius and elsewhere (in Louisiana, USA, for example).

The roots of the debates about creolisation date quite a few years back. It was in the 1940s that anthropologist Melville Herskovits called for the survival of African cultures as the norm in New World regions. He challenged the idea that the ruptures of the Middle Passage had produced a near-total cultural *tabula rasa* in the environments of the colonial plantation societies.[32] The term 'creolisation' emerged to nuance this premise; it derived from linguistic theories emphasising that African syntaxes and vocabularies were persistent in the European-lexifier creole languages.[33] Social scientists and humanists drew on these studies when analysing cultural and religious processes, arguing that creolisation was a form of hybridity, generating new blends.[34] Languages, religious customs, legal traditions, cultural folkways of all kinds, racial/ethnic categories, and even people (through human genetics) have been examined as by-products of creolisation. More recently, anthropologists have moved beyond the prevailing hypothesis of creolisation as hybridity; Jacqueline Knörr, for example, suggests the term 'indigenisation' as a dissolution of boundaries accompanied by the recontextualisation of cultures and diasporic forms of ethnic belonging to new environments.[35]

Historical scholarship relating to the Mascarenes has responded to these trends. Robert Chaudenson considered creolisation as uniform and unidirectional, dooming cultural entities to European assimilation.[36] Moving beyond this view, Megan Vaughan sees the ongoing merging of multiple influences and the naturalisation of cultural identities in their new context of Mauritius.[37] In the same way, Françoise Vergès provides further nuances by suggesting that cultural resilience to the rupture of forced migration and slavery occurred despite creolisation.[38] Scholars in general, however, have continued to consider that creolisation naturally and inevitably includes some variant of dominant European culture in its end product.[39]

The notion of creolisation is quite often either misunderstood or overused to the point that the concept has lost much of its explanatory power. Currently, it seems to be undergoing a kind of *re*-mystification. In decolonial literature, it is used to describe forced assimilation of culture in slave societies by which African heritage and culture must be forcefully

[32] Herskovits, *Myth of the Negro Past*. See also Gershenhorn, *Melville J. Herskovits*.
[33] Mintz and Price, *Anthropological Approach*. [34] McGuire, *Lived Religion*.
[35] Knörr, 'Contemporary Creoleness'.
[36] Chaudenson, *Creolization of Language and Culture*.
[37] Vaughan, *Creating the Creole Island*.
[38] Vergès, 'Creolization and the Maison des Civilisations'.
[39] Larson, 'Enslaved Malagasy'; Alpers, 'Becoming "Mozambique"'.

abandoned and rejected.[40] Not only does that belief ignore the nuances of creolisation as a process that recent scholarship has identified; it also dismisses the agency of non-Europeans, particularly enslaved people, and denies the existence of any possible blending of non-European traditions. *Creolised Science* responds to these considerations by reasserting the agency of enslaved people and focusing on the importance of non-European intellectual inputs within the context of the often violent European colonialism.

Plant Knowledge as Portal into Knowledge Production

This book proposes an unbounded, nuanced, and contextualised approach to the constitutive and transformational processes of knowledge production about plants in eighteenth-century Mauritius, where European knowledge was not necessarily the dominant form of knowledge. Various traditions from all over the world were interacting to produce new, sometimes unexpected, results. Thus, I do not consider European frameworks and the metropolitan market as an endpoint of knowledge production. By insisting on local knowledge production in an island space, *Creolised Science* examines the (sometimes fractured) process through which knowledge deriving from all over the world was created by the people actually interacting in the Indo-Pacific. The knowledge traditions of the people of Mauritius, a creole island in the making, did not necessarily correspond to a pre-existing hierarchy. A deeper look at knowledge practice, then, allows us to revisit the dynamics of knowledge production and colonial practice more generally. By focusing upon creolisation via a praxeological approach[41] ('knowing how' instead of 'knowing that'), I have been able to retrieve critical instances of knowledge production, different paths of knowledge, ruptures, and moments of incoherence of cultivation, along with the contested nature of the knowledge that disappeared from the public records.

Located at the intersections of several distinct Indo-Pacific cultures, Mauritius connects local practices (cross-cultural knowledge production) with global flows (scale, movement, and disruptions of knowledge).[42] I use the term 'local knowledge' to clarify who the knowledge providers

[40] Casimir, *Haitians*.

[41] Freist, *Diskurse – Körper – Artefakte*; Brendecke, *Praktiken der Frühen Neuzeit*; Chateauraynaud and Cohen, *Histoires pragmatiques*; Freist, 'Historische Praxeologie Als Mikro-Historie'; Cerutti, 'Histoire pragmatique'; Lepetit, *Les formes de l'expérience*. On natural history as a practice, see, for instance, Terral, *Catching Nature in the Act*; Mariss, *Johann Reinhold Forster*; Hünniger, 'Nets, Labels and Boards'; Endersby, *Imperial Nature*. See further Manning and Rood, *Global Scientific Practice*.

[42] See also, for instance, Kroupa, 'Georg Joseph Kamel'.

were, specifying cultural origins and names of providers where available in the sources. But I do not use 'local knowledge' to contrast it to Western science, implying a lesser knowledge.[43] Nor do I present 'local knowledge' as something in opposition to 'global science'. Instead, I stress that the local is integral within the global. But also, instead of 'global' I use the term 'cross-cultural' in order to emphasise that 'globality' should be understood as a 'locally situated production'.[44] I take up recent lively debates in the scholarship on science and empire,[45] global science,[46] environmental history, natural history (with its subfield, colonial botany),[47] material culture, oceanic histories, and islands as places of knowledge production.[48] In particular, I follow recent calls in the history of global science to reveal the importance of local knowledge negotiated in diverse and highly complex cross-cultural situations.[49] Appealing for the use of 'cross-contextualization', Sujit Sivasundaram argued that:

> by stretching the category of science across cultural perspectives and across different genres of recording and thinking, it is possible to appreciate the distinctive features of the history of science in the Pacific and to take indigenous knowledges seriously.[50]

While this monograph is by no means the first to interrogate botanical experimentation in the Mascarenes, it is the first that attempts to centre the contributions of non-experts, including a wide range of subaltern actors – and enslaved people in particular – within Mauritius's botanical past. Colonial botany is commonly defined as plant transfer schemes intended to increase European knowledge, often by exploiting non-European

[43] As Cañizares-Esguerra points out, this has been the case even since the recovery of non-Western Indigenous science in the 1990s ('Iberian Colonial Science', 64–70).

[44] Seth, 'Putting Knowledge in Its Place', 380. See further Ophir and Shapin, 'Place of Knowledge'; Golinski, *Making Natural Knowledge*; Jasanoff, *States of Knowledge*.

[45] Delbourgo and Dew, *Science and Empire*.

[46] On methodological reflections on global science, see Sivasundaram, 'Sciences and the Global'; Roberts, 'Situating Science in Global History'. On French responses to global history, see Maurel, *Manuel d'histoire globale* and on global science in particular, Raj and Sibum, *Histoire des sciences et des savoirs*. See also Wood, *Archipelago of Justice* and Peabody, *Madeleine's Children*.

[47] Schiebinger and Swan, *Colonial Botany*; Batsaki et al., *Botany of Empire*; Drayton, *Nature's Government*; Schiebinger, *Plants and Empire*; Barrera-Osorio, *Experiencing Nature*; Spary, *Utopia's Garden*; Blais and Markovits, 'Introduction' and the whole special issue of *Revue d'histoire moderne et contemporaine* 66 (2019) on 'Le commerce des plantes: Empires, réseaux marchands et consommation (XVIe–XXe siècle)'; Jardine et al., *Cultures of Natural History*; Curry et al., *Worlds of Natural History*.

[48] Grove, Green Imperialism; Sivasundaram, *Islanded*; Gómez, *Experiential Caribbean*; Kroupa, Mawson and Brixius, 'Science and Islands'.

[49] Schaffer, *Brokered World*; special issue on 'Global Science', *ISIS* 101 (2010); special issue on 'Entangled Histories', *Colonial Latin American Review* 26 (2017).

[50] Sivasundaram, 'Science', 239.

cultures: in short, 'colonial science par excellence'.[51] But the history of plant knowledge has proved particularly useful, through its various colonial contexts, for the writing of cross-cultural histories of knowledge production.[52] Colonial botany has helped to reveal both the overlap of exchange, commerce, and science, and second, the nuanced human interactions connecting histories of scientific practice.[53] By examining the botanical networks shaping the conditions and outcomes of cross-cultural exchanges, recent studies have linked various parts of the globe while providing extensive details of local encounters.[54] Nevertheless, these same scholars have all too often relied on the migration of such knowledge back to European printed texts as their endpoint. In so doing, they ignore the actual practice of knowledge both in local contexts (whence the knowledge originated) and in the new contexts. The methods and techniques applied in the new environments and new sociocultural contexts requires the reworking of the extant knowledge, and this merits further study. In accordance with recent critiques, I suggest that it would be particularly fruitful to explore the ways in which European interpretations of non-European plants, applied knowledge, and environmental conditions impacted the cultivation and the usage of plants. When knowledge was only partly transmitted to the island, the cultivators, both European and non-European, had to fill the gaps with their own knowledge and experience. That experience produced not only new horticultural knowledge but also a recognition of the limits of the existing European science.

In order to critically analyse both social and intellectual power relations in science, we must look at disruptions rather than any presumed smooth circulation of knowledge.[55] For instance, as I show in Chapter 6, while South Asian classification systems helped the practical implementation of knowledge of cultivation and reproduction of plants on Mauritius, they did not necessarily result in success. Both European and non-European

[51] Easterby-Smith, 'Recalcitrant Seeds', 216; Drayton, *Nature's Government*; Schiebinger, *Plants and Empire*; Schiebinger and Swan, *Colonial Botany*; Batsaki et al., *Botany of Empire*.

[52] Parsons, *Not-So-New World*; Boumediene, *La colonisation du savoir*; Bellégo, *Enraciner l'empire*; Schiebinger, *Plants and Empire*; Schiebinger, *Secret Cures of Slaves*; Kroupa, 'Georg Joseph Kamel'; Yoo, 'Wars and Wonders'; Menon, 'What's in a Name?'. For natural history in colonial contexts, see Winterbottom, *Hybrid Knowledge*; Delbourgo, *Collecting the World*; Curry et al., *Worlds of Natural History*; Chakrabarti, 'Networks of Medicine'; MacLeod, 'Introduction'.

[53] Schiebinger and Swan, *Colonial Botany*; Batsaki et al., *Botany of Empire*; Cook, *Matters of Exchange*; Schiebinger, *Plants and Empire*.

[54] Schaffer et al., *Brokered World*; Boumediene, *La Colonisation du Savoir*; Raj, *Relocating Modern Science*; Safier, *Measuring the New World*; Sivasundaram, *Islanded*; Winterbottom, *Hybrid Knowledge*.

[55] For this call, see Fan, 'Global Turn'.

theoretical knowledge faced a reality test when contemporaries tried to put knowledge into action. It is incorrect to assume that particular techniques would have travelled from France to Mauritius and vice versa. Instead, methods were developed independently in different parts of the world. Europeans possessed only rudimentary knowledge about the natural world outside Europe, and the 'exotic' was frequently constructed from insufficient information.[56]

I examine knowledge production primarily in the context of the cultivation of 'useful' plants: foodstuffs, animal fodder, medicines, and/or for economic profit. On the one hand this included the augmentation of agriculture for the settlers' supplies, and on the other, the cultivation of spices as a way of securing new monopolies and trade. In relation to both, I would like 'useful knowledge' to be understood as the practical understanding and knowledge application of the ways in which plants should be cultivated and propagated, and this was indeed an eighteenth-century fashion in all botanical collecting projects.[57] Botany and its numerous sub-branches supported each other. Practical knowledge formed the substance of agriculture, which proved botany to be useful – above all in the colonies for cultivation techniques, handling, and preparation methods for both subsistence and commerce.[58]

'Useful knowledge' about the tropical natural world was based on the skills and knowledge traditions ('applied knowledge') of those who practised it. Practice-based knowledge traditions were of prime importance. Here, knowledge of plants should be seen on two levels: (1) the location, climate, soil, and cultivation techniques, and (2) the processing of plants for various purposes (food, remedies, and materials). The explorers would garner local expertise, hoping to learn from it, because they knew that local knowledge was crucial in the ways that plants were cultivated and processed. Nevertheless, as the transmission and practical application of knowledge was very complex, it would often be of little practical use in the end. Even though scholars have shed light on the significance of embodied knowledge, we do not know much about what happened when the plant material, the knowledge, and the planters all arrived at different times at a site of knowledge practice, such as a colonial garden.[59] As this book will show, there were huge practical challenges in the collection, transmission, and ultimately application of the knowledge.

[56] For examples of confusion and misinterpretation in other contexts, see Boumediene, *La colonisation du savoir*, 185–214, 191–4, 207–13. See also Lawrence, 'Assembling the Dodo'.

[57] Easterby-Smith, 'Botany as Useful Knowledge', 286. See also Berg, 'Useful Knowledge'.

[58] Bourde, *Influence of England*, 10.

[59] On embodied knowledge, see in particular Smith, From Lived Experience to the Written Word.

These considerations should also be understood as a response to economic theories that seek to understand 'usefulness' for Western economic 'growth' and 'progress'.[60] Instead, it is more appropriate to think in terms of the interactions between the human actors, the environment (plants, animals, climate, water, soil) and the material culture (in this case the seeds and plants, and the tools needed for their cultivation and processing).[61] Scholars have stressed that it is more useful to explore the entanglements of practices, instruments, and imaginaries within a spatiotemporally situated realm of activity.[62] *Creolised Science* thus makes reference to the concept of 'learning by doing' and the social, cultural, and environmental spaces in which knowledge was collected, adapted, and created.[63] Knowledge may have been wrongly transmitted and/or wrongly practised and meshed with natural conditions in the field. Above all, knowledge about the cultivation and preparation of plants remained scant, fragile, and unsettled throughout the period under review.

Against the Archival Seed[64]

The European archival sources used in this book contain a plethora of information on Indigenous knowledges. In addition, my research focused on many subaltern historical actors. To that end I drew primarily on the accounts created by the knowledge recipients and knowledge producers in the colonies themselves (including governmental, secret, and private correspondence, private travel journals, notes, drafts, instructions, official reports such as registers and yearly statistics, periodicals, memoirs, botanical examinations, and printed travel accounts). In light of the cross-cultural nature of my work, my research is based on extensive investigation into underexplored archival material in five languages (French, English, Spanish, Dutch, and Latin), connecting events taking place in locations from Paris to China.[65] Of prime importance is the series C *Correspondance à l'arrivé* of the *Archives nationales d'outre-mer* with its sub-series *Bourbon* (C3), *Ile de France* (C4), and *Madagascar* (C5), containing copies of letters from the governors and the administration.

[60] Mokyr, *Gifts of Athena*.
[61] See also Easterby-Smith, 'Recalcitrant Seeds'. See also Klein and Spary, *Materials and Expertise*; Craciun and Schaffer, *Material Cultures*.
[62] Roberts, 'Practicing Oeconomy', 2; Berg, 'Genesis of "Useful Knowledge"', 127; Roberts et al., *Mindful Hand*; Smith, *Body of the Artisan*; Halleux, *Le Savoir de La Main*.
[63] Arrow, 'Economic Implications'. See Berg, 'Genesis of "Useful Knowledge"', 127.
[64] Of course, I make reference to Stoler, using 'seed' as a pun in the context of plants. Stoler, Along the Archival Grain.
[65] On calls for such an approach, see Subrahmanyam, *Explorations in Connected History*; Conrad, *What Is Global History?*

I have combined the knowledge I gained from French colonial archives with that from scientific institutions such as the Muséum nationale d'histoire naturelle (the former Parisian Jardin du Roi) and the Académie royale des sciences, the manuscript department and cartography and map department of the Bibliothèque nationale de France, several departmental archives including those of Réunion and Chartres, local archives in Mauritius, the Bibliothèque Sainte-Geneviève, the National Archives in Kew, the British Library manuscript department, and to some extent Spanish and Dutch archival, and Dutch printed, sources related to the French contacts with the Philippines and the Moluccas. Many of the sources used relate to the intendant of Mauritius, Pierre Poivre (1719–86), who features consistently throughout the book. This is partly a reflection of his significance in managing the various projects discussed, but also due to the availability of sources about him and his activities.[66] But that does not mean that Poivre is the main focus of this book.[67] Rather, my pursuit of his activities allowed me to trace various forms of knowledge and to develop a deeper understanding of their origin and adaption in relation to Mauritius. In the very same sources (often nameless), subaltern historical actors in the background can be accessed; their experiences and their perspectives merit further research.[68]

During my first weeks of archival research, I swiftly realised that these sources are very rich and often (even though unwittingly, sometimes) allowed me to trace Indigenous knowledge outside its use for the purposes of the French establishment. Thus, reading sources against a colonial approach (in Europe as the endpoint), I began instead to try to understand knowledge within its cultural background and the ways in which it was recontextualised within the island. Practical knowledge for everyday use (making objects, gardening, food preparation, cosmetic products), but also for medical purposes, can be examined more deeply because the descriptions are detailed. Also, Indigenous names for plants or objects can be traced and interpreted; they often describe purposes, origins, a way of growing, or circumstances, as I reveal in Chapters 2, 4, and 6.

Although the sources used are European, that does not mean that they need to be viewed from a European perspective. Even though much of the

[66] See also www.pierre-poivre.fr (last accessed 28 August 2020), which reflects the great number of sources related to Poivre, and indeed provides numerous transcriptions of archival material.

[67] On Poivre's biography, see Malleret, *Pierre Poivre*. On biographical approaches to the history of science, see also Terrall, 'Biography as Cultural History'.

[68] For creative methodological strategies to follow when it comes to revealing the perspectives of enslaved people in the archives, see Ferrer, *Freedom's Mirror*; Turner, *Contested Bodies*; Peabody, *Madeleine's Children*; White, *Voices of the Enslaved*; Berry, *Price for Their Pound of Flesh*.

plant knowledge explored in this book was expressly aimed at sustaining (and controlling) the island's populations, this does not mean that we should use the information revealed to see maintenance or control as the point of the story. Instead, we can trace the different actors and influences of knowledge creation – or, say, creolisation – while considering the full array of contributors of knowledge and their motivations. Here, we need to ask what happened when actors *thought* they knew enough but on turning their knowledge into action rapidly realised that they did not. This can be understood only by looking at knowledge as a practice. In so doing, my findings reinforce the calls in the discipline to cross-contextualise science while providing key evidence regarding interactions between the non-Europeans and the Europeans, and the role played by the enslaved people in the construction of plant knowledge.

Creolised Science

In its six chapters, this book retells the story of the creation of plant knowledge and its practices within, and in relation to, Mauritius; it follows that knowledge in its various forms from the plants' countries of origin to the practical implications of plant knowledge within the island. Here, we can see a wide range of sociocultural interaction between local populations outside the island, enslaved people in the island, and Europeans, both beyond and on the island. My exploration of these details meant that I encountered several more challenging Eurocentric narratives: the presumed superiority of European science, and the focus on the Parisian acclimatisation garden in particular. Hence, the themes of the six chapters pursue interlocking questions of political, social, cultural, material, and environmental aspects of knowledge production as a process. Each chapter represents a coherent part of my thematic approach, but it is when they come together as a whole that they sustain my wider claim that knowledge in Mauritius was being creolised.

Chapter 1 introduces the geopolitical and scientific–colonial context of eighteenth-century Mauritius. It argues against the master narrative of institutionalised French science centred on Paris. By setting out the hidden dynamics of empire, this chapter provides a detailed discussion that explains why Mauritius was primarily a stopping-off point in the period under review rather than an island of commerce. It examines the various experiments in colonial autonomy undertaken on Mauritius between the 1760s and the 1780s, including the complex alternatives relating to the agents who tried to build networks through alliances with local actors and Indigenous populations in the Indo-Pacific region.

Lastly, this chapter spotlights the use of enslaved people in various projects in the island.

Each chapter focuses distinctly on the roles and activities of local actors embedded within the wider Indo-Pacific encounters, environments, and practices on or in relation to Mauritius. But now, instead of the usual academic practice of summarising chapters one by one, I intend to explain the purpose of each chapter in relation to a different chapter with a similar approach; their division was created by geographical necessity but also by context-based content and sub-argument. On the one hand, I explore encounters in the plants' native countries *outside the island* (Chapters 2 and 5) and on the other, I analyse explicitly the plants' arrival and employment *within the context of the island* (Chapters 3, 4, and 6). Even though I decided to make this geographical distinction for the collection (outside the island) and practical application of plant knowledge (inside it), all the chapters are connected via the plants' imagined purposes.

These five chapters are grouped according to two basic forms of plant knowledge: Chapters 2 and 3 relate to staple crops, focusing on the idea of making the island self-sufficient and feeding its population via the augmentation of agriculture. The context of Chapters 5 and 6 was the efforts to introduce cash crops (mainly nutmeg and cloves), that is, the project of growing spices on the island, while asking what happened when the actors thought they knew of, or they found they faced, practical agricultural challenges.

But while these chapters are divided into food crops and cash crops, they are closely linked as well. Cross-cultural encounters outside the island form the heart of analysis for Chapters 2 and 5, in that they both shed light on the exploration and collection of plants and related knowledge about them, which was made possible via the assistance of local populations in Madagascar and in numerous regions of Asia. The actors in question drew on several diverse types of plant knowledge, communication systems, languages, and negotiation practices, often leading to cultural misunderstandings. Cross-cultural connections were essential, and merchants, brokers, and subaltern actors were indispensable for the acquisition of plants and seeds. Local rulers, meanwhile, hoped to arouse European interest in their goods for their own purposes. Looking at those Eurasian encounters, Chapters 2 and 5 lead us away from Eurocentric narratives in sociocultural terms. Ultimately, in Chapter 5 in particular, I apply a European transnational perspective, particularly through the lens of French–Spanish collaboration in South East Asia.

While Chapters 2 and 5 focus on sociocultural interactions, Chapters 3 and 6 focus on practical approaches to the collection, transportation, and acclimatisation of the plants introduced to Mauritius. These chapters highlight the practical significance of knowledge about plants in relation to their cultivation by Malagasy and several Asian communities. Both chapters seek to understand how settlers, labourers, and enslaved people tried to cultivate foreign, newly introduced staple crops, such as rice, root vegetables, and fruit trees, plus spices: nutmeg and cloves. Emphasising the importance of non-European knowledge and the limits of European knowledge, I look at the interplay of knowledge between its practical implementation and environmental factors. Both chapters reveal that the cultivation techniques were difficult to implement and would often lead to the failure of a crop. Until at least the 1780s the sketchy and uncertain knowledge of the French actors remained fragmented, and that led to miscalculations and the eventual failure of the project to establish a spice trade. Exploring the reasons for the ruptures of knowledge and the failures, Chapter 6 in particular sets out the slow and asymmetrical processes of plant knowledge in the making.

In its central position, Chapter 4, with its focus on enslaved people, brings together all the aspects discussed in Chapters 2, 3, 5, and 6. It highlights the importance of the enslaved people in Mauritius, both for their labour and as sources of plant knowledge. Making an important contribution to the history of slavery and natural history, it serves as a link between the chapters on staple crops (Chapters 2 and 3) and those on commercial crops (Chapters 5 and 6), because it elaborates on both types of crops in relation to slavery. In particular, Chapter 4 reveals the disconnects of knowledge circulation. It seeks to explain what happened when new and unknown crops were introduced and knowledge of their cultivation or preparation techniques was lacking or faulty. Lastly, this chapter focuses on the Bengali enslaved gardener, Charles Rama. His knowledge of cultivation earned him praise from French actors, and he was later freed because of it. In the same way, I examine the work of the enslaved gardener Hilaire, who initiated and tested the new grafting methods that were adopted by European-trained naturalists in the island. These two cases not only highlight the importance of the enslaved people's knowledge, but more importantly, they reveal the shortcomings of European plant knowledge within the creolising processes.

In brief, in examining the construction of natural knowledge in a colonial context, *Creolised Science* shows how the hierarchy of knowledge claims and their practical implementation during cultivation can be studied through a perspective that takes the focus off European science and instead values Asian and African knowledge. In order to achieve this goal, I explore the ongoing process of knowledge creation in the insular space of Mauritius, which is an arguably unique case study for the analysis of plant knowledge as creolisation.

1 The Limits of French Colonial Visions and Science

Navigating the *Comte d'Argenson* through the vast tracts of the Indian Ocean, the astronomer Alexandre-Gui Pingré remarked in his journal on 15 March 1761: 'Today we had to discuss a question: it was the matter of whether or not a stay in the Bastille was preferable to the one on our ship.'[1]

You might well, then, be able to imagine his heartfelt relief when he finally, for the first time in many months, set foot on terra firma in Port Louis, the main harbour of Mauritius.[2] Many explorers, merchants, and travellers would have experienced a similar feeling of relief after months out of sight of land, and to them Mauritius was a welcome place of rest and refreshment. This was not the only reason, however, why that small island, 700 miles off the east coast of Madagascar, had become a topic of interest to the Europeans.

In this chapter I set out to explain the geographical, geopolitical, commercial, social, scientific, and administrative context in which I have anchored this study. To that purpose, this chapter locates Mauritius within the French scientific–colonial strategies, and reflects on the island as an insular space with maritime tensions bound up in its insularity. Further, this chapter forms an introduction to the historical development of Mauritius throughout much of the eighteenth century, from the start of the French settlement there in 1721 until the start of the French Revolution in 1789.

Broadly, the French encounters with Mauritius during the Ancien Régime can be divided into two periods. The first, that of the Compagnie française pour le commerce des Indes orientales (CIO or French East India Company) lasted from the earliest French settlement of the island in 1721 until the end of the Seven Years War in 1763, during which time the island was mainly used as a *relâche* (port of call) on the shipping routes between Europe and Asia, around the Cape of Good Hope. The second period, defined by the French royal administration, began after the Peace of Paris

[1] Pingré, *Voyage à Rodrigue*, 90. [2] Ibid., 127.

in 1763, when the Mascarene Islands were given to the French Crown, to be put into the hands of the Ministry of the Navy and the Colonies in 1766.

First, I look at the early French settlement of Mauritius, followed by the impact of Anglo-French rivalry in the Indian Ocean and the politics of the CIO. Third, linking local management and utilitarian botany, I examine the lack of scientific institutional interest in the Mascarenes until the last quarter of the eighteenth century. Before closing this chapter with a discussion on colonial life and provision more generally, the penultimate section is mainly concerned with the local management of the island before and after the Seven Years War, focusing on the role of the intendant and naturalist Pierre Poivre. I reveal how the metropolitan French aims clashed with the local conditions on the island, and look at the ongoing tussle between local administrators. Then, lastly, I elaborate on colonial life, with its introduction of enslaved people, its productivity, and its provisions. As the latter were connected to networks in the south-west Indian Ocean, I have included a discussion of the island's dependence on the food networks in the region. In short, the overall aim of this chapter is to explore the role of Mauritius as an island colony within the French imperial strategies, the colonial botany, and the Indo-Pacific world more widely.

Isle de France: A Second France in the Indian Ocean?

Lying in the Indian Ocean on the major commercial shipping routes connecting the West Indies to the East Indies, thus in a strategic location between the Atlantic and the Pacific, Mauritius became of great interest to the various competing European colonising powers in terms of geopolitics, communication, and trade. Although traders from the Mediterranean and Arabian regions had been criss-crossing the Indian Ocean for more than 2,000 years, the Mascarene Islands had apparently remained undiscovered owing to their location well south of those traders' regular routes. There is evidence that Arab travellers came across the islands in 1300, but there is no known indication of any earlier landings there, by any proto-Malagasy people or any from South East Asia.[3] Sailors from the Indian and Arabic worlds must, however, surely have discovered the islands long before any European set foot on them. According to Islamic studies scholar Shawkat Toorawa, Mauritius did not play any significant role in the Arabic world, but scholarship has tried to embed the island in the wider Islamic context.[4]

[3] Cheke and Hume, *Lost Land of the Dodo*, 21.

[4] Wâq al-wâq 'Fabulous, Fabular, Indian Ocean (?) Islands'. Cited in Vaughan, *Creating the Creole Island*, 3.

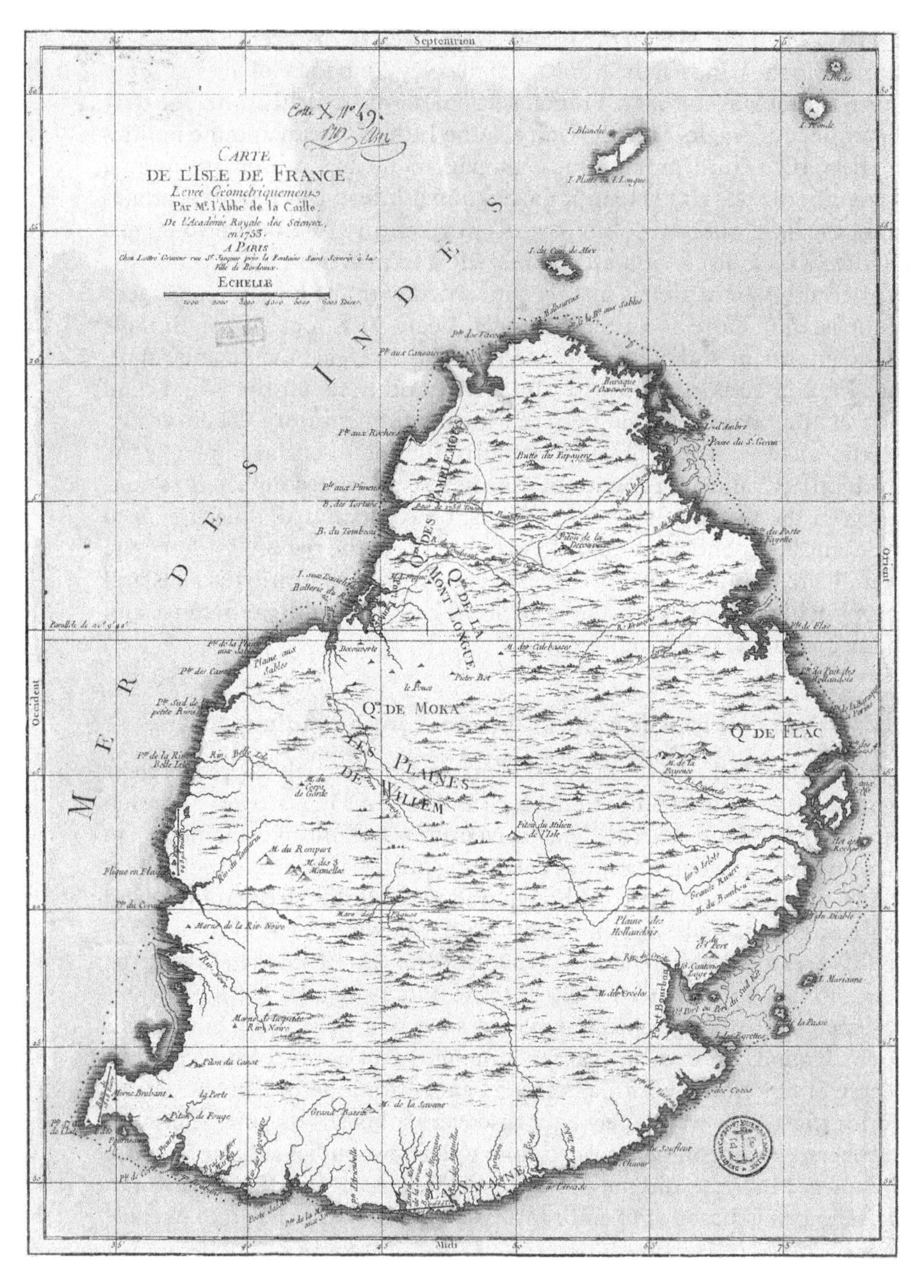

Figure 1.1 Map of Mauritius, by Abbé de la Caille, 1753, Bibliothèque nationale de France, https://gallica.bnf.fr/ark:/12148/btv1b8595773n

As mentioned earlier, though, the strategic location of Mauritius became more and more evident during its French colonisation, particularly after the Seven Years War.

The French were not the first Europeans to colonise the Mascarene Islands. It was probably the Portuguese who were the first of the European traders to drop anchor there, off the coast of Bourbon in 1510, and then off Mauritius, in 1516.[5] Almost a century later, in 1598, the Dutch landed on the latter island, and their fear of rivalry by both the French and the British led to their colonisation of the island in 1638, naming it Mauritius in honour of Prince Maurits van Nassau.[6] Yet the Verenigde Oost-Indische Compagnie (the VOC, Dutch East India Company), whose headquarters were in South East Asia, saw the island only as a minor outpost, so under Dutch rule the island had little effect on European imperial strategies.[7]

The French, however, had had their eye on the Mascarene Islands since 1664, the founding of the CIO under the finance minister, Jean-Baptiste Colbert (1618–83). Under Colbert, Africa, including Madagascar, had become an object of colonial interest, so expeditions to that island were undertaken.[8] The French settlement of Mauritius, then, cannot be appreciated without taking into account the French attempt to establish a colony in Madagascar. This enterprise would meet with success in the second half of the eighteenth century,[9] but it had initially been the island of Bourbon that attracted the French traders, because of its fertility; there, after unsuccessful attempts to domesticate cloves (1681) and pepper (1702), the CIO tried to make its fortune with the cultivation of coffee.[10] The first coffee plants arrived in Bourbon in 1715.

The Dutch, having abandoned Mauritius, 1715 was the year, too, when France formally took possession of the island, renaming it Isle de France. Six years later, in 1721 – surprisingly late, since the French knew that the English governor of St Helena had been flirting with the idea of resettling its inhabitants in Mauritius – the French administration did take an interest in its island and attempt to do something with it.[11] The interest of the French was mainly aroused by its two good harbours, a desirable feature lacked by neighbouring Bourbon.

[5] Regarding the study of known early voyages in the Indian Ocean, see North-Coombes, *La découverte des Mascareignes*.

[6] Toussaint, *Histoire des Iles Mascareignes*, 24.

[7] Roos, 'Dutch as Globalisers', 7–16.

[8] Steiner, *Colberts Afrika*.

[9] Tricoire, *The Colonial Dream*; Foury, *Maudave et la colonisation de Madagascar*.

[10] Toussaint, *Histoire des Isles Mascareignes*, 38.

[11] Grove, 'Conserving Eden', 328; Ly Tio Fane-Pineo, *Île de France*, 283–4, 285–6.

Planting a Colony within Indo-Pacific Expansion

The shift in French imperial politics towards an increasingly global trade inspired by the mercantilist economy gave new significance to the CIO and its strategies, and these were tested in the Mascarene Islands. It is incorrect to think of the CIO as a single entity; from its foundation in 1666, the company changed radically three times.[12] The most crucial period was in 1769, in the aftermath of the Seven Years War, when the CIO was dissolved due to the post-war introduction of free trade. It is also incorrect to think of private interests in relation to the CIO; although it was involved in the cultivation of Mauritius from 1721 to 1764, the relationship remained a formal one. Between 1669 and 1763, the *Secrétaires d'état à la marine* directed the colonies.

French royal power in Mauritius was shared between three established institutions: (1) the governor, responsible for military affairs ; (2) the intendant, responsible for the civil administration; and (3) the *conseil supérieur*, in theory the equivalent of the parliament and sovereign courts.[13] In a contemporary report, the astronomer Alexandre Guy Pingré (1711–96) explains that the *conseil supérieur* consisted of five or six *conseilleurs* (advice-givers) under a public prosecutor with the governor as president, but he also states that laws were commonly violated.[14] In short, even during the presence of the CIO the power remained with the monarchy.[15]

The Ministry of the Navy and the Colonies formed one of the main elements of the royal government, the political and bureaucratic organisation centred on the king. Located in Versailles, the Ministry sought to coordinate, instruct, and correspond with the royally appointed governors and intendants in the colonies. The Ministry did not itself consist of scientific experts, but it financially lobbied scientists and the navy.

From 1739 to 1782, Duhamel du Monceau (1700–82) was the navy's inspector general. Duhamel was also a naturalist working in collaboration with the director of the Parisian Jardin du Roi (King's garden), Georges-Louis Leclerc de Buffon (1707–88). According to Richard Grove, Duhamel stimulated Poivre's views on agriculture and environmental awareness.[16] The scientific ties became closer when Duhamel engaged actively with the Académie des Sciences and the Société Royale

[12] On the CIO and its development, see Haudrère, *Les compagnies des Indes orientales*; Manning, *Fortunes à Faire*.
[13] Marion, *Dictionnaire des institutions de la France aux XVIIe et XVIIIe siècles*, 155.
[14] Pingré, *Voyage à Rodrigue*, 223–4. [15] Ibid., 223.
[16] Grove, *Green Imperialism*, 168.

d'Agriculture.[17] Having instructed botanists to collect plants, he received some specimens which he passed to the Jardin du Roi,[18] which was part of La Maison du Roi and the offices of the Bâtiments du Roi, the ministry of culture and art.[19]

Despite its origins as a strictly commercial enterprise, the CIO developed over the course of the eighteenth century into a political instrument of conquest. For the trading companies, however, moneymaking was part of the imperial strategies, and although Western Europe was far from poor, the continent 'did lack many things for which consumers developed a taste and in which entrepreneurs saw a profit'.[20] The French trading companies were closely related to the royal government, and were even run by government insiders.[21] On the foundation of the CIO, the nomination of its directors had to be approved by the king, which meant that the Crown decided its political stance. The king's ministers defined both the limits of the CIO and its privileges: its monopoly of commerce in the colony; exclusive commercial traffic between the colony and metropolitan France; certain exemptions from customs duty on imported and exported commodities; and the recruitment of subaltern agents.[22] In return, the CIO was obliged to bring colonisers to new lands.

In contravention of the CIO's motto, '*florebo quocumque ferar*' ('I will flourish wherever I am brought') Mauritius was deforested. The intense deforestation and clearing of the island wrecked its ecological system. 'Sustainability' was barely beginning to emerge as a concept within eighteenth-century Europe, and was thus not at that point on the CIO's agenda.[23] While early Dutch records describe Mauritius as covered with trees and being a 'very highly goodly and pleasant land, full of green & fruitful vallies [*sic*]', the natural environment changed drastically over the course of the eighteenth century.[24] The French astronomer Guillaume Joseph Hyacinthe Jean-Baptiste Le Gentil de la Galaisière (1725–92; Le Gentil for short), who charted the Indian Ocean, stayed for a while in

[17] Technically, the Société fell under the Ministère des Affaires Etrangères et du Contrôle Général. See Passy, *Histoire de la Société nationale d'agriculture de France*, vol. 1.

[18] See, for instance, BCMNHN, MS 1327, 'Journal des Envois des Plantes', n.d.

[19] On the institutional–administrative structure of scientific networks, see Regourd, 'Les lieux de savoir et d'expertise colonial à Paris au XVIIIe siècle'.

[20] Aldrich and McKenzie, 'Why Colonialism?', 4.

[21] Haudrère, *La Compagnie Française des Indes au XVIIIe siècle*, 147–53.

[22] Ly Tio Fane-Pineo, *Île de France*, 13.

[23] On the concept of sustainability in historical perspective, see Warde, *Invention of Sustainability*, and Hay, *Companion to Environmental Thought*. See also Fressoz, 'Les politiques de la nature au début de la révolution'.

[24] Anon., *True Report of the Gainefull, Prosperous and Speedy Voiage to Iaua in the East Indies*. This is the earliest account by the Dutch to be translated into English (Grove, *Green Imperialism*).

Mauritius in the 1760s and on observing that the trees were not growing back, forecast that if the CIO's policy continued the island would soon become a desert.[25]

In the years of the early settlements, Mauritius was by no means the CIO's new 'pearl'; its commercial hub was the (then) French colony of Pondicherry (now Puducherry) on the south-eastern coast of India, the Coromandel coast. In fact, Pondicherry was the only well-established French possession in the East Indies, and it needed the support of a naval base less distant than France, as neither was it very active, nor did it possess ports of sufficient quality.[26] So for the CIO's purposes, Mauritius seemed like a far better option as a base for its ships. In other words, the harbours in Mauritius were needed for refreshment and as a naval base on the long sea route between Europe and South Asia.[27] The initial idea was that Mauritius would serve as a port of call. No sooner said than done; the administration's offices promptly moved from Réunion to the new possession.[28] Réunion was to serve as the granary of the Mascarene Islands – but as it produced insufficient food for the two islands, basic necessities had to be imported from the wider Indian Ocean world, mainly the Cape of Good Hope and Madagascar.

Next, with the idea of bringing a population to the island, the CIO imported a large number of enslaved people from the African coasts and Madagascar, and offered tempting concessions – free of charge – to any Frenchman who settled and worked in the new island colony. Especially during the first decades of the CIO's governance, settlers were lured by the promise of building a life there without having to pay anything back to the Crown.[29] From the early 1720s, settlers were transferred from Réunion to Mauritius, but an outbreak of leprosy in 1729 caused a hiatus in that scheme.[30] The first French families from France to seek their fortune in Mauritius arrived on *Le Bourbon* on 1 August 1728.[31] But the consequences of such rapid settlement were disastrous; the island lacked food and basic necessities, leading to a famine, and supplies had to be imported from Madagascar.[32] Meanwhile Mauritius's remaining

[25] Le Gentil de la Galaisière, *Voyage dans les mers de l'Inde*.
[26] See Toussaint, *Histoire des Îles Mascareignes*, 47.
[27] For life on board during these long journeys, including water supply, food, and other difficulties, see Girault-Fruet, *Les voyageurs*, 71–98.
[28] The fact that Mauritius became the centre of operations after 1735 is also reflected in a lack of records relating to Bourbon in the second half of the eighteenth century.
[29] Brouard, *History of Woods and Forests in Mauritius*, 10.
[30] Ly Tio Fane-Pineo, *Île de France*, 289. On leprosy and other sicknesses in Bourbon during the CIO's administration, see ADR C 936–9.
[31] Brouard, *History of Woods and Forests in Mauritius*, 10; Ly Tio Fane-Pineo, *Île de France*, 101.
[32] Ly Tio Fane-Pineo, *Île de France*, 94.

resources were being rapidly depleted by extensive hunting and fishing. The Director of the Colonies in the East, Pierre-Christophe Lenoir (1683–1743), had reported in September 1726 that hunting there had had to be prohibited by law.[33] He nonetheless recommended keeping Mauritius because of its ports and the high quality of its soil. Once cultivated as he envisaged it, the island would be able to supply not only its inhabitants but also sailors seeking refreshment.

But the conflicting views between the local governor and the far-away CIO on how to use the island were counterproductive. While the governor was aiming to create a base and settler home indispensable to the CIO, he also had to face the fact that the island's other task was to supply French ships with food, water, and medical care. But Mauritius was not productive enough to achieve this, and the CIO directors were harshly critical about that. Worse, the CIO's monopolistic policies did not allow the vicious circle to be broken. The heavy financial investment required to build up a proper depot and improve the fortifications was a constant thorn in the CIO's side.

The recommendations made by Lenoir and the CIO were ambiguous and contradictory. While in 1727 the new inhabitants were encouraged to cultivate coffee and edible goods, such as rice and wheat, as well as pepper and other spices, instructions dated 1729 enjoined them to abandon coffee and focus on food crops instead.[34] A journal by Captain Jonchée de la Goleterie relating to his voyage to the Indies (October 1727–April 1729) provides insights into the island's produce: it grew wheat, rice, maize, beans, vegetables, sugar cane, pineapple, figs, oranges, lemons, and generally all fruit that grew in the Americas, as well as grapes, peaches, figs, and mulberries from Europe – and 10,000 square feet of coffee.[35] But as the inhabitants lacked fundamental agricultural know-how their plants would not grow well enough, if at all, in the insufficiently prepared soil. The discontent of the colonists grew, and their opinion of the CIO deteriorated every day.[36] In addition, as the captain remarked, the enslaved people could not be fed sufficiently, so were likely to rise in revolt.[37] The food shortages continued throughout the eighteenth

[33] Brouard, *History of Woods and Forests in Mauritius*, 10–11.
[34] Lougnon, *Correspondance du Conseil supérieur de Bourbon*, vol. 1, 61.
[35] 'Extrait du journal du vaisseau Le Mars, Capitaine Jonchée de la Goleterie durant son voyage aux Indes commencé le 27 octobre 1727 et terminé en avril 1729', printed in Ly Tio Fane-Pineo, *Île de France*, 287–90.
[36] This is what Haudrère observes in the *mémoire* of Philibert Orry, dated 28 July 1730, AN B/3/302 (*La Compagnie Française des Indes au XVIIIe siècle*).
[37] 'Extrait du journal du vaisseau Le Mars, Capitaine Jonchée de la Goleterie durant son voyage aux Indes commencé le 27 octobre 1727 et terminé en avril 1729', printed in Ly Tio Fane-Pineo, *Île de France*, 287–90.

century, in part because of the overpopulation of enslaved people there, who suffered terribly from hunger and died in great numbers. The local administration had to explore new ways to secure food on the island by increasing agriculture, which required more enslaved people to work the grounds, plant, harvest, and process food material in the first instance.

It was only in the mid-1730s that things started to improve, albeit slowly. During the years 1735–46, the governor, Bertrand-François Mahé de La Bourdonnais (1699–1753), employed two helpful strategies. Lissa Roberts observes that he first had to improve the island's infrastructure and organise the available knowledge as well as reinforcing the natural and human resources, and second, he had to reconcile the sometimes ambivalent visions of the CIO and the Ministry of the Navy and the Colonies.[38] The clash between the governor and the CIO administration illustrates the difficulties inherent within local management, especially as the governor was dependent on influential officials in institutions in France and India. Although these connections may have helped La Bourdonnais make a career in the networks of the CIO, they ended the moment the different patronage networks pulled the strings in different directions. La Bourdonnais's governorship ended abruptly, with him being arrested, accused of insubordination to the CIO and the Crown. He was taken back to France, where he died a few years later, a broken man.[39]

Despite his unfortunate end, La Bourdonnais was celebrated by his contemporaries as the founding father of the colony. However, we read that the economist Pierre-Samuel Dupont de Nemours (1739–1817) observed that what La Bourdonnais as a governor had tried to establish in Mauritius was rapidly destroyed by his CIO successor due to consistently poor administration.[40] The Scottish geographer Alexander Dalrymple (1737–1808) commented on La Bourdonnais's work in the late 1750s:

> The sooner to improve Mauritius, the French Company gave the inhabitants of Bourbon great encouragement to transport themselves thither. M. de la [Bourdonnais] was sent from France in quality of governor of both islands. A man who was every way qualified for executing the great designs the Company had formed, of great experience and a good seaman, merchant and mechanic. It is to this gentleman's great capacity, and indefatigable industry that the French Company owns all the advantages they now enjoy from one of the most flourishing colony and best ports in all India.[41]

[38] Roberts, 'Le Centre de Toutes Choses', 323. [39] Ibid., 324.
[40] Dupont de Nemours, *Notice sur la vie de M. Poivre*, 36.
[41] BL, Add. MS 33765, f. 2v, 'Geographical Collections of Alexander Dalrymple'.

Indeed, agricultural activities were increasingly encouraged, in particular, rice and wheat, and sugar cane.[42] Another important crop was manioc, originating from Brazil, mainly used to feed the ever-increasing population of enslaved people. La Bourdonnais experimented with drought-resistant plants and mulberry trees (for domestic silk production).[43] In contrast to contemporary accounts, however, he did not achieve a breakthrough, nor did his governorship stimulate explorations or new discoveries. Yet he had laid the foundation for the future of the island: Mauritius's experimental gardens were to serve as a setting for the acclimatisation of imported plants. The first acclimatisation garden, Monplaisir in the Pamplemousses area, close to Port Louis, was established by La Bourdonnais, and it later served as the administrators' residence on the island (see Figure 3.2).[44]

While the War of the Austrian Succession (1740–8) was being played out, French botanists, together with their political patrons, were pushing for an economic use of plants and forests in the service of the CIO's wealth. Plants, and spices in particular, became more and more an object of interest, and became associated with the business of economic botany. Captain Gennes de la Chancelière observed in 1735, the year of La Bourdonnais's arrival:

> Since it is not enough, in order to make a colony flourish, for it to have an abundance of food and refreshment, if, moreover, it does not provide some particular production that can be used for a solid trade, and compensate the owners for the great expenses that one is obliged to make to maintain it, it would be advisable, in my opinion, for the good of the service, to encourage the inhabitants to cultivate the trees and plants that can be used for trade, as much for the particular utility as for the interest of the Company.[45]

The idea was to challenge the Dutch spice trade of the Moluccas (now the Maluku Islands), as not only could Mauritius become a port of call on the new routes to Asia but also it could be turned into an important island of commerce.[46] Gennes de la Chancelière suggested the planting of olive trees – olive oil being rare and expensive in the European colonies – plus cacao, pepper, indigo, sugar cane, and other crops cultivated in the Americas.[47] Further, he made the point that producing sugar was not the best way of making a profit, as it was already a common product in Europe. According to him, one should rather experiment a little with the

[42] See the *mémoire* by Père Cossigny's son: Cossigny de Palma, *Mémoire sur la fabrication des eaux de vie de sucres*. For an illustration of the big sugar mill introduced by La Bourdonnais in 1744, see Piat, *L'île Maurice*, 66–7.

[43] Roberts, 'Le Centre de Toutes Choses', 323. [44] Grove, *Green Imperialism*, 175.

[45] Gennes de la Chancelière, 'Observations sur les Isles de Rodrigue et de France en mars 1735', 223–4.

[46] Ibid., 228, 236. [47] Ibid., 223–5.

market and look at what flavours the inhabitants developed. He hoped to achieve this first by having all sorts of cereals sent from France, together with olive trees, and by importing coffee from Réunion, mulberry trees from China, and cotton and pepper. Second, he sought to provide for each inhabitant one or two enslaved people in order to work the fields; third, he intended to oblige every new arrival in Mauritius to plant food crops and a certain quantity of spices.[48]

Ideas about the acquisition of spices had been circulating for some time. In an anonymous *mémoire* about the commercial relations between Europe and China from 1733 to 1735, the author reflects on the opportunities in China, where cinnamon of better quality than that in Ceylon could be found (however, according to the *mémoire*, this was a rumour and needed to be confirmed) and also nutmeg and cloves, although these could not compete with the 'Malay crop' (the crop from the Moluccas).[49] It was only in 1748 that the CIO agreed to finance a project to acclimatise spices in Mauritius, spearheaded by Pierre Poivre, a novice at the time. It is important to underline that the spice project was not an idea that Poivre alone developed, as indeed he himself – self-promoting though he was – highlighted in his writings, both unpublished and published.

Although La Bourdonnais had pushed for the greater and proper usage of the north-west part of Mauritius, both to accommodate troops for the defence of Pondicherry and to provide a living for the island's inhabitants, the CIO obdurately insisted on using the island as a mere stopover to the East Indies.[50] Even though the CIO directors agreed in principle with the advantages of the island being an entrepôt for French vessels, they were more concerned about the huge expenditure that would be needed to build a proper depot and upgrade the fortifications. They decided that the investment would be too high for too low a return and the CIO directors declared that 'the Company will never think about forming a depot in Mauritius'.[51] La Bourdonnais's vision of an improved lifestyle on the island was disrupted by the Company's reluctance to create a parallel French society in the East Indies. He explained in his long *mémoire* that Mauritius did not offer the same advantages as Réunion in terms either of the number of inhabitants or for the plantation of coffee, and that it was

[48] Ibid., 226.

[49] BnF, NAF 9347, f. 19r, 'Commerce européen avec la Chine' (1733–1735)'.

[50] On La Bourdonnais in the service for the CIO, see Haudrère, *La Bourdonnais*. On Cossigny's work and his struggle with the governors, see MA, OA 97, 'Papiers concernant M. de Cossigny'.

[51] Ly Tio Fane-Pineo, *Île de France*, 305–6.

therefore necessary to support living and working on the island colony.[52] La Bourdonnais then wrote that even if the CIO had expected the colony only to perform its task as a provisioning station, Mauritius was greatly in need of inhabitants committed to the island and to their lives there. Only with a society committed to its living space, he argued, was the island sustainable.

He did not abandon his plans, but instead tried to convince the French institutions. In October 1740, on the outbreak of the War of the Austrian Succession, he presented his plans for a campaign in the Indian Ocean, in which he set out the current state of the colonies in the East Indies and explained the need to fortify Mauritius.[53] The conflict between La Bourdonnais and the general commander of the East Indies, Joseph François Dupleix (1697–1763), reveals that this was not merely a dispute between two high officials, but was actually a reflection of the conflicting views on French expansion in the Indian Ocean. At the end of the day, it was the Ministry of the Navy and the Colonies that was most likely to sympathise with this proposal.[54] While the French islands with their strategic position remained important to the Ministry of the Navy and the Colonies, the CIO preferred to aim for the short-term exploitation of the island's natural resources, and to have forced migrants taken to the island as enslaved people. To the CIO, Mauritius was still primarily a means of defending Pondicherry.

Although from 1746 the situation of Mauritius improved, under La Bourdonnais's successor Pierre Félix Barthèlemy David (1710–95; nominated 1746–53), the island was still unable to prove its worth to the CIO. David's successor, René Magon de la Villebague (1722–78; in office 1754–9), compared the establishment of Mauritius to that of the French Antilles, concluding:

> The reasons for our misfortunes are the same as the ones in Saint-Domingue and Martinique in the first years of their establishment. The Company, which oppressed us, is a dismemberment of those to whom Louis XIV gave exclusive privileges for their commerce. The principal managers and their protégés made immense fortunes at the expense of the inhabitants. (...) The screams of the miserable colonists eventually reached the throne.[55]

[52] Mahé de La Bourdonnais, *Mémoire Historiques de B. F. Mahé de La Bourdonnais*, 14. A draft of this *mémoire* can be found as a manuscript in the BnF: NAF 9344, ff. 197r–239r.

[53] 'Plan de Campagne dans l'Ocean Indien présenté par La Bourdonnais au Ministre de la Marine en date du 17 octobre 1740', printed in Ly Tio Fane-Pineo, *Île de France*, 308–13.

[54] Toussaint, *Histoire des Îles Mascareignes*, 51.

[55] ANOM, C/4/19, f. 199r, 'Mémoire de Magon' for Poivre and Dumas, 14 November 1767.

Apparently, the CIO ignored the screams of the miserable colonists, but saw the geographical advantage of the island because, as a contemporary *mémoire* states: 'Isle de France is the gate to all Indies'.[56] Furthermore, in 1770 the Abbé Raynal observed:

> Somewhat remote from the ordinary routes, [Mauritius] is the surest of the secrets of armaments. Those who imagine it closer to our continent do not see that it would consequently be impossible to transport oneself to the Coromandel coast in two months or more to the most distant gulf, a priceless advantage for people who do not have any port in India.[57]

Sure enough, the growing importance of Mauritius's ports attracted other European powers; in 1756 Dalrymple remarked: 'I am much afraid the port [of Mauritius] will give [the French] greater advantage over [the British] in India, than we at present seem sensible of.'[58] Although the main Anglo-French disputes took place in South Asia, where the trading companies were well established, this rivalry was felt in the trading networks throughout the Indo-Pacific regions, and definitely also in Mauritius, a strategically located island base and naval station for the CIO.

Despite the island's value as a *relâche* on the way to the East Indies, a persistent problem was its isolation. In the middle of the Indian Ocean as it was, its distance from Europe, the demands and the ignorance of the CIO, and the indifference of the government all combined to reduce its value.[59] The fact that it was not self-sufficient had two consequences. On the one hand, it was extremely vulnerable in the event of a blockade, while on the other the lack of provisions made it impossible to supply the French troops adequately.[60]

The journals of the governor Magon reveal the struggle to manage the island in wartime, its biggest problem over the course of the eighteenth century: this exacerbated the scarcity of the food and the mortality rate of the enslaved people rose even higher.[61] The CIO had failed to build the storage the French forces so desperately needed in wartime. When they were sent to the Indian Ocean for the first time, in order to defend Pondicherry, Mauritius was unable to supply the men with the food and

[56] ANOM, C/4/153/1, Anon. 'Mémoire sur l'Isle de France', n.d., unpaginated.

[57] Raynal, *Histoire de deux Indes* (1770), vol. 2, 141. The quotation refers to the 1770 edition and was deleted in the 1780 edition since action was taken regarding the fortifications.

[58] BL, Add. MS 33765, f. 12r, 'Geographical Collections of Alexander Dalrymple'.

[59] Bourde de la Rogerie, *Les bretons aux îles de France et de Bourbon*, 9.

[60] These are the two major problems that occur throughout the correspondence between the CIO director Dupleix and Mauritius's new governor Lozier-Bouvet in 1753, see ANOM, C/4/7, 'Correspondence Générale', 1751–3.

[61] ANOM, C/4/9, 'Correspondence Générale', 1755–7.

the materials needed to repair and restore their ships.[62] Throughout the eighteenth century, there was great disappointment about the western Indian Ocean colonies: a recommendation to discard the Mascarenes, particularly in view of its bad weather conditions, was made more than once.[63] However, as underlined in a *mémoire* from 1752, the French were in dire need of a port of call on the long route to India, and Mauritius was the only feasible option.[64]

Lacking Botanical Interest

The Parisian scientific institutions were fairly indifferent about Mauritius. The Parisian Académie royale des sciences, created in 1666, was patronised by the state, so it had little freedom for manoeuvre.[65] It was a state-supported institution which possessed a restricted yet heterogeneous membership, and was governed by a strict hierarchy.[66] Although you might expect it to have received a large number of letters from the Indian Ocean colonies during the first half of the eighteenth century, only a few concerning the Mascarenes have survived (although the precise number sent cannot be ascertained).

It was only in the 1730s that the French royal Academy of Science took an interest in Mauritius; this was after the engineer Jean-François Charpentier de Cossigny (1690–1780) had undertaken experiments on the weather conditions there, about which he corresponded with the Academy's member Antoine Ferchault de Réaumur (1683–1757).[67] Even so, he was not in Mauritius in the service of the Academy, but was working for the CIO.[68]

Cossigny's example illustrates the double function of individuals attached to the CIO and scientific institutions, confirmed by the case of Poivre. He, as a CIO agent, wished to make use of the Parisian Botanical Garden in order to confirm his findings on the nutmeg. According to a report on his activities in the Indo-Pacific region, he had, during his first expedition to the Moluccas in 1753, instructed CIO officials to send specimens of nutmeg to the Parisian garden in order to prove the

[62] Toussaint, *Histoire des Isles Mascareignes*, 61.
[63] Vaughan, *Creating the Creole Island*, 37.
[64] ANOM, DFC IV, Anon. 'Mémoire' 10, piece 38, June 1752. Vaughan also makes a reference to his said memoir (*Creating the Creole Island*, 37).
[65] Marion, *Dictionnaire des institutions*, 3, entry 'Académie'; Raj, '18th-Century Pacific Voyages of Discovery', 85–6.
[66] For statistics, see McClellan, 'Académie Royale des Sciences' and *Specialist Control*; Crosland, *Science under Control*.
[67] See *HARS* and *MARS* from the early 1730s.
[68] Lacroix, *Notice historique sur les membres et correspondants de l'Académie des sciences*, 7.

originality of the plants.[69] As he explained in a report, he had been asked to send a *mémoire* to France about the manner of their cultivation, together with twelve long and ten round nutmegs, all still in their shells. His findings were to be examined by the *contrôleur général des finances*, Jean Baptiste Machault d'Arnouville (1701–94), and the Parisian garden.[70] It is, however, unclear whether the CIO actually sent these specimens from Mauritius to Europe or, if they did, whether the plants arrived in Paris.

In 1754, while in the service of the CIO, Poivre was also named as a correspondent of Réaumur.[71] At that point the director of the Academy's library, Chrétien Guillaume de Lamoignon de Malesherbes (1721–97; in office 1750–63), became very interested in Poivre's explorations of Asian natural goods.[72] After Réaumur's death in 1757, Poivre was named as a correspondent of the botanist Antoine de Jussieu (1686–1758) on 10 December 1757, but Jussieu died a year later.[73] Any further direct attachment of Poivre to the Academy is unknown.

The Jussieu family was one of the main families in the field of French botany, occupying important seats in the Académie des Sciences and the Jardin du Roi, and later the Muséum d'histoire naturelle for one and a half centuries (1712–1853). The first of this 'botanical dynasty' were the two brothers Antoine and Bernard (1699–1777), followed by their nephew Antoine-Laurent (1748–1836) and his son Adrien-Laurent-Henri (1797–1853). Although most members of the Jussieu family did not travel abroad themselves, they were interested in botanical activities overseas, particularly in the last third of the eighteenth century.[74] During his lifetime, Antoine de Jussieu was an active correspondent: fifty-six of his letters between 1724 and 1729 are about the French colonies in the Atlantic and Indian oceans. However, there is no evidence of a botanical correspondence between him and Poivre.[75]

[69] BnF, NAF 9377, 'Rapport de la mission du sieur Poivre', f. 59r. [70] Ibid.

[71] ADS 69J 67/26, Poivre to Réaumur, 18 March 1754. See also *MARS* 1733, 417–37; *MARS* 1734, 553–63; *MARS* 1735, 545–76; Dupont de Nemours, *Notice sur la vie de M. Poivre*, 33.

[72] BCMNHN, MS 1765, 'Extraits de quelques conservations avec M. Poivre'. On Malesherbes, see Shaw, *Problems and Policies*.

[73] See ADS, PV 76 (1757), 643.

[74] Bernard de Jussieu also corresponded with botanists overseas, such as Michel Adanson in Senegal. See Lacroix, *Notice historique sur les cinq de Jussieu*, 27 and *L'Académie des sciences et l'étude de la France d'Outre-Mer de la fin du XVIIè siècle au début du XXè*. Yet, the surviving correspondence in the botanical archives is sketchy and by no means as systematic as it became in the last two decades of the eighteenth century under head gardener André Thouin. On André Thouin's correspondence networks, see Spary, *Utopia's Garden*.

[75] Lacroix, *Notice historique sur les cinq de Jussieu*, 12; on the role of the Academy in relation to botany in the seventeenth century, see Stroup, *Company of Scientists*, 65–168.

It is clear, though, that Antoine de Jussieu was involved in the first scientific description of Réunion coffee. In the early eighteenth century, after the Treaty of Utrecht (1713), coffee specimens were sent to the Jardin du Roi, where he examined the plants and submitted an official report on them to the Academy. Later, in 1721, specimens were sent directly from Réunion to Jussieu, and he reported to the CIO on the coffee's (poor, as it turned out) quality.[76] Although his motives were economic, Réunion coffee was not a popular product in France and in December 1725 the CIO stopped the administration from sending it to Europe.[77] Minutes of a letter to the CIO, dated 14 August 1722, also reveal that Antoine de Jussieu received information about plants and drugs in Réunion, especially the juice of aloe vera, turmeric, and 'féhénate', probably a kind of lemongrass.[78] Possibly soon afterwards, he wrote a *mémoire* on Réunion in which he stressed once again his interest in plants for nourishment and pharmacy, especially spices such as turmeric, cinnamon, and cloves, plus coffee, and several kinds of timber.[79]

By the 1720s Antoine de Jussieu was already flirting with the idea of strengthening his Indian Ocean correspondence network. He proposed that one could do 'with Bourbon as with Madagascar', and asked for reports and general descriptions of cultivation expertise and know-how.[80] As an unpublished treatise from probably 1729 suggests, he felt a strong need to correspond with men who had confidence in the colonies, and he felt that these men, in return, should rely on information about cultivation from those who knew it best: locals, meaning local inhabitants with plant knowledge.[81] He underlined that he was particularly interested in the plants and seeds of food crops. He argued that since useful plants would also be of interest to the Crown, such a project, and the ties with these '*amateurs de l'histoire naturelle*' was likely to receive funding.[82] Jussieu also saw that it was necessary to learn from people in other places such as Greece, Turkey, and the East Indies with regard to how the plants were

[76] See *HARS* (1713), 43–4. See also BCMNHN, MS 357, 'Papiers provenant de Louis-Guillaume Lemonnier (1717–1799) et en partie de sa main', piece 3: 'Observations sur la culture du Café recueilles par M. Deforges-Bouché, l'intendant du roi en 1720'.

[77] Lougnon, *Correspondance du Conseil supérieur de Bourbon*, vol. 1. See also on coffee introduction in France, Spary, *Eating the Enlightenment*.

[78] Lacroix, *Notice historique sur les cinq de Jussieu*, 17, 19.

[79] BCMNHN Jus 3, f. 93v, 'Pour l'isle de Bourbon'. See also BCMNHN 1140/9bis, f. 3r, 'Lettres concernant des drogues envoyées de L'isle Bourbon'.

[80] BCMNHN Jus 3, f. 93v, 'Pour l'isle de Bourbon'.

[81] BCMNHN Jus 3, ff. 95r–105v, 'Des Avantages que nous pouvons tirer d'un commerce littéraire avec les botanistes étrangers'. See also the scientific correspondence and notes of his brother Bernard, BCMNHN Jus 6 and Jus 7.

[82] BCMNHN Jus 3, f. 97r, 'Des Avantages que nous pouvons tirer d'un commerce littéraire avec les botanistes étrangers'.

cultivated and used in those countries. Plants from India and Turkey, such as raisins, beans, and red truffles, Jussieu concluded, were by then common on French tables and would often form the literal fruits of correspondence.[83]

During the first half of the eighteenth century, however, it was *not* Mauritius's flora that attracted the Academy's or the Garden's interest, but the island's mineralogy, astronomy, and meteorology. Astronomy was a matter in which the Academy took a great interest throughout the second half of the eighteenth century. But while the Academy worked closely with the astronomers sent by the observatory to look for new routes in and around the Indian Ocean[84], the ties between the Mascarenes and the Academy or the Parisian garden were not very strong. Only in the last three decades of the eighteenth century did the Academy receive items of interest from the Indian Ocean colonies, as revealed by the Academy's minutes: a coconut from the Indian Ocean, a report on dwarfs and minerals from Madagascar, and observations from Mauritius and new islands off Madagascar's coast.[85]

It was only after the dissolution of the CIO in 1764 that the networks were strengthened, and Parisian naturalists then made increasing use of the Ministry of the Navy and the Colonies to transmit their tropical plants.[86] At the same time the Jardin du Roi's professor of botany, Louis Guillaume Le Monnier (1717–99), successor to Antoine de Jussieu from 1758, started to establish a network of correspondence on natural curiosities.[87] Indeed, and as Poivre's own writings show, Poivre attempted to correspond with Le Monnier and Bernard de Jussieu during his travels in Cochinchina[88] in the 1740s and 1750s.[89] During those years, Poivre spent four months in Madagascar as well, where he documented several plants.[90]

Only in the late 1760s, when the agricultural activity of the island had started increasing, did Le Monnier become interested in the useful plants

[83] Ibid., ff. 97v–98r.

[84] Rochon, *Voyage a Madagascar et aux Indes Orientales*; BnF, NAF 9345, ff. 287r–294r, 'Nouvelles routes de Isles de France et Bourbon aux Indes dans l'une et l'autre Monsoon'.

[85] See ADS, PV 89 (1770). See also ADS, PV of 1769, 1772, and 1773 on nutmeg and cloves.

[86] Spary, *Utopia's Garden*, 123; Schiebinger and Swan, 'Introduction', 5; for the English case, see Brockway, *Science and Colonial Expansion.*

[87] See BCMHN, MS 357.

[88] 'Cochinchina' is the historical term for the region comprising southern Vietnam and parts of eastern Cambodia.

[89] 'Voyages de Pierre Poivre de 1748 Jusqu'à 1757', 49, 65, 66, 72, 76, 82. Poivre's manuscript was printed by Henri Cordier almost 200 years later. See also Laissus, 'Note sur les manuscrits de Pierre Poivre', 42.

[90] BCMNHNm MS 1269.

of the Mascarene Islands. The detailed information he received show that he must have had a correspondent in that part of the world, this turns out to have been one De Reine, who presented the *Catalogues de graines d'arbres et d'abustes* of the Mascarene Islands to Le Monnier in 1768.[91] The said *Catalogues* contain not only a report on plants and trees growing in Mauritius, but also specific instructions for the Abbé René Galloys.[92] The Abbé was a *protégé* of Le Monnier, and, having been given the title '*naturaliste du roi et du Jardin royal*', had been sent to assist Poivre with the introduction of useful plants and birds into Mauritius.[93]

The correspondence between Le Monnier and Poivre of between 1771 and 1772 is sketchy, with only three letters from Poivre to Le Monnier. In a letter dated 25 August 1771, in addition to the attempts to source plant material in the Moluccas under Jean Mathieu Simon Provost (1728–76), commissaire de la Marine, another point catches the eye: Poivre mentions the seeds which he had sent to the Jardin du Roi and that he was happy to hear they were sprouting.[94] Although he does not specify the type of seeds, this does prove that Le Monnier must have received specimens from Poivre earlier. In the following year, Poivre wrote that he was sending more boxes with plant material to Paris on the *Duc de la Vrillière* – unfortunately, however, without specifying what types of plant they were.[95] It also remains unknown whether he sent seeds or live plants, since his term for a transportation box was '*caisse*', '*boîte*', or '*barigue*' ('case', 'box', or 'barrique'/'barrel'), none of which were exclusively used for one or the other.[96]

The overall knowledge and specimen exchange between the Mascarenes and the Jardin du Roi remained limited, and it was not until the late 1770s, or even the 1780s, that the Parisian garden set up an established correspondence network with botanists in the colonies. While the acclimatisation of tropical plants became more and more prominent and of great economic interest, the Jardin du Roi, with its

91 Lanux junior refers to him as a 'officier chez le Roy a Versailles', BSG, MS 2551, f. 59v, Lanux to Pingré, 15 August 1763. For the *Catalogues*, see BCMNHN, MS 357, file 15.

92 BCMNHN, MS 357, file 15, 'De Reine, Catalogues de graines d'arbres et d'abustes', 1768. See also the list of plants from China, August 1766, and BCMNHN, MS 357, file 14, 'Catalogue de graines de plantes de la Chine. Liste des graines envoyées de Pekin 1778'.

93 On this mission, see Malleret, 'Pierre Poivre, L'abbé Galloys et l'introduction d'espèces botaniques et d'oiseaux de Chine à l'Île Maurice', 117–30. Some letters from Galloys to Le Monnier are also printed in Laissus, 'Note sur les manuscrits de Pierre Poivre', 45–56.

94 Galloys to Le Monnier, 25 August 1771, printed in Laissus, 'Note sur les manuscrits de Pierre Poivre', 47.

95 Poivre to Le Monnier, 1 April 1772, ibid., 48–9.

96 For instance, see on the methods of packing: BCMNHN, MS 47, 'Mission de Joseph Martin à l'Mauritius'.

laboratories and experiments, was used as a nursery for plants and objects that produced millions of livres for the French kingdom in the second half of the eighteenth century. It was, however, not until the early 1770s that the Jardin du Roi, under the intendant Buffon and André Thouin (1747–1824, the head gardener from 1764 to 1793), established correspondence networks recognised by the state, justified on the basis that they attracted economic wealth.[97]

Plans, Plants, and Management after the Seven Years War

For France, the Seven Years War was a disaster: Louisburg and Gorée fell in 1758, Québec and Guadeloupe in 1759, Montréal in 1760, Pondicherry in 1761, and Martinique in 1762. The Treaty of Paris, signed on 10 February 1763, humiliated Louis XV as few monarchs had been humiliated hitherto.[98] Although France managed to retain its Caribbean colonies, its territorial downsizing in the Atlantic had a serious effect on the financial crisis that the French monarchy was undergoing. The response to this was the French colonial administration's attempts to compensate for the loss of its possessions, leading to more ambitious and stronger control over the remaining colonies and their inhabitants. This is also reflected in the colonial archives: they became an instrument of imagined control and a repository of imperial consciousness.[99]

This crucial point overlapped with colonial innovations and traditional approaches.[100] The rivalry between the great European powers turned in new directions, European interest in trade and imperial advantage moving towards the Pacific; this meshed with the promotion of knowledge and resulted in increasing contact with the Pacific world.[101]

The global dimension of the Seven Years War and its aftermath had a strong impact on the development of Mauritius in the second half of the eighteenth century.[102] France's defeat by the British triggered an outburst of French colonial ambitions and schemes. In the East Indies, Mauritius was France's principal remaining colony, and throughout the eighteenth century it was the only stable French colony, that is, left unoccupied by British troops. In order to reinforce the links between

[97] Spary, *Utopia's Garden*. [98] Hodson, *Acadian Diaspora*, 79.

[99] Houllemare, 'La fabrique des archives coloniales', 8. See also Regourd, 'Capitale savante, capitale coloniale', 121–51; Ruggiu, 'India and the Reshaping of the French Colonial Policy', 25–43.

[100] Røge, 'Natural Order of Empire', 32.

[101] Ly-Tio-Fane, 'Pierre Poivre et l'expansion française dans l'Indo-Pacifique', 80–8.

[102] On the global dimension of the Seven Years War, see especially Baugh, *Global Seven Years War*; De Bruyn and Regan, *Culture of the Seven Years' War*; Gascoigne, *Encountering the Pacific*, particularly for the legacy of the Seven Years War, part II, ch. 5.

Versailles and the Indian Ocean colonies, the French king – or rather his minister – declared that the CIO would return the Mascarenes to the Crown in August 1764. On payment of the sum of 7,625,348 livres the islands were placed under the direct control of the Ministre de la marine et des colonies (Minister of the Navy and the Colonies).[103] The Ministry designed an *Ordonnance du Roi* from 25 September 1766, which proclaimed:

His Majesty wishes to regulate everything which concerns the general and particular administration of the Isles de France and Bourbon, not only by report of the government of these islands but also by report of the distribution of justice.[104]

At the beginning of the *époque royale* (the decades of royal power), the most important French actors as far as Mauritius was concerned were Étienne-François de Choiseul (1719–85), who managed French politics via his roles as Minister of the Navy and the Colonies (1761–6) and Minister of Foreign Affairs (1766–70), and his cousin César Gabriel de Choiseul-Praslin (1712–85; Praslin for short), Foreign Minister (1761–6), Minister of the Navy and the Colonies (1766–70) and a future honorary member of the Académie des Sciences.[105] The importance of Choiseul and Praslin to the Mauritius project was well known to contemporary observers. In 1772, a Mr Colpoys (who was probably Admiral Sir John Colpoys (1742–1821), of the Royal Navy), explained that the new governor of Mauritius planned to fortify and reinforce the island. Later, however, after Choiseul and Praslin had left their posts, this idea was not supported by Versailles.[106]

Choiseul and Praslin looked for new possibilities for expansion in order to compensate for France's serious losses in the Atlantic. They tried, but failed, to regain control of the posts lost in western Africa, Senegal, and Gambia. So French Guiana had to serve as a base for the army and the French Antilles, an idea which Étienne-François Turgot, governor of

[103] Toussaint, *Histoire des Îles Mascareignes*, 64.

[104] MA Vol Z3B7, No. 29 'Ordonnance du Roi concernant le gouvernement civil des iles de France et de Bourbon, 25 septembre 1766', printed in Napal, *Les Constitutions de l'Ile Maurice*, 30.

[105] All information and dates are taken from Marion, *Dictionnaire des institutions*, 503–4. I refer to Choiseul-Praslin as 'Praslin' in order to avoid confusion with his cousin Choiseul. As the two cousins exchanged their ministry between 1761 and 1770 the situation is sometimes very confusing, particularly in letters, since 'Choiseul' could refer to either of the men. Even the secondary literature on the French administration of the Mascarene Islands is often misleading. Most letters from Poivre to Versailles between 1767 and 1770 are addressed to the 'Duc de Praslin'. Until the nomination of Pierre Étienne Bourgeois de Boynes (1718–83) as Minister of the Navy and the Colonies in April 1771, the Abbé Joseph Marie Terray (1715–78) held the post from late 1770.

[106] BL IOR/H/111, f. 142v, 'Mr. Colpoy's Remarks on the Mauritius', 25 September 1772.

Guiana, strongly supported.[107] For this programme, Choiseul suggested an expedition to seek out Guiana's natural resources to the botanist Michel Adanson (1727–1806), who had explored Senegal between 1749 and 1753 and later became an important figure at the Académie des Sciences. But this project would only be undertaken later, by the pharmacist Jean Baptiste Christophore Fusée-Aublet (1720–78).[108] All in all, the Guiana project was badly planned and ended in disaster. Choiseul felt he had to make up for his mistakes;[109] new projects were to be considered and new imperial venues for resources that could bring new wealth to the debt-burdened kingdom explored. After the French navy had been defeated, the Canadian colonies had been lost, and the South American projects had failed, hopes were pinned on the small island of Mauritius. Choiseul had ambitious plans that were not shared among other governmental officials, because – as so many of them argued – Mauritius was too far from the French possessions in India.[110]

The *Mémoire du Roi*, instructions for Mauritius's new local administration, drafted by Choiseul and enacted in 1766, provide further insights into the imagined and visionary importance of the Mascarene Islands to Choiseul.[111] Three major points become clear. First and foremost, the islands were to serve as a place of commerce and exchange for the spice trade. Trade along the coasts of Africa was said to be varied, abundant, and highly profitable. So as a base for the spice trade, Mauritius would be important not only for the trade with Europe, but also – and to a much greater extent – for the trade with Asia; there, as the *mémoire* underlines, the consumption of spice was much higher. Second, the Mascarene Islands were important *relâches* for ships' companies to rest and refresh en route to the eastern Indian Ocean. According to the *mémoire*, if France had not owned the islands, its ships would have had to stop in either Brazil or the Cape of Good Hope, which would have meant relying on foreign governments. So, the Asian territories would have been much harder to access, making it difficult to renew food and other resources. Third – the most important factor – the *mémoire* explains the advantageous

[107] Quoted in Pluchon, 'Choiseul et Vergennes', 228.

[108] Lacroix, *Michel Adanson*, 15; Adanson, *Histoire naturelle du Sénégal*. After his return from Africa, Adanson was invited by Bernard de Jussieu to live in his house, where he remained for ten years. This also explains the intimate use of Bernard's taxonomic procedure, which Adanson justified (Morton, *History of Botanical Science*, 302).

[109] Regourd, 'Kourou', 233–54; on the colonisation of Guiana and the struggle for a French America, see Godfroy, *Kourou*.

[110] Reussner, 'L'Ile de France', 218.

[111] ANOM, C/4/17, esp. ff. 3r–4v, 'Mémoire du Roi pour server d'instructions aux Sieurs Dumas, Commandant general, et Poivre, Commissaire general de la Marine, faisant function d'Intendant aux Isles de France et de Bourbon', 28 November 1766.

geopolitical position of Mauritius in light of the French possessions in India. Choiseul was convinced that through its possession of the island the French could dominate the entire region; the island being located centrally, it could most certainly serve as a base to defend the French territories as well as to attacking foreign colonial possessions.[112] According to the *mémoire*, if the island were to be used optimally its ports, its population, its salubrity, and its strategic location would enable the French in the Indian Ocean to equip, feed, and move its troops.[113]

Choiseul's wishful thinking was to remain, however, just that. Among government officials in Versailles, there was no homogeneous idea of how to use Mauritius. Richard Grove suggests that Choiseul's cousin, the Duc de Praslin in his role as Minister of the Navy and the Colonies, was more interested in the establishment of an agricultural island according to physiocratic ideals than in establishing a focal point and naval repair base in the middle of the Indian Ocean.[114]

In official documents, Praslin sugar-coated the situation on the island, focusing on a hypothetical geopolitical importance of Mauritius and its overall advantages. He created his imagined reality in order to retain royal financial support. The Parisians were suspicious, however.[115] Although after 1766 Praslin pushed for the establishment of Mauritius, he could not act independently by sending money or troops. The game of balancing visions, creating reality, and the support of financial donors can thus be observed both in the colony and in France itself. In other words, politics was symbolically present in the colonies. The tension between Choiseul and the government, and the nomination of Poivre as intendant of the Mascarene Islands by Choiseul, created multiple identities. The battle fought for the economic use and the political role of Mauritius, a battle fought between the varying, and conflicting, agendas of state officials, continued throughout the *époque royale*. The battle was fought not only between Versailles and the local administration but also locally, between the governor of the island and its intendant.

The local administration of Mauritius was divided into the governor-general, who had the military authority, and the intendant, concerned with the economic and financial questions. In 1766, Jean-Daniel Dumas (1721–94) and Pierre Poivre were sent to Mauritius as governor and

[112] See also on the instructions and Choiseul's plans, Ly Tio Fane, 'Pierre Poivre et l'expansion française dans l'Indo-Pacifique', 460.

[113] ANOM, C/4/17, 'Mémoire du Roi pour server d'instructions aux Sieurs Dumas, Commandant general, et Poivre, Commissaire general de la Marine, faisant function d'Intendant aux Isles de France et de Bourbon', 28 November 1766.

[114] Grove, 'Conserving Eden', 332–4.

[115] Above all, Choiseul was also replaced by De Boynes in 1770 as *secrétaire d'État à la Marine*, who stayed in the post until 1774.

intendant respectively, to clear up the mess that the CIO's political and commercial programme had left on the island. Besides general compensation for the CIO's short-sighted usage of the island, the new royal administration had to stabilise the colony's social infrastructure and bring it under productive control. The detailed regulations of the French administration and their visions of order in the colonies were far from consistent with colonial life. And, above all, the conflicting views of the governor and the intendant on how to run the island were seriously counterproductive. During the years of Poivre's directorship (1767–72), the governor-general of Mauritius changed three times; between 1767 and 1769, Jean-Daniel Dumas (who was in fact replaced by Jean Guillaume Steinauer, as acting governor-general, from 1768 onwards), was succeeded by François Julien du Dresnay, chevalier Desroches (1719–86) (Desroches for short, in office 1768–72), who was in turn succeeded by Charles-Henri-Louis d'Arsac de Ternay (1723–80, in office 1772–6).

Dumas and Poivre's relationship was complicated enough already, embedded as it was in conflicting views and conflicting personalities. The governor, Dumas, focusing on his military projects, did not share Poivre's ambitious plans for the establishment of a settler community on the island.[116] The fact that Poivre's initial clove and nutmeg acclimatisation project had failed added fuel to the fire, and in 1768, after the conflict had been brought to the minister's attention, the quarrel between governor and intendant ended in Dumas's precipitate departure, hence Steinauer becoming acting governor.[117] Poivre's visions for agricultural engagement clashed, too, with the ideas of the next governor, Desroches, relating to the island's usage. Like Dumas, Desroches was concerned with military issues rather than food crops, let alone Poivre's obsession with spices. Desroches was at least, however, interested in the continuing production of timber (for ship construction) and agreed to regulations on the conservation of the forests (printed in 1769) and the encouragement of agriculture in order to service the incoming troops.[118] Both Desroches and Poivre, even with their different aims, were keen to improve the situation and to compensate for the CIO's disastrous policy.

[116] See, for instance, AD Loire FGB 15 J 16 and 15 J 17.

[117] Toussaint, *Histoire des Îles Mascareignes*, 70. For the conflict between Poivre and Desroches, see also Malleret, *Pierre Poivre*, ch. 8.

[118] 'Reglement economique sur le défrichement des terres et la conservation des bois de l'Mauritius', 15 November 1769, signed by Poivre and Desroches (Bouton, *Sur le décroissement des forêts à Maurice*, 13).

Still sugar-coating the problems, Poivre explained in a letter to the minister, dated 3 April 1769:

In order to fulfil your expectations of the colony, I especially encouraged its cultivation . . . . You have seen by my previous accounts that your island nowadays can feed itself.[119]

This was a huge – and disastrous – miscalculation, particularly striking as Poivre went on to explain that the island's potential was by no means maximised and the majority of the land was still unused. As Lissa Roberts points out:

local management strategies and practices would have to mesh with a different governance process as colonial policy, appointments and administration, officially made in Paris, became a matter of negotiation with officials claiming to represent state interests or with metropolitan patrons whose influence had to be acquired and maintained.[120]

Although the geopolitical importance of the island was growing, its progress and maintenance as such remained limited. If we are to believe Dupont, Poivre's contemporary biographer,[121] when Poivre arrived as intendant in 1767 he found the Mascarenes in a state of almost total devastation; cultivation, commerce, fortifications, everything had been neglected.[122] Although Dupont asserts (as does Poivre himself in his writings), that Poivre managed to rebuild the entire island, the sources I have seen that do not relate to either Poivre's person or official reports from the royal government do *not* confirm this picture. That absence seems more reliable, in that Poivre would never have been able to reconstruct the island within an intendancy lasting only about five years; he might have been able to lay the foundations for the future recovery of the island, but he would certainly have been unable to reconstruct Eden.

The Abbé Raynal summarised the impact of conflicting views as follows:

Mauritius was for a long time much more occupied by the imagination of its possessors than their industry. They are married to the speculations on the use that could be made of it. There are those who want it to be a depot where all the goods from Asia can be stored. These must be transported on boats from these countries and then be loaded on French vessels.[123] . . . But the island lacked

[119] ANOM, C4/25, ff. 40r–42v, Poivre to Praslin, 3 April 1769.
[120] Roberts, 'Le Centre de Toutes Choses', 327.
[121] Later, Dupont would become the second husband of Poivre's wife Françoise.
[122] Dupont de Nemours, *Notice sur la vie de M. Poivre*, 37.
[123] These conflicting views were also closely observed by the Abbé Raynal, see book IV, chs. 31–2 of the critical rerelease of the 1783 version (*Histoire de deux Indes* (1780), vol. 1, 541).

vessels and money, it has no objects of export, no means of import. For all these reasons, the enterprise was unlucky and the island remained nothing more than a purely agricultural establishment.[124]

The Ministry of the Navy and the Colonies had to make financial sacrifices for the Mascarene Islands. According to a calculation of February 1766, the annual expenses came to 768,750 livres for Mauritius and 450,000 livres for Réunion, a total of 1,218,750 livres, an enormous amount of money to invest in islands that were far from self-sufficient.[125] A document from the Ministry of the Navy and the Colonies listing the debts for 1774 reveals that unlike the American and Indian colonies, the Mascarenes were very expensive, costing almost as much as the general navy, the Atlantic colonies, and Pondicherry put together.[126] In 1780, the Abbé Raynal estimated the annual expenses of Mauritius as being up to 8,000,000 livres.[127] Yet the French government – possibly motivated by pride combined with the honeyed tongues of those with vested interests – continued to support its islands, albeit at enormous expenses.

The destiny of Mauritius was a matter of the negotiations and personal patronage of the high officials who supported the maintenance of the colonial islands. While Minister Praslin remained convinced of the geopolitical potential of the island and felt sure that a new approach to Australia would be possible only with the help of Mauritius, the efforts made by the Ministry of the Navy and the Colonies to establish a self-sustaining colony on the island remained limited.[128] For example, in 1770, when Mauritius needed Choiseul's and Praslin's support the most, the cousins were demoted from their ministerial positions. This was at a time when Madagascar was increasingly becoming an object of interest, and the naturalist Philibert Commerson (1727–73) was assigned to explore its natural resources.[129] While Poivre was hoping to make greater use of Madagascar over and above its existing role in the supplies of food and enslaved people for Mauritius, government officials in both the Indian Ocean and Versailles were expressing concerns that Madagascar's resources might become exhausted.[130] At the same time, the officer Louis Laurent Fayd'herbe de Maudave (1725–77) and Maurice Benyowsky (1746–86), the self-proclaimed king of Madagascar, who together enjoyed the support of the French government for their colonising

[124] Ibid.
[125] AN MAR/3JJ/353, 'Etat de la depense des Mauritius et de Bourbon pendant un an suivant les pièces déposés au Bureau des Indes', n.d.
[126] AN MAR/1JJ/111, doss. 17, 'Dettes du Département de la Marine', 25 November 1774.
[127] Raynal, *Histoire de deux Indes* (1780), vol. 1, 541.
[128] NAF 9408, ff. 49r–50v.
[129] See BCMNHN, MS 887; MS 888; MS 279; MS 280; MS 281.
[130] Røge, 'La Clef de Commerce'.

efforts, made new attempts to create a permanent French colony, based on Fort Dauphin in Madagascar.[131] As Lissa Roberts correctly observes, these were more than unfortunate circumstances for the prosperity of Mauritius.[132]

The seizure of Madagascar threatened Mauritius; this is clearly indicated in the *mémoire* by Maudave, who in 1767 had proposed taking possession of Madagascar. He argued that Mauritius had nothing to offer to either the Asian or the European market, and could not provision the sailors and troops landing in Port Louis. Only with a proper settlement in Madagascar could the new colony be secured.[133] But Poivre sent a letter to the minister, deeply concerned by the likelihood that the financing of the project in Madagascar could lead to the abandonment of the projects in Mauritius: he begged for more efforts to be made in relation to Mauritius.[134] Finally in February 1770, the expenses of Maudave's projects in Fort Dauphin created such problems for Mauritius that Governor Desroches too pleaded for their abandonment; they were ruining his island colony.[135] The French colonies were competing with each other, not only within the 'Malagasy triangle' (Madagascar, Mauritius, Réunion) but also with the colonies in the Atlantic (French Guiana in particular). The competition with the Atlantic colonies was intensified by individual actors (and old rivals) who tried to cash in on the vulnerabilities of both Mauritius and its intendant. Lastly, the botanist Fusée-Aublet, a bitter rival of Poivre's, asserted in his *Histoire des plantes de la Guiane françoise* (1775) that the best conditions for the acclimatisation of nutmeg could be found in Guiana, not Mauritius.[136]

Networks of Supply and Enslaved People

Throughout the eighteenth century, the settlers in Mauritius had to face overwhelming problems because of the lack of supplies and the lack of produce cultivated there. Although the island's potential was well recognised by the CIO and later by the royal government, Mauritius was of marginal value for sixty years, particularly in comparison with the French Antilles and their huge profits from the sugar industry.[137] In addition to

[131] On colonial attempts in Madagascar, see Tricoire, *The Colonial Dream*; Foury, *Maudave et la colonisation de Madagascar*.
[132] Roberts, 'Le Centre de Toutes Choses', 329.
[133] Maudave's *mémoire*, printed in Rochon, *Voyage a Madagascar et aux Indes Orientales*, vol. 1, 91–8.
[134] ANOM, C/5A/2, no. 54, Poivre to Praslin, 29 July 1768.
[135] AQ, Cote 12, Desroches's letters, ff. 251v–259r, Desroches to Praslin, 1 February 1770.
[136] Fusée-Aublet, *Histoire des plantes de la Guiane françoise*, vol. 2, 87.
[137] See also Raynal's observations about the state of the island. Raynal, *Histoire des deux Indes*, vol. 1, book IV, ch. 32. The quote refers to the 1770 edition, see p. 719 of the 2010

the supplies of enslaved people and food from Madagascar, imports from other parts of the world such as the Cape of Good Hope, Batavia, and even Guiana were indispensable to the survival of those on Mauritius. The maritime exchange networks were reinforced, exposing both the dependency of the island on its contacts with the French maritime empire and its marginal status.

As mentioned earlier, in wartime of course the situation became even worse. In 1770, during Poivre's directorship, the Anglo-French rivalry in the Indian Ocean seemed to be verging on open war once again, and Mauritius had to be prepared to accommodate numerous French troops. As Minister Praslin could send neither money nor basic necessities, Poivre approached the Dutch VOC at the Cape.[138] As the wartime need for provisions increased so greatly, the supplies from Bourbon, Madagascar, and the African east coast were insufficient, and this required the creation of new networks with the Dutch possessions on the Cape, in Ceylon and Batavia, and with the French Atlantic. For instance, in January 1771, express vessels were sent to the African mainland in order to load up with foodstuffs.[139] But 'express' does not mean 'immediate': the sailing time from the Cape to Mauritius was about three weeks, from Bombay five, and from Madras and Bengal more than seven.[140]

Mauritius's dependency in terms of supply was evident to contemporary observers. Alexander Dalrymple remarked in the 1750s that the island did not produce half enough provisions for the maintenance of its inhabitants. Thus, ships were sent to Bengal and Pondicherry for beef and rice, and above all to Bourbon and Madagascar, trading gunpowder and shot, armies, strong liquor, and coarse clothing in exchange for food.[141] About two decades later, in 1772, the British officer Colpoys reported to Madras that Mauritius was still having to send for food supplies, particularly after the cyclones in March that had destroyed an entire crop and caused considerable damage all over the island. Because of food shortages,

edition. This passage was absent in the edition from 1780, which is the basis of the 2010 critical reprint.

[138] Boullée, 'Voyages de Poivre (2° article)', 60.

[139] According to a report from 25 September 1772, 'Mr. Colpoy's Remarks on the Mauritius, 25 September 1772', BL IOR/H/111, f. 135v.

[140] BL, Add. MS 34450, f. 241r, 'Memorandum respecting the reduction of the Mauritius or Isle of France in the East Indies, London', 14 April 1793. Signed by Adm. Diron.

[141] BL, Add. MS 33765, f. 12r, 'Geographical Collections of Alexander Dalrymple'. See also BL, MSS EUR/ORME OV 4, ff. 78v–79r; Richard Smith, 'An Account of the Island of Bourbon in 1763'. On the connection between Madagascar and Mauritius, see also BnF, NAF 9347, ff. 193r–205v, 'Extrait d'un mémoire intitulé, Iles de France et de Bourbon' (1770–1). As the correspondence of Bourbon with Mauritius clearly reveals, the latter was supplied on a regular basis with maize, coffee, honey, and wax, see ADR C 1573 and 1575 for the years 1748 and 1753, C 1577 and 1579 for the year 1767.

sheep and cattle had to be imported, milk was 1 livre a bottle, and there was no butter 'to be had for love or money'. Fruit and vegetables were:

> extremely scare, beef a livre per pound, and fowl or duck five *livres*[142] ... from what I saw myself I can very easily believe it, besides the heavy complaints which everybody ... made of it, are very sufficient proofs of its poverty.[143]

At the Cape of Good Hope, too, the miserable situation of the French in Mauritius was well known, as confirmed by the Swedish botanist Carl Peter Thunberg in 1773.[144] He concluded that because of their desperate need for supplies, it was:

> the French who most enriched the Cape merchants, as, on account of the credit they took, they were obliged to pay more than others, and at the same time had occasion for a greater quantity of merchandise, not only for their ships, but also for their garrison in the *Isle of France*.[145]

Thunberg was quite right in his observations. In the initial years of La Bourdonnais's governance, the links with the Dutch Cape colony were tightened when he proposed offering wine and wheat in exchange for coffee and ebony. This proposition, however, only materialised during the Seven Years War with the permanent installation of a French agent at the Cape, Jean-Joseph Amat.[146] Amat was also instructed by Poivre to buy wine and strong liquor – indispensable for the hospitals and the incoming troops – and ship them to Port Louis.[147]

The majority of the sources reveal the fact that without the supplies from the surrounding islands and the African mainland, the settlement on Mauritius would soon have been lost without trace. Over the course of the eighteenth century, in addition to the arrival of the many troops and the threat of warfare, it was the natural factors that made island life less than pleasant. In addition to the general food shortages, catastrophes such as cyclones and plagues of locusts destroyed the crops, and famines were frequent.

Instead of focusing on the supplies for the island's inhabitants, the policy was to import more and more enslaved people, who of course needed food, just like every individual on the island. So, the island needed more natural

142 BL IOR/H/111, ff. 131v–132r, 'Mr. Colpoy's Remarks on the Mauritius, 25 September 1772'.

143 Ibid, f. 132r. 144 Hansen, *Linnaeus Apostles*, vol. 6, 101. 145 Ibid., 101–2.

146 I thank Rafaël Thiébaut for pointing me to Amat. On the correspondence between the French agents stationed at the Cape and Mauritius (1772–89), see MA, OA 34A; OA 34 B; OA36A.

147 ANOM, C/4/23, f. 71r, Amat to Duc de Praslin, 4 November 1768. See also ANOM, C/ 5B/1, Amat to Duc de Praslin, 16 January 1769, and ANOM, E/4, 'Instructions au Sr Amat pour sa mission au cap de Bonne-Espérance', 6 October 1768.

resources. In a report from 1764, the number of white inhabitants was 1,883; there were 589 Indians and free Black people, and 15,022 enslaved people, a total of 17,494.[148] Another document on the population of Mauritius in 1766, listing the total number at 20,000, of whom 5,000 were white and 15,000 Black, stated that 'the cultivation of one Black person feeds five persons. 4,000 Black cultivators thus feed the colony.'[149] Poivre wrote in his journal upon his arrival in Mauritius in August 1767:

> In the 10 years of my absence, into this colony they imported a great number of slaves during the war and since the peace, either for a particular commerce, the fraudulent commerce of the Company's officers, or Portuguese ships from Mozambique; however, the population has increased amongst these miserable people since this time.[150]

Relying on the estimate made by the Mauritian archivist Harold Adolphe, the population of enslaved people expanded dramatically and was nine (!) times higher than that of white colonists: in 1766, there were 1,998 whites and 18,100 enslaved people and in 1788, 4,457 whites and 37,915 Black people.[151] The number of enslaved people rose almost exponentially: between 1766 and 1788 their number doubled.

Only recently has it been revealed that the Mascarene Islands played an important role in the slave trade in both the Indian Ocean and the Atlantic world. Between 1768 and 1789, 236 slave voyages were undertaken involving the Mascarene Islands.[152] These voyages started out mostly from Madagascar and Mozambique, but also from the Swahili coast, including Kilwa, Lindi, Mafia, Mombasa, Mongale, Mouttage, and Zanzibar, and also, but to a lesser extent, India. The estimated embarkations of involuntary Malagasy migrants to the Mascarene Islands between 1670 and 1831 come to 136,000, of which about 40 per cent were enslaved people originating from Madagascar's east coast.[153] These significant numbers provide a powerful insight into the contribution of Mauritius to the global slave trade.

Slaving voyages reflect very clearly the explicit link between French mercantile interests and the slave trade. While it remains uncertain quite

[148] AN, MAR/3JJ/353, 'Récapitulation du nombre des habitants de l'isle de france au 30 7. bre 1764'.
[149] AN MAR/3JJ/353, 'Population de Isle de France', 1766.
[150] AD du Finistère, Brest, Fonds des Roches du Dresnay, Cote 1E 3 39, 'Journal de l'intendant Poivre du 5 au 13 août 1767', entry from 6 August 1767, transcribed by Jean-Paul Morel, www.pierre-poivre.fr/doc-67-8-mois-b (last accessed 7 April 2015).
[151] Adolphe, *Les Archives démographiques de l'Ile Maurice*, 6; see also Toussaint, *Histoire des îles Mascareignes*, 75.
[152] I am relying here on the work by Richard Allen. For the exact numbers, see Allen, 'Constant Demand of the French', 50.
[153] Pier Larson makes these estimations. See Larson, 'Enslaved Malagasy', 459–61.

how profitable any specific slaving voyage would have been, the selling of enslaved people to the Mascarene Islands was clearly a highly lucrative trade. Richard Allen suggests that in the late 1760s and 1770s, in comparison to the slave prices along the Malagasy and East African coasts, 'slaves could sell in the Mascarenes for twice their original purchase price, if not more'.[154] Why was that the case? It could be concluded that because of the increasing number of island activities, the local demand for labour was growing so rapidly. Slave traders made a fortune out of the enslaved people taken to Mauritius, and from the 1770s onwards the island served as a staging point for slaving expeditions.

The earliest known such enterprise took place in 1772, when *La Digue* set sail from Mauritius to Mozambique in order to gather human resources for Saint-Domingue.[155] Ever since its settlement by the French Mauritius had served as a port of call not just for naval vessels but also for merchant vessels, including slave traders, and this was particularly so between the 1770s and the early 1790s, with an increase in the number of Atlantic-bound slave ships. There is a lack of detailed information on the exact numbers of vessels, but it is known that between 1776 and 1790 eight voyages were to some extent attached to Mauritius, transporting more than 3,800 enslaved people from Angola and Cabinda to the French Americas.[156]

The importation of the huge numbers of enslaved people has been an understudied and underestimated factor in the island's struggle for food supplies. More and more enslaved people were imported from other parts of the world, and some of them would have escaped and joined those who had escaped earlier. The number of runaways would have increased with every new slaver calling into Mauritius. In addition, more and more soldiers were sent to Mauritius – and not only in wartime. More soldiers meant an even greater demand for food, and this meant an even greater need for additional supplies from Madagascar, Bourbon, and the Cape. In short, the French Indian Ocean colonies cannot be seen in isolation from the wider French Empire, and from the inter-imperial relations with other Europeans overseas.[157]

The administration was running in circles: in order to increase the cultivation of wheat and rice, more enslaved people were introduced, but since these newcomers also needed food and the island was not self-sustaining, the island had to rely heavily on its food networks in the

[154] Allen, 'Constant Demand of the French'. See also Filliot, *La Traite des esclaves vers les Mascareignes au XVIIIe siècle.*

[155] Allen, 'Constant Demand of the French'. For statistics, see Allen, *Slaves, Freedmen, and Indentured Laborers*; Régent, *La France et ses esclaves.*

[156] Allen, 'Constant Demand of the French', 62.

[157] This point is made by Allen, *European Slave Trading*, 22.

region. Yes, a consequence of more enslaved people was indeed an increase in food production, but the concomitant increase in agricultural activities demanded yet more enslaved people. The administration was unable to break this cycle. Even when, after the French Revolution, Mauritius became the seat of the French government in the East Indies, the island still remained desperately dependent on supplies from the African mainland and Madagascar.[158]

Conclusions

The purpose of this chapter was to locate eighteenth-century Mauritius within the highly complex situation of entanglements between the mainland French and colonial actors, which led to the island becoming a test site for experiments in colonial autonomy from the 1760s to the 1780s. Mauritius was a very expensive colony to run. It was managed by the CIO until 1763 and then, in the aftermath of the Seven Years War (and the dissolution of the CIO), was purchased by the French Crown. The main period under consideration is the aftermath of the Seven Years War, with a particular focus on Poivre's time as intendant between 1767 and 1772. This chapter has, furthermore, set the scene for inserting the history of science into the existing histories of eighteenth-century Mauritius and the extent (or rather, lack) of interest in the Mascarene Islands in the French scientific institutions.

There is now a sizable literature on the practices of early modern empires, emphasising decentred views.[159] The Iberian empires were part of what has been considered a 'polycentric monarchy'[160] and they were in fact created by stakeholders[161] while a centralised information system did not immediately lead to more efficient government control.[162] This chapter has focused on that decentred view. All the factors discussed here (the poor fortifications, the dependency on cross-oceanic networks for food, the lack of scientific interest, the massive imports of enslaved people, and the consequences of that policy on the island's environment and infrastructure) went hand in hand with the rivalries between the individual actors and between the French colonies themselves. In Mauritius, as elsewhere, there were ongoing tussles for supremacy between different merchant groups and rival Crown administrators,

[158] BL, MSS EUR/ORME OV 325, f. 9r and BL, MS EUR 707/7c, 'Memoir of the French settlements in the East Indies & Plans for reducing them under the British Government', 9 July 1793, unpaginated. See also BL, MSS EUR D 707/4, Admiral Diron, 'Plan of the Reduction of the French settlement in the East Indies', London, 9 March 1793.

[159] For the history of science in particular, see László et al., *Negotiating Knowledge*.

[160] Cardim et al., *Polycentric Monarchies*.

[161] Grafe and Irigoin, 'Stakeholder Empire'.

[162] Brendecke, *Empirical Empire*.

both in their relations with one another and with the ministers in France. The interplay of people and sites must be revealed in order to create an understanding of the conflicting ideas about what the purpose of the island colony should have been. These ideas related to its potential geographical significance, its symbolic significance and the commercial profit that it could potentially provide for France. Conflicting ideas were often tossed back and forth between the island's two local rulers – the one civic (the intendant) and the other military (the governor). The internal divisions on the island over its management and policy appear to have created assumptions in the historiography that such interests would be divided between policymakers in the metropole and those in the colony, where each has been assumed to represent a unified view.

But in fact, actors working in Mauritius had to try to balance the contradictory instructions from the French state officials and implement them within their own visions which, in turn, were so often destroyed by local circumstances. The acclimatisation of plants of both economic and nutritive value was part of the strategic plan to make the island of Mauritius less vulnerable and less dependent on resources imported from all over the world, though mainly from the Indo-Pacific region. It was the very aim of being independent of the foreign provision of food and basic necessities that resulted in exactly the strategies of acclimatisation which this book brings to light.

The tensions on the island because of its maritime location made it vulnerable and marginal; the local administration tried to overcome this by increasing the cultivation of domestically produced livestock. These problems were not solved, however, because there was a lack of food production and provision, exacerbated by the enormous number of enslaved people imported against their will into the island. Indeed, as mentioned earlier, there was a deep twofold relation between the huge slave population and food crops. On the one hand, enslaved people had been purchased to work the island, but their high number of course became a problem because of the lack of provisions. On the other hand, the slave trade was inevitably linked to the travel and cultivation of certain food crops, and the lack of food would eventually be overcome by the augmentation of the agricultural activities, as I explore in depth in Chapters 2 and 3.

While food crops were to be domesticated in order to feed the island's population, spices were to be planted as cash crops, as set out in more detail in Chapters 5 and 6. Those two projects coexisted, merged, and clashed on the island, leading to tensions because of the conflicting ideas about what the purposes of the island colony should be. These ideas were related to its potential geographical significance, its symbolic significance and the financial profit it could potentially make for France. The attempts to acclimatise the two kinds of crops cannot be regarded in isolation since they intermeshed, just as their collection went hand in hand with the collection of other plants in the Indo-Pacific and beyond.

Overall, the island relied heavily on imported labour (predominantly comprised of enslaved people). Mauritius not only needed to become self-sufficient but also to be able to supply provisions to the many ships that stopped off en route between South and East Asia and Europe. The inability to produce enough food was exacerbated in particular during wartime, as the island could easily be blockaded; Mauritius was a very fragile, primarily geopolitical, project. Though Réunion was (and still is today) much more fertile than Mauritius, it lacked good enough ports to accommodate the royal vessels. Because, however, the Dutch had established two ports in Mauritius before abandoning it, the island had become of particular interest to the geopolitical and imperial projects of the French Crown. The idea was that vessels could anchor at Mauritius while being supplied from Bourbon, a plan that worked well in theory but not in practice; Réunion could not produce enough for two islands, especially not when troops stopped over.[163] In late 1767, Poivre ended a letter to Réunion's director Crémont about the misery of Mauritius and the lack of food, as follows: 'Adieu, my dear Ordonnateur, I do not wish to keep going on about the poverty of this island, of which you have enough of your own.'[164] The imperial–geopolitical project of Mauritius failed in the long run: the desperate cultivation activities could not break the vicious circle created by the lack of provisions.

Most importantly, this chapter has illustrated the underlying reasons for the subsequent cultural and botanical diversity of the island. Likewise, as I have underlined, the island lacked plants that might be used as human foodstuffs or as animal fodder, or that could be cultivated for economic profit. Thus, both the human population and the flora and fauna that might support that population had to be imported and acclimatised. As the following chapters examine in greater detail, the arrival of the flora was not coordinated with that of the humans who might possess the relevant knowledge about how to cultivate them. The circumstances meshed with the environmental factors and the many discrete approaches to handling natural materials both on and beyond the island, creating an extraordinary diversity as the years went by.

[163] ANOM, C3/12–15. See also BSG, MS 1085.
[164] AD Réunion, 12 C, Poivre to Crémont, 16 November 1767.

2 The Acquisition of Knowledge and Plants, from Madagascar to China

This chapter is about the search for the plants intended to make Mauritius more resilient, sustainable, and self-sufficient in its agricultural productivity. I examine the exploration, collection, and understanding of the plants and the related practical knowledge of their cultivation and utilisation in the Indo-Pacific. Mauritius itself had barely anything to offer in terms of food for either humans or livestock, since no staple crops were native to the island. So, the potential food crops had to be collected from elsewhere. French naturalists, commercial agents, engineers, and naval officers were sent to several parts of the world in order to gather useful plants and the relevant specialised knowledge.

The common, and very simplistic, narrative about Mauritius is usually based on the search for plants of economic value: nutmeg and cloves.[1] Yet for the local administration it was actually much more important to introduce and cultivate food crops and other plants for subsistence, in order to turn the island into a self-sustaining colony. That aim, however, was never achieved, and that might explain why in the historiography the role of agriculture has been largely overlooked.

As mentioned in Chapter 1, under the governorship of La Bourdonnais several types of staple crops had been introduced to the island from the mid-1730s onwards. It was only from 1766 onwards, however, when Poivre was present, that cultivation itself would play a key role in the hope of making the island more self-sufficient. His dual role as intendant and agronomist led him to push for the import and acclimatisation of spices, and this clashed with the agricultural projects vital to the islanders' subsistence, and the everyday use of plants, as Chapter 3 elaborates. Above all, Poivre's physiocratic ideas relating to reliance on nature as a food source could only be put into action once enough plants had been collected from around the Indo-Pacific and acclimatised in Mauritius.[2]

[1] Ly-Tio-Fane, *Mauritius and the Spice Trade*, 2 vols.; de Fels, *Pierre Poivre*; Le Gouic, 'Pierre Poivre et les épices'; Piat, *L'île Maurice*.

[2] On physiocratic discourse and improvement, see also Lacour, *La République naturaliste*, 91–6. On the different agendas of physiocracy and cameralism as well as political economy

This chapter, then, is primarily concerned with what happened outside the island: the collection of plants suitable for food crops, and the related knowledge embedded in cross-cultural interactions.

In Mauritius and elsewhere, plant transfer was clearly more than an intellectual exchange; the Europeans were attempting to make economic use of the plants that had been transferred. The naturalists on eighteenth-century Mauritius in particular were acutely aware of the value of useful plants; Bernardin, for instance, exclaimed: 'The gift of a useful plant seems to me more precious than the discovery of a gold mine and a monument which lasts longer than a pyramid.'[3] The physician-naturalist Commerson, who explored the South-West Indian Ocean littoral for its flora, was convinced that experience was the main criterion for identifying the relevant plants. He declared:

> In a completely unknown country, it suffices an experienced glance to pronounce: here, we have an edible fruit; there, a vegetable good for cooking; there again starchy foods out of which we could make bread. Here, it will be a remedy similar to another already well-known; there, finally, a poison of which we should keep our hands off.[4]

The 'experienced glance' was not, however, sufficient if you wanted to avoid the risk of poisoning yourself. So, you had to rely on the expertise of those who lived in countries hitherto barely known to Europeans.

This chapter examines the various actors involved in plant acquisition, operating as they did in both directions across the Indian Ocean. It highlights the geographical regions in the Indo-Pacific and beyond where the plants, and the knowledge relating to them, were found. The main regions concerned were India, Cochinchina, China, Madagascar, Mozambique, and the islands 'beyond the Ganges'[5] – the Maluku Islands. Analysing in-depth encounters on the ground, I examine the local agency of a range of Indo-Pacific actors on a broader scale, both regional and cross-cultural. I consider intermediaries in different localities and with different knowledge traditions, arguing that local knowledge was negotiated in diverse and highly complex cross-cultural situations.

Whereas in the Indian, Chinese, and Arab-Islamic world, and particularly in the African context, studies in global science are emerging, the extent to which Europeans relied on local knowledge and *how* they did so

and agronomy, see Jones, *Agricultural Enlightenment*, 14–31; Stockland, 'Policing the Oeconomy of Nature'.

[3] Bernardin, 'Voyages à l'Isle de France', vol. 1, 62.

[4] Commerson, 'Sommaire d'observations d'histoire naturelle', 132.

[5] ANOM, C/4/22, Poivre to Praslin, n.d. [1768?]. See Provost's list of collected plants, MA OA 116.

is still unclear.[6] Elizabeth Green Musselman suggests that in the African region the Europeans relied much more on local knowledge because the region had not been heavily colonised prior to the mid-eighteenth century.[7] Then in eighteenth-century Madagascar and South East Asia, most of the plants were unknown to Europeans, so they had to rely even more on knowledge provided by the people who were familiar with them.[8] As this chapter shows, the search for plant-based practice knowledge was fundamentally shaped by the activities of agents from a wide range of backgrounds – but these tended to be missionaries, marine officers, and local mediators rather than naturalists trained in plant knowledge. A close examination of these encounters will help us to understand how people communicated about plants and named them; mediation and knowledge transmission are closely linked to the questions of language, naming, and negotiation practices.[9] Here, know-how depended on implicit or tacit knowledge between individuals, which when observed consciously would result in descriptions. Thus, overt communication about a plant was the key to documenting, identifying, and using it. Yet, when plant knowledge was turned – to use historian Harold Cook's term – into a 'matter of fact', the sources of the knowledge were erased from the resultant public records, as were the identities of the knowledge providers.[10]

I have set this chapter out in several steps, in order to help clarify the entanglements of the various encounters. First, I examine the agents in the Indo-Pacific, ranging from the European engineers to the Mandarin mediators; this section highlights the human diversity and the complexity of the networks that spread across the Indo-Pacific. Second, I explore human encounters in Madagascar, Cochinchina, and China raising, the question of language, negotiation, and diplomacy, closely linked to the third aspect: naming, identifying, and documenting unknown plants. I show that when the French tried to make sense of plant knowledge, they made use of their own experience, comparing it to, and enriching it with, local knowledge. Here, instead of explicit knowledge, it was tacit knowledge that – once it had been observed and described by Europeans – played such an important role, because of the language barriers. From there, I look at the types of knowledge and the types of plants collected,

[6] Musselman, 'Plant Knowledge at the Cape', 367–8. Musselman also shows examples of Africans implementing European agricultural knowledge in their own lands.

[7] Musselman, 'Indigenous Knowledge and Contact Zones', 32.

[8] Victor Savage puts forward a similar argument with regard to cartography and plant knowledge ('Southeast Asia's Indigenous Knowledge', 257). For cross-cultural science in East Asia, see Fan, *British Naturalists in Qing China*.

[9] On the importance of language during negotiations across different contexts, see Couto and Péquignot, *Les langues de la négociation*.

[10] Cook, 'Global Economies and Local Knowledge', 100–18.

arguing that the search for plants was fundamentally driven by utilitarian botany. Finally, the chapter analyses practical global rice cultivation in relation to French agrarian thinking.

Non-Naturalists

The key elements of the plant-related endeavours in the service of Mauritius were located in many different parts of the world. The actors in the eighteenth-century Indo-Pacific, working without any outside help, had to adapt to many new situations, including their interactions with local populations. Mauritius was left to struggle with its self-organisation, without barely any assistance from France. As the naturalist Pierre Sonnerat (1748–1814), a relative of Poivre's, observed, the vegetables and fruits growing in Mauritius were not native to the island but had come, together with numerous travellers, from China, the Indies, the Cape of Good Hope, and Europe.[11]

Poivre, meanwhile, in his role as intendant and agronomist, employed actors of every kind wherever and whenever he could. As explained by the astronomer Alexis-Marie de Rochon (1741–1817), who imported food crops and useful plants from Madagascar to Mauritius in 1768, Poivre never missed an opportunity to instruct captains and officers to bring live plants 'as souvenirs' from their voyages.[12] He would initiate the collection of useful plants by asking for personal favours and by engaging naturalist voyagers and captains. The horticultural project of Mauritius was built on the informal relations inherent in both local cultures and between other Europeans in the region. Poivre skilfully used the channels that the government had been unwilling or unable to provide, in terms of both finance and ships.

The case of Commerson, who in 1770–1 explored Madagascar to study its flora, is a good example of the powerful and complex state relations built on local negotiations. His biographer, Antoine Cap, suggests that Poivre had been given the task of persuading Commerson, who was in the region because he was the botanist on Louis-Antoine de Bougainville's circumnavigation, to explore Mauritius and Madagascar thoroughly, specifically because of the increasing exploitation of Madagascar's resources, as I explained in Chapter 1. Commerson, in a letter to a friend of his, the astronomer Lalande, explained that Poivre had aroused his interest in exploring the natural world and had then persuaded him to

[11] Sonnerat, *Voyage aux Indes orientales et a La Chine*, vol. 2, 85.
[12] Rochon, *Voyages à Madagascar*, vol. 1.

explore Madagascar.[13] Very probably, however, Poivre had seen a chance to promote his own policies by using Commerson as a weapon to counter those in France who did not share his ideas of building an agricultural island state.[14] Both Bougainville and his expedition doctor, the powerful Académie member Pierre-Isaac Poissonnier (1720–98), supported Poivre's idea, and together they ensured that Commerson's stay was justified.[15]

Although the official reason for Commerson's stay in Mauritius was for him to gather knowledge of plants for nourishment and medicine, he was in fact more involved in undertaking a survey of the flora of the Mascarene Islands and Madagascar, where he discovered more than 3,000 species and sixty new plants.[16] He engaged deeply with the flora of Mauritius and built up a herbarium. The hospitals needed remedies, especially after the failure of the garden at Le Réduit (see Figure 3.1). Rather than Praslin, it was Poivre who designed Commerson's mission to explore northern Madagascar before Fort Dauphin was abandoned at the end of 1770.[17] By disguising Commerson's real purpose – designed by Poivre – on the islands, the two were playing a subtle game whose risks Commerson was well aware of.[18] In return, he tried to make use of Poivre's channels by, for instance, asking the naval officer Simon Provost to collect some live plants for him, while Provost, in turn, was instructed by Poivre to collect spice plants.[19] This shows how intricately the networks were linked by personal favours and informal requests.

As mentioned earlier, in those botanical networks the types of naturalist varied significantly; some of them were missionaries, and many were trained as physicians or apothecaries. The trading companies, the Crown, or the scientific academies sent most of them.[20] Only after 1770 – that is to say, after the period I am focusing on – did a new type of a state naturalist begin to appear: specialists, funded by the Crown, whose remit was to gather botanical knowledge and specimens from overseas. These men would sometimes stay at their stations in the colonies for quite a long time.[21]

[13] Cap, *Philibert Commerson*, 173, 16. Commerson to Lalande, 18 April 1771.
[14] Grove, *Green imperialism*, 217.
[15] ANOM E/89, Poissonnier to Praslin, 7 April 1770, f. 8.
[16] Société de l'histoire de l'Ile Maurice, *Dictionnaire de biographie mauricienne*, vol. 1, 116, entry on Commerson. On Commerson's archive, see Béltrán, 'Scribal Practices'.
[17] Undated letter, printed in Cap, *Philibert Commerson*, 106.
[18] Commerson to Lemonnier, 1 May 1772, ibid., 163.
[19] Commerson to Cossigny, 19 April 1770, ibid., letter 3.
[20] Schiebinger and Swan, 'Introduction', 10.
[21] McClellan and Regourd, *Colonial Machine*, 307.

Notably, in the case of Mauritius it is necessary to rethink the very nature of a trained botanist. Not even Poivre was trained: he was merely a self-proclaimed botanist who relied on a network consisting of engineers (such as Bernardin), some actors trained in medicine (such as Commerson), naval officers (such as Provost), missionaries (such as Galloys), and astronomers (such as Rochon) who were active in the region. In brief, nearly all, if not all, of the eighteenth-century agents in Mauritius were amateurs who relied on the complex networks of anonymous providers of knowledge and materials.

In the various countries, the naturalists would look for various types of plants, depending on their status and their uses. In China, for instance, Poivre, familiar with local products because of his travels in East Asia in the 1740s and 1750s, was able to give detailed instructions to the Abbé Galloys, who was sent to that part of the world in 1766. Commerson, Rochon, and Galloys were not the only individuals Poivre employed. His nephew, Pierre Sonnerat, was attached to Commerson, and joined him on several expeditions to be trained in botany and drawing. Sonnerat acquired his drawing skills from the engineer Paul Philippe Sauguin de Jossigny (1750–1827), who accompanied Commerson during his voyages as his draughtsman.[22]

Sonnerat, born of a middle-class Lyon family and son of Poivre's first cousin Benoîte Poivre, was soon introduced to the adventurous world of his uncle in Mauritius, in that Commerson's arrival there in 1768 was a turning point in Sonnerat's career.[23] Sources reveal that Commerson and Sonnerat worked together for three years in the Mascarenes and Madagascar. However, some other contemporary evidence, such as letters by Marie-François Eloy de Beauvais (1743–1815), veterinarian in Mauritius, do not make an intensive collaboration between the two seem very plausible.[24] Although Sonnerat claimed that he accompanied Commerson on his expeditions to the Mascarenes and Madagascar, chronological and contemporary evidence seem to disprove this assertion.[25] In the documents related to Commerson's exploration of Réunion, it is not Sonnerat but Jossigny who is mentioned as his companion.[26] However, other documents related to Commerson

[22] ANOM, E 231, Jossigny's file, ff. 102, 116.

[23] Ly-Tio-Fane, *Pierre Sonnerat*, 7. According to Commerson's unpublished *mémoires*, the exact date of his arrival was 25 August 1768. See BCMNHN, MS 888, f. 1r.

[24] See on Beauvais, Heintzman, 'Cabinet of the Ordinary'; Dalbine, *Un vétérinaire sous les tropiques*.

[25] DSB, vol. 11, 536.

[26] Ly-Tio-Fane, *Pierre Sonnerat*, 10; Laissus, 'Catalogue des manuscrits de Philibert Commerson'. There is a reference to Jean-Baptiste Lislet-Geoffroy as Commerson's companion.

include observations apparently made by Sonnerat on the flora of Réunion.[27]

Bernardin, who explored Mauritius's natural resources between 1768 and 1770, had been trained as a military engineer. Malcolm Cook suggests that Bernardin started to see geographical locations with the eyes of a skilled observer.[28] However, as he had never been trained as a naturalist it seems inappropriate to consider him as one.[29] In fact, in his famous *Voyage à l'Isle de France* he clearly states that he did not consider himself a naturalist, adding that he did not know anything about botany and that he simply described things as he perceived them.[30] This was, of course, also a way of self-fashioning.

In a nutshell, very few actors in Mauritius's botanical network had actually had any training whatsoever in the knowledge of plants. So, when the Europeans arrived in each new world of unknown flora they had to rely on their local informants. Language was of course a serious barrier, so interpreters or go-betweens were indispensable. For example, when the naval officer Provost was given the mission of acquiring spices in 1768, he also gathered seeds and live plants suitable for nutrition, timber in Kedah (Malaysia), such as the oil tree (the '*arbre Bray sec*') and fruits: *yangomme*, *quiampeda*, mangosteen, bergamot, and others. He most probably achieved this with the help of his local go-between, the dervish Hadé Hachem, who also helped him buy spices. In Aceh (see Figure 5.1) which was also used as a transit hub by the French during their spice quests of 1768–70, Provost loaded his ships with the mangosteen, the rambutan, and the langsat, as well as *mamplan*, *manquian*, *moulin*, *fatoul*, and *sanasse* – species I have been unable to identify.

In the same way, Poivre, during his travels in Asia in the 1740s, established several personal relations with local go-betweens, such as a man whom he called 'On doï' or 'the little Mandarin'.[31] A report on Poivre's activities in Cochinchina provides more insights into On doï: he was the uncle of the second wife of Cai An-Tin, a man who held an important mediating position when the CIO official Friell had arrived to Cochinchina in 1743.[32] It appears that On doï had been of 'excellent service' since the first day of Poivre's arrival, and that he had been the only

[27] BCMNHN, MS 1904, XIII. 'Description des plantes observées dans le voyage au volcan de Bourbon', which also contains 'Nottes de quelques plantes observées dans l'isle de Luçon pas Mr. Sonnerat' (7 folios). The exact date is not given.

[28] Cook, *Bernardin de Saint-Pierre*, 36.

[29] This is stressed by Crestey, 'Bernardin de Saint-Pierre', 354.

[30] Bernardin, 'Voyage à l'Isle de France', 63.

[31] Poivre's journal, entry from 30 January 1750, ANOM C/1/2, ff. 161–215, 'Journal d'un voyage à la Cochinchine'.

[32] Launay, 'Rélation de la Persécution de Cochinchine en 1750 par Mgr Lefebvre'.

person who had assisted the Europeans with true interest and politeness.[33] That is why Poivre, prior to his return to Mauritius, sought to offer On doï a gift in the name of the CIO .

Another example is Poivre's interpreter, 'Miguel' or 'Michel' to the French, whose true name I could not trace.[34] He was a young Cochinchinese who had previously worked as an interpreter for the Portuguese. However, Manuel Matthaus, captain of the Portuguese vessel *Saint-Louis*, had discouraged Poivre from engaging him; Miguel had a bad reputation, was known for his multiple identities – and, while asking for great sums for his efforts, had also tried to play the Portuguese in Macao against the French in Pondicherry before then seeking his fortune with the British.[35] Clearly, he had tried to take advantage of the shifting presence of the Europeans in the region, seeking new alliances to achieve his personal goals.

Despite the captain's warnings, Poivre, hopeful of the advantages that hiring Miguel might well give him, took him on. Poivre reported that he had given Miguel some lessons in sketching, and that Miguel had used his new skills to introduce himself to the king via a 'Black servant', a 'great favourite of the prince', who, interested in improving his art skills, had taken Miguel on as his protégé, facilitating his entry to the palace.[36] Poivre, after making promises to Miguel prior to departing, sent a letter to him in Chinese, albeit using the Latin alphabet, or so he claimed, then revisited Cochinchina in a second attempt to make good there, at which point he was received at court with open arms. So, it was through Miguel that Poivre was eventually welcomed by the king.

The interactions between the Indigenous and local populations and the Europeans were far from a one-way process, for two main reasons. First, they were determined by the willingness of these populations to communicate and share their knowledge. Second, the local populations naturally sought to make a profit from their interactions with the Europeans. This is illustrated by a comment on trade in Cochinchina made by a French missionary, M. d'Azemat. Upon Poivre's arrival in Cochinchina on 30 August 1749, probably anchoring in Faifo (present-day Hội An, Vietnam), he was sent directly to Père Azemat, who warned Poivre against the idea of setting up commercial relations with the Cochinchinese because, or so Azemat claimed, they pursued only their own interests while making empty promises.[37]

[33] Ibid. [34] ANOM C/1/2, ff. 161–215, 'Journal d'un voyage à la Cochinchine'.

[35] Launay, 'Rélation de la Persécution de Cochinchine en 1750 par Mgr Lefebvre'.

[36] Ibid. See also the correspondence between Lefebvre and Poivre printed in the same volume, 323–32.

[37] ANOM C/1/2, ff. 161–215, 'Journal d'un voyage à la Cochinchine depuis le 29 aoust 1749, jour de notre arrivée, jusqu'au 11 février 1750'; Launay, 'Rélation de la Persécution de Cochinchine en 1750 par Mgr Lefebvre'.

While this section has made clear the wide variety of agents in many of the locations involved in the process of plant knowledge making, the following section will examine some actual encounters between the various actors, linked to the question of diplomacy and communication.

Diplomacy, Language, Negotiation, Misunderstandings

How did French-speaking actors communicate and negotiate with non-French-speaking ones? Let us explore this question through various examples, starting with French-Malagasy encounters. On Madagascar, the French actors would approach local populations and vice versa, as illustrated in Commerson's botanical exploration of Madagascar. He had received orders from the governor, Desroches, and Poivre to explore Madagascar for its natural history, and he was to be accompanied by Jossigny as his draughtsman.[38] Commerson added in his *mémoire* that Poivre had assured him he would receive the best possible equipment for his journey.[39]

But this equipment was clearly not going to be enough; he would need local help. When he anchored in Madagascar in September 1768, he was welcomed by a Malagasy chief, Dian Mananzar, who presented fruits and three bullocks, and offered to provide assistance in the exploration of the big island.[40] While in the maritime world the presentation of newcomers with refreshments and water would usually herald a trade relationship,[41] this specific deal was set up in a specific way. First of all, glasses were clinked. The Malagasies offered 'une grande tasse pleine d'Eau de vie' ('a big cup filled with liquor'), which everyone drank from;[42] Dian Mananzar was apparently a jolly fellow who enjoyed his liquor. Then one night, on 7 October 1768, the two had dinner and the chief's mood became mellower and mellower thanks to 'the help of a little wine and eau de vie' as Commerson commented.[43] Then, not for the first time, the chief offered Commerson his wife, an offer Commerson reported he

[38] BCMNHN, MS 887, doss. II, f. 44. On the eighteenth-century practice of preparing and engraving objects, see Mariss, *'World of New Things'*, 230–70. On Jossigny, see also Bour, 'Paul Philippe Sanguin de Jossigny', 415–48; and his file in the colonial personal series ANOM E 231.

[39] BCMNHN, MS 887, doss. II, f. 44r.

[40] BCMNHN, MS 888, f. 3, 'Mémoire pour servir a l'histoire naturelle & politique de la grande isle de Madagascar anciennenent appelée isle de St Laurent par les Portugais qui la decouvrirent les premiers l'an'.

[41] Bertrand, 'Spirited Transactions', 45.

[42] BCMNHN, MS 888, ff. 3–4, 'Mémoire pour servir a l'histoire naturelle & politique de la grande isle de Madagascar anciennenent appelée isle de St Laurent par les Portugais qui la decouvrirent les premiers l'an'.

[43] Ibid.

refused, concluding that these people had a seemingly 'bizarre custom' of offering strangers their women and girls.[44]

Dian Mananzar was a businessman who wanted to arrange a profitable deal. For the exploration of the island, he offered to hire his enslaved people to the French. Since Commerson and his crew considered his price too high, Dian Mananzar had a new idea: to order the enslaved people of Malagasy slave owners to assist the French explorers. But the next day only three out of the promised two hundred turned up.

This episode merits some clarification. First, according to Maudave's *mémoire* on the project of colonising Madagascar, its labourers were not like the enslaved people in Mauritius, where 'all the work in this colony is conducted by slaves'.[45] Second, Dian Mananzar, hoping to make a fortune, had asked for a high price for his own enslaved people. Third, his people did not obey his orders, refusing to provide their labourers for nothing. Later, in a different part of the island, some Malagasies set up their own individual arrangements – not with enslaved people but with animals, selling ten bullocks to Commerson.[46] The Malagasies were very keen on private trade with the French.

During his exploration of the island, Commerson encountered more chiefs who offered practical assistance as well as help with plant knowledge. In a small village he met a chief, Ramainon, who spoke French quite well and who was, according to Commerson 'more civilised, more honest' than the others he had met.[47] Commerson also received information about useful plants, especially about the cultivation and the value of the timber which grew in the northern parts of the island.[48] In his 'Notte de quelques plantes et fruits de Madagascar et de leurs propriétés' ('Note on some Malagasy plants and fruits and their properties'), relying on the chief Ramainon, the Frenchman referred to a certain herb which was not good for cattle, and listed other native names for plants.[49] These examples show that the exploration of Madagascar's natural world was highly dependent on sociocultural interaction, local expertise, and the willingness of the Indigenous people to give such information.

[44] He had also encountered this custom in the Pacific during Bougainville's circumnavigation.
[45] Maudave's *mémoire*, printed in Alexis-Marie Rochon, *Nouveau Voyage à la Mer du Sud*, vol. 1, 91–8.
[46] BCMNHN, MS 888, f. 4r, 'Mémoire pour servir a l'histoire naturelle & politique de la grande isle de Madagascar anciennenent appelée isle de St Laurent par les Portugais qui la decouvrirent les premiers l'an'.
[47] Ibid, f. 11v. Entry from 30 September 1768.
[48] Ibid, f. 47r. Entry from 10 November 1768.
[49] Ibid., f. 11v, 'Notte de quelques plantes et fruits de Madagascar et de leurs propriétés', 1769.

It was a matter of local negotiation practices, friendly encounters, diplomacy, and, above all, language. Some Malagasies spoke a little French, which enabled Commerson, who could not speak Malagasy, to communicate with them. So, it was the Malagasies' knowledge of French that enabled them to communicate with him. Successful encounters were also facilitated by go-betweens, such as Commerson's interpreter Antoine Renard, whom he called 'René'.[50]

The importance of learning foreign languages becomes clear on examining French–Asian collaboration at about the same time. French agents sought to negotiate with the Maluku sailors who frequently entered the ports of Kedah (in today's Malaysia).[51] Poivre indicated in his secret instructions to the French naval officer Provost that in Malay nutmeg was called *pala* and cloves *digne* – words that he must have picked up during his own travels in the 1740s and 1750s, or perhaps in Rumphius's *The Ambonese Herbal* (1741).[52] In modern Malay, *pala* does indeed mean nutmeg, but cloves is translated as *cengkih*. *Digne*, meanwhile, was probably the French way of spelling *chenge(h)*, *chengkeh*, or *chenke(h)*, an earlier form of today's *cengkeh*, cloves. In early modern Dutch the spelling was *tjengke*.[53] In the language of the Bugis (a South East Asian people), cloves is *imokimong*: this was the word used in Celebes and thus used by Dinck Poudony, a Bugi pirate, who was supposed to collect nutmeg and cloves for the French.[54] Thanks to his ability to speak both Malay and Dutch, Provost could interact with many islanders directly and could, therefore, more easily acquire information about the spice plants as well as the Dutch vessels omnipresent in and around the Moluccas. A 'French–Malay' vocabulary list, which Provost shared with his collaborator Cordé, contained various useful phrases such as '*N'y a-t-il aucune nouvelle de quelque vaisseau?*' ('Isn't there any news of shipping?'), translated into the Malay '*Tida Gabar daron derry Darang capal*'.[55]

[50] Ibid., ff. 11r, 38v, 44v, 'Mémoire pour servir a l'histoire naturelle'.
[51] ANOM, C/4/22, letter no. 147, Poivre to Praslin only, 18 December 1768.
[52] I explore the French use of Rumphius's *Ambonese Herbal* in greater detail in Chapter 6.
[53] My thanks to Peter Boomgaard for pointing me to these different spellings and distinct vocabularies.
[54] ANOM, C/4/22, ff. 119r–121v, 'Instructions secrètes pour Mrs. Trémigon, lieutenant des vaisseaux du Roi, commandant la corvette Le Vigilant et Provost subrécargue sur ledit bâtiment pour le voyage de Quéda et d'Achem', 4 February 1768. See also the instruction on how to distinguish the plants: ANOM, C/4/22, f. 127r, Poivre, 'Description abrégée du muscadier et du géroflier pour servir à mettre les Srs Trémigon et Provost dans le cas de n'être pas trompés dans le choix des plants de ces deux espèces d'arbres'.
[55] ANOM, C/4/29, f. 245r, Provost, 'Instruction secrète pour M. Cordé commandant la corvette du Roi le Nécessaire, au cas de séparation', 25 December 1771.

Islanders would often present refreshments upon a vessel's arrival. When Provost reported on this hospitality, Governor Desroches forwarded an extract of Provost's journal to Versailles in July 1770. In it Provost had stated that although the Malays were said to be the most malicious people on earth, he had instead met with humanity, friendship, and reliability.[56] The tactful, even friendly, relationship between the French actors and the Malay-speaking locals in the region can also be illustrated through the 'French–Malay' vocabulary list; '*mon ami*' ('my friend') was translated as '*mohabat*'.[57] There are indeed some scholarly works which mention these friendly relations, such as Vanessa Smith's study on eighteenth-century Pacific explorations.[58] Cultural and scientific diplomacy clearly played a significant role in the natural sciences. This has also been shown by Simon Schaffer in the context of scientific instruments and the English in China, and Sarah Easterby-Smith in the context of diplomacy and plants as gift exchange.[59]

In brief, communication was not only a question of the spoken word and the ability, or lack of it, to speak and understand a foreign language. It was more deeply linked to cultural diplomacy and ways of gaining trust. Only then did the communication about plants, as detailed in the next section, become possible.

Communicating about Plants

How did actors identify plants, and what names were employed? What naming and identification strategies were applied? The difficulties attendant upon the use of native names of plants have been discussed in works which touch on that issue. For instance, in *The Names of Plants*, David Gledhill provides a sustained discussion and historical account of the significant changes and problems of plant nomenclature in terms of botany and agriculture.[60] Anthropologists in particular have focused on the differences between European and non-European modes of classification, in particular Scott Atran in *The Cognitive Foundations of Natural History*, in which he explored the

[56] ANOM, C/4/26, ff. 168r–179r, Desroches to Praslin, 22 July 1770.
[57] ANOM, C/4/29, f. 245r, Provost, 'Instruction secrète pour M. Cordé commandant la corvette du Roi le Nécessaire, au cas de séparation', 25 December 1771.
[58] Smith, *Intimate Strangers*. On intercultural exchanges in early modern South East Asia, see also Alberts and Irving, *Intercultural Exchange*; and in relation to natural knowledge in particular Sargent, 'Global Trade and Local Knowledge', 144–60.
[59] Schaffer, 'Instruments and Cargo', 217–45; Easterby-Smith, 'On Diplomacy and Botanical Gifts'.
[60] Gledhill, *Names of Plants*.

emergence and development of natural history from common conceptions of folk biology, and Clifford Geertz in his essay, 'Common Sense as a Cultural System'.[61] The difficulties of using native names have not been the subject of any profound study, except for Geoff Bil's recent research on Maori and Polynesian plant names for Pacific botany.[62]

As far as historians go, in the context of the British botanist Joseph Hooker, Jim Endersby's study on the practices of Victorian science touched on the naming difficulties, although primarily with respect to the gap between the European and the colonial contexts.[63] Minakshi Menon explored Sir William Jones's strategies to create culturally specific plant descriptions in India, leading to a method falling somewhere between Linnaean plant–object terminology and local knowledge of plants.[64] Alexandra Cook, looking at the presence and absence of 'local names', and Chinese names in particular within the Linnaean system, tested the thesis of linguistic imperialism. She concluded that while Linnaeus had used generic names of diverse origins, he had misidentified the habitats of certain Chinese plants.[65]

Historian of science Šebestián Kroupa explored several strategies in naming and describing a plant undertaken by the Jesuit pharmacist Georg Joseph Kamel, stationed in Manila at the turn of the eighteenth century. Kroupa highlighted that these strategies either drew on a comparison with a commonly known plant, hence making a reference to a certain author, or adopted a well-established reference. If these strategies failed, Kroupa argued, Kamel would record the native name followed by a simple description.[66] Here, Kroupa suggests that the use of local names indicates two important things. First, the inclusion suggests direct contact with local populations, in order to bolster the authority and credibility of Kamel's account. Second:

> synonomata presented a useful comparative tool, which scholars could employ as a common reference in communication or to trace identical plants across different works.[67]

Kamel also included names in different languages, so that they could be of more general use. While it was important to know, trade with, and use the

[61] Atran, *Cognitive Foundations*; Geertz, 'Common Sense as a Cultural System', 5–26.
[62] Bil, *Indexing the Indigenous*. [63] Endersby, *Imperial Nature*.
[64] Menon, 'What's in a Name?'. [65] Cook, 'Linnaeus and Chinese Plants'.
[66] Kroupa, 'Georg Joseph Kamel', 11–12, unpublished manuscript, by permission of the author. I thank Šebestián for providing a copy of his unpublished work.
[67] Ibid., 12.

local names during encounters with local populations, Kamel's primary aim was to establish a classification system to:

> appropriate local plants into a European body of knowledge and thus convey a representation which would enable his correspondents to better imagine and understand the unfamiliar Philippine flora.[68]

Other cases show that some Europeans would transliterate Indigenous names into the Latin script. Historian of science Huiyi Wu, for example, made the point that in the context of the Jesuit translation of texts related to Chinese plants, the eighteenth-century botanists found it more useful to transliterate plants' names than translate them. This practice was exemplified by André Thouin, head gardener of the Jardin du Roi, who instructed missionaries in China to follow that strategy. The transliterations, as Wu observes, thus referencing Joseph Needham, would tend to be 'taxonomical in quality', because a plant's name quite often related to a linguistic component referring to a more general category, that is, a particular type of species.[69]

Transliteration of names was indeed a common strategy, as confirmed by an example relating to Tamil names. Commerson also sought to collect useful trees and fruits from India; these were provided by Joseph-François Charpentier de Cossigny de Palma (1786–1809), son of the engineer Cossigny, who included some 'Malabar' (Tamil) names, and other plants whose names are written in French, including sandalwood, orange, ginger, lemon, jasmine, durian, white betel, and the palm tree.[70] When Cossigny sent specimens from Pondicherry to Mauritius, he noted: 'The names written in Malabar characters are the names of the country, they are transliterated in our European characters.'[71] To the catalogue, Commerson added explanations of some Malabar words frequently used to name plants.[72] The strategy of identifying and naming plants

[68] Ibid. See also Müller-Wille, 'Collection and Collation', 541–62; Koerner, *Linnaeus*.

[69] Wu, 'Entre curiosité et utilité', 187–8.

[70] BCMNHN, MS 277, file 2, piece 4, 'Notte des plantes, arbres & arbustes qu'on peut tirer de l'Inde pour l'Isle de France', n.d., possibly these were communicated to Commerson by Cossigny, piece 5 consists of a similar list of plants from India: 'Notte des Plantes et Graines à faire venir de la coste Coromandel et du Bengale'. More species are listed, but the observations are missing. 'Malabar' referred to the entire Tamil region in southern India.

[71] BCMNHN, MS 1319, 'Catalogue des plantes des environs de Pondichéry, que m'a envoyé M. de Cossigny, officier au régiment d'Aunis, en 1769'. See also BCMNHN, MS 277, VI, 'Observations sur la qualité des bois des Isles de France et de Bourbon communiqués par Mr Cossigny et le nommé Bichet ancien charpentier. Nottes pour M. de Commerson', n.d. These observations were required by Commerson (Commerson to Cossigny, 22 September 1770, printed in Cap, *Philibert Commerson*, 134–6).

[72] BCMNHN, MS 1319, 'Catalogue des plantes des environs de Pondichéry, que m'a envoyé M. de Cossigny, officier au régiment d'Aunis, en 1769'.

was thus a dual one: local names were transliterated to facilitate the communication of them; and the explanation of the meaning of a name made the plant in question more accessible to Europeans.

The importance of Indigenous plant names to identification can be traced across the Indo-Pacific, from Madagascar to East Asia. For the documentation and identification of an unfamiliar plant, it was necessary both to know its name and to make comparisons with the familiar. The latter was important to the identification and understanding of the plant, in order to locate it within a known epistemological system. This can be shown through the example of the ravenala or traveller's palm, whose appearance Commerson compared to the banana tree.[73]

When it was impossible for Commerson to make a comparison with exotic plants, he would make use of plants familiar to a European understanding. This strategy can be observed regarding the so-called *Voua-Sohî* which he described as 'the size of an olive' or the 'Tambouroun angats', depicted as greener and bigger than those of India, noting that the Malagasies 'do not use [them] ... like the Indians'.[74] His observations contain not only native names but also descriptions of the plants and their uses: for instance, the *Voua Faha*, which could replace rice in times of famine, the *Voua Sourindi*, which had a similar scent to nutmeg; the *Voua-n-ïou* (the coconut palm tree), and the *Diti-Voua-Zin*, which could be used to make gum or rubber.[75] The Malagasy expression *Vohaniho* means 'fruit of Nio', which derives from the Malay and Polynesian word *Niu* or *Nihu*: the palm tree had probably arrived on the island with the Hindus and Arabs long before the European colonisers came.[76]

The astronomer Alexis Rochon, who in 1768 was sent to collect useful plants in Madagascar, used strategies similar to Commerson's. Rochon collected numerous useful plants and listed them under their Malagasy names, most with a brief description of their appearance and usage. These were attached to his travel accounts, published in 1791 as *Description de plusieurs Arbres, Arbustes et Plantes qui croissent dans la partie du Nord de Madagascar, et que j'ai apporté à l'île de France, à la fin de l'année 1768.*[77] Rochon's list included useful plants, trees, spices, food crops, timber

[73] BCMNHN, MS 887, III, 'Voua – Raven âla', 'Voyage de Madagascar, en 1770, concernant les dattes et les observations d'histoire naturelle faittes dans le sud de cette isle & principalement aux environs du Fort Dauphin par Sr. Philibert de Commerson'. These notes refer to Commerson's second voyage to Madagascar. On the traveller's palm, see Allorge-Boiteau and Allorge, *Faune et flore de Madagascar*, 94.

[74] BCMNHN, MS 888, 'Notte de quelques plantes et fruits de Madagascar et de leurs propriétés, 1769. 'J'y joins une notice de quelques oiseaux, quadrupèdes et reptiles de la mesme isle par M. de V[algny]'; and 'Notte de quelques plantes et fruits de Madagascar'.

[75] Ibid. [76] Perrier de la Bathie, 'Les Plantes introduites à Madagascar', 833.

[77] Rochon, *Voyages à Madagascar*, vol. 1, 272–88.

(such as the *Fouraha*), cosmetics (such as the *Harame*, from which Malagasy women produced a moisturiser), poisons (such as a huge quantity of the fruit of the *Tangeum*), dyeing (such as the roots of the *Hounits*, from which the Malagasies made a nice red shade), and 'pleasant flowers' of all different kinds.[78] So, it was clearly not just plants of everyday value that were to be introduced, but also plants for pleasure. Similarly, Commerson found and researched plant sources of food and rubber, along with several palm trees. The eighteenth century was a 'period when the consumption of exotic goods was generally rising throughout Europe', and consumption was rising in the colonies, too.[79] There, the broader economic debate over the value of certain plants – particularly in Mauritius, where barely any plants were of economic value or indeed everyday value to island life – was crucial in the search for local resources.

While Poivre sent Commerson and Father Rochon to Madagascar, he sent Père Galloys on a similar mission to East Asia. On his way to China in 1768, Galloys was to call at Malacca and other ports in Malaysia in order to collect offshoots and young live plants of several fruits and timber, including the mango, from which Poivre hoped to receive 'the best fruit which there is on earth', the durian, which had a revolting smell but tasted delicious (as is still the case today), and two sorts of guava, one of which was the 'Malay rose apple', which Poivre called the 'jambon malaca'.[80] He was interested, too, in the 'champaca' (the magnolia tree) and a plant called 'lani-lani'. I have not been able to identify this; but as *la ni* in contemporary Malay means 'right now' (or for that matter 'some time soonish, depending . . .') it's entirely possible that Poivre, on asking for its name, misinterpreted the response.

It also seems that despite the instructions Galloys had received from Poivre in 1768, he did not jot down the Chinese phonetic transcriptions for the vegetables but instead provided brief translations and some notes on colour and appearance. There is evidence that Poivre, meanwhile, during his earlier travels in China in the 1740s and 1750s, had followed the former practice and expected that his collaborators would use a similar method. Although he did not do this for every plant, he must have paid particular attention to the various teas and their names, writing them down with their phonetic transcriptions: Tchao-Tchong, Haissuen, Bouy, Pékao, and all teas of the Fo-kien province.[81]

As this section reveals, the plants' native names were important: if they were to be accessed subsequently, their names had to be given to

[78] Ibid., 274–7. [79] Spary, 'Peaches which the Patriarchs Lacked', 15.
[80] ANOM, C/4/18, ff. 16–21, 16v, 'Instructions pour Mr L'abbé Galloys', n.d.
[81] Ibid., f. 17r.

the local people. Indigenous knowledge was important not only in the first encounter but also for the future; it was important to know the local name in order to gather more information, as in the cases of Rochon and Galloys.

Documenting Knowledge for Everyday Usage

Throughout the eighteenth century, Asian and Malagasy farming practices were important sources of knowledge and materials for Mauritian agriculture. Contemporaries, such as Carl Linnaeus, urged travellers to gather useful plants according to the information given by the people whose lands they were exploring.[82] Similarly, the French botanist Antoine de Jussieu emphasised that naturalists should include the local use of remedies in their reports.[83] The authorities in Mauritius were more concerned, though, about utilitarian botany than classification. The French made 'useful knowledge' of mainly everyday value, which they drew from many local, even Indigenous, informants in the Indo-Pacific world. For example, as mentioned earlier, Commerson, on exploring the natural world of Madagascar in the late 1760s and early 1770s, encountered native knowledge and names, listing thirty useful plants and about the same number of birds and other animals. He concluded in his *Summary of the Observations on Natural History*:

> Plants make up the largest part of natural history, and the most multiplied in individuals on the surface of the earth. We will nevertheless see everywhere that each of their parts will give way to dietetic, economic, mechanical and medicinal uses, and consequently become objects of consumption and trade.[84]

This section provides details of the dual strategy, mentioned earlier, that the European naturalists would use to interpret and document knowledge about plants and their usage: while they would rely on their own knowledge and experience while examining the unknown, they would also document local practices in order to learn from them, reproduce them, and apply them. As the Spanish naturalist Francisco Noroña (who contributed to the plant catalogue of Mauritius's Jardin des Pamplemousses in 1787) remarked, Europeans, 'perfect imitators', were good at adopting the ways of preparing and using plants and crops. Here, Noroña makes references to the usage of tea from the Chinese and Japanese, coffee from the Arabs and Turks, cacao or chocolate from the Americans, and opium

[82] Drayton, *Nature's Government*, 73.
[83] BCMNHN, Jus 3, ff. 95v–105r, 'Des Avantages que nous pouvons tirer d'un commerce littéraire avec les botanistes étrangers', n.d.
[84] Commerson, 'Sommaire d'observations d'histoire naturelle', 132.

from the Malays and Persians.[85] This process – that the Europeans imitated local customs and their means of preparation – is fundamental to my argument. The Europeans would adopt, apply, and practise the new methods once they had encountered and documented them.

Let's start with Asian plants and the related knowledge. Poivre, travelling in the 1740s and 1750s around what is now peninsular Malaysia, and possibly also on the islands of Sumatra and Borneo, observed the sago palm or breadfruit tree. This tree he considered 'a gift of nature' which in part made up for the lack of grain in the area: it grew naturally in the woods, did not require labour, and propagated itself.[86] Once a tree had ripened, he documented, the Malays cut it up into sections, quartered them, scooped out the mealy mass, extracted the fibre from it, and dried the remaining paste. When it was ready to eat, it was very nourishing and could be preserved for years. He explained that the 'Indians' not only used the paste as food but also as a remedy against stomach ache.[87] He was interested, too, in the 'rotang de Malaisie' (rattan), which by the late 1740s would have been cultivated in the colonial garden at Le Réduit, and in the 'choux des Moluques (arbre)', which I have not been able to identify.[88]

During the 1750s, Poivre had already encountered Chinese knowledge on his travels. His observations became the basis for the detailed instructions which he gave to Galloys, one of the most important figures in his network in China.[89] Galloys's expedition was designed to collect the plants categorised as useful, based on observations Poivre had made during his earlier travels, published in *Voyages d'un Philosophe*; in the next section I will examine these in greater detail. Galloys's mission was focused on plants for food crops. Poivre, who had interacted with locals in China himself, asked him to write a detailed report on all plants he gathered, including the Chinese name for every one, together with its price, the agricultural methods it required, and the domestic animals that might be needed for its agriculture in Mauritius.[90] Poivre was well aware of the fact that for the success of his mission Galloys needed one thing in particular: local Chinese people with whom he could collaborate.[91]

[85] BCMNHN, MS 43, doc. VIII; Pinar García, *El Sueño de las Especias*, 249.

[86] Poivre, *Mémoires d'un botaniste et explorateur*, 74–7.

[87] Ibid., 75; see also Rouillard and Guého, *Les plantes et leur histoire à l'Ile Maurice*, 569.

[88] ANOM, C/4/18, ff. 16r–21r, 16v, Poivre, 'Instructions pour Mr L'abbé Galloys'.

[89] Poivre, 'Observations sur le décreusement de la soie sans savon', transcribed in Michel Dürr, 'Quatre Inédits de Pierre Poivre', 210–32. Poivre read it at the Academy in Lyons in August 1762. On Galloys's mission, see in particular Malleret, 'Pierre Poivre'.

[90] ANOM, C/4/18, ff. 16r–21r, 'Instructions pour Mr L'abbé Galloys', n.d.; and ANOM, E 197, Galloys's file, f. 2. On 'useful' Chinese plants and their acclimatisation in France, see in particular Dumoulin-Genest, 'L'introduction et l'acclimatation des plantes chinoises en France au XVIIIe siècle'.

[91] ANOM, C/4/18, ff. 16r–21r, 16v, 'Instructions pour Mr L'abbé Galloys'.

Galloys received detailed instructions; he was to investigate:

all useful trees, all nutritive grains, and the seeds of fruit trees and vegetables whose cultivation could be attempted with some experience of success in our Isles de France and de Bourbon.[92]

As in his instructions, wheat and rice are listed first and second, this indicates that those food crops had priority and were greatly needed in the Mascarenes. Galloys was to collect all different kinds of fresh wheat 'ready to sow', a quintal of each to test, and different kinds of fresh rice (50 pounds of each kind ready to sow), and in particular a type of rice suitable for dry soil.[93] Poivre was clearly planning to test the cultivation of different kinds of rice in the Mauritian climate. For the same purpose, Galloys was to gather 'an assortment of all [kinds of] cereals and vegetables', guava plants, citrus fruits such as the orange, the lemon, the tangerine, the grapefruit, the kaffir lime, plus all sorts of litchi plants (of different seasons), the fig (because, as Poivre explained, the fruit was excellent when dried), the jujube (the Chinese red date), the peach, the pear, the 'raspberry-strawberry' (which resembled the European strawberry in taste), the Chinese walnut and chestnut, and the seeds of fodder plants such as vetch (for cattle, goats, and sheep).[94] These examples confirm the importance of fruit trees, but also the importance of fodder crops.

In a letter to Minister Praslin, Poivre explained that Galloys would have to travel all the way to Beijing and the northern provinces, which would require time but would result in his sourcing good tea, mulberry trees, the Chinese wax tree, and many other species of excellent quality.[95] In these northern provinces of China, Galloys was also to gather plants for making rubber, resin, and oil (the Chinese lacquer tree) and the 'tree whose fruit make the nice Chinese wood oil', which Louis Malleret identified as *aleurites fordii*, or the tung oil tree.[96] In addition, Galloys was told to bring the Chinese tallow tree (*triadica sebifera*) (whole trees and lots of seeds) and other oil and wax trees, such as the benzoin tree (*benjoin*) native to Sumatra.[97]

Poivre also elaborated on how coconuts were used for the production of cooking oil, 'despite the unpleasant taste', and described the liquor

[92] Ibid., f. 16r. [93] Ibid., ff. 16v–17r.

[94] BCMNHN, MS 277, file 2, piece 9, 'L'orangerie de Chine', October 1769; ANOM, C/4/18, f. 17r, 'Instructions pour Mr L'abbé Galloys'. See also BCMNHN, MS 1385, 'Graines potageres venus de Chine des Jardins de l'empereur par Galois', 1770.

[95] ANOM, E/197, f. 3r, Poivre to Praslin, 6 March 1767.

[96] Malleret, 'Pierre Poivre', 121.

[97] ANOM, C/4/18, ff. 16r–21r, 18r, 'Instructions pour Mr L'abbé Galloys'. See also BCMNHN, MS 280 vol. 1, and ADR 4 J 65, 'Dissertation sur le benjoin par François Beauvais'.

named 'soury'.[98] For the latter, he observed that during the season of young nuts, no bigger than 'ordinary French nuts' (walnuts?) which appeared immediately after the flower was over, the local people would cut a slit in the stalk of the shoot, about 7 or 8 inches from the trunk of the tree, placing an earthen vessel below it to collect the juice, covering it with some cloth in order to keep air out, as that would make the juice bitter. He also mentioned oily seeds cultivated on the Malabar Coast, including sesame, *gergelin*, and ricin (or *palma christi*). In this part of India, Poivre explained, *palma christi* oil was used as a remedy for children's diseases, and had a different effect from that of the ricin used in France.[99] These examples underline the Mauritian need for crops that would produce everyday goods: wood of course, but also rubber and glue, not forgetting medical remedies.

In addition to China and India, an important source of knowledge and material was, as mentioned earlier, Madagascar, which Poivre had also visited during his earlier travels.[100] Its trees, used for both construction and materials, were of particular interest. Poivre himself, during a four-month stay in Madagascar in 1756 on the way back to France, researched useful plants and palm trees, making detailed notes about their appearance and their usage by the Malagasies. For instance, in the description of the raffia, which consists of a brief general discussion followed by a separate description and examination of its utility, he documented the practices of Malagasy women and how they produced fibre from the leaves (a process still followed to this day).[101]

Similarly, Poivre made observations on the traveller's palm or ravenala ('Raven-Ale'), the ravensara ('Raven-Tsarap'), and trees such as the *Hin-Tchi* (the most common tree around Foulpointe in eastern Madagascar), the *Labeau*, *Harougan*, *Ramontchi*, and *Fouraa*, the latter being 'after teak the best timber in the Indies for the construction of vessels ... the Malagasies prefer it ... to all other kinds of timber'.[102] Malagasy ravensara, whose essential oil is used today in medicine, was then used as a spice; the Malagasies used its leaves. (This practice was the origin of the term 'ravensara', meaning 'good leaf'.)[103] While Sonnerat refers to

[98] ANOM, C/4/18, ff. 16r–21r, 18r, 'Instructions pour Mr L'abbé Galloys'. [99] Ibid.

[100] He also took specimens with him back to France, many of which are kept in the herbarium of the Muséum national d'Histoire Naturelle. The herbarium recently digitised an entire collection of vascular plants, including the Herbier d'Antoine Laurent de Jussieu (including Poivre's herbarium), see https://science.mnhn.fr/institution/mnhn/collection/p/item/search (last accessed 18 April 2020).

[101] BCMNHN, MS 1265, 'Le Rafiâ', n.d.

[102] BCMNHN, MS 1265; Laissus, 'Note sur les manuscrits de Pierre Poivre', 42. For an illustration of the raffia fruit, see BCMNHN, 279, plate 29.

[103] BCMNHN, MS 1265, 'Raven-tsara'. There were different spellings of the plant, including, Ravent-sara, Raven-sara, Ravend-sara.

only one species of ravensara in his published work, Poivre observed that the Malagasies distinguished several kinds.[104] Poivre noted that the fruit of the ravensara had the same scent as cloves; the inhabitants of the Mascarenes used it in their curries.[105] Given that Poivre mentioned the inhabitants of Mauritius and Bourbon making use of ravensara, the tree had probably been introduced there much earlier, maybe via Malagasy forced immigration.

As far as trees are concerned, it was the coconut palm that drew attention in France, as a useful and exotic curiosity. Commerson's manuscripts contain hundreds of descriptions and drawings of plants, including palm trees, fruit trees, and local 'average' flora.[106] He planned to turn his findings about the water coconut native to the Seychelles (*cocos de mer, Lodoicea Callipyge)* into a printed account under the auspices of the Académie des Sciences. His fascinating work on palm trees consists of forty-eight plates, thirteen of which give clear visual and textual details about the water coconut; and its leaves, bark, and fruit (the nut) are discussed. This palm was first recorded by a certain Barré, who joined Marion-Dufresne during an expedition to the Seychelles in 1768. Commerson was probably able to examine the tree because Father Rochon, who was sent to the archipelago in 1769, was instructed to report on the botanical discovery; he would probably have taken a live plant back with him to Mauritius.

For the botanical survey of the flora of the South West Indian Ocean, Commerson had been in touch with the botanist Bernard de Jussieu, writing to him in 1770:

> Regarding the water coconut, I promised you a complete history of the palm tree that carries it. I have made all the sketches and all the natural cuts ... and one cannot see any better execution. I reserved all this to be offered to the Academy.[107]

Commerson attempted to classify the flora of both Madagascar and the Mascarenes, hoping to put an overload of information into some sort of a logical order, but never completed it due to his early death.[108] He passed away in 1773 and although the manuscript, entitled 'Monographica de Cocô Maritima dictâ & aliis nonnullis Palmarum Generibus' (apparently part of *Palmarium Volumen In quo isquedum desiderata Palmarum arborum*

[104] Sonnerat, *Voyage aux Indes orientales et a La Chine*. Sonnerat may have used the information provided by Commerson, with whom he had explored the region. See also BCMHN MS 279–81.

[105] BCMNHN, MS 1265, 'Raven-tsara', n.d.

[106] BCMNHN, MS 887, doss. II, f. 44. During his botanical missions in Madagascar and the Mascarenes, he was accompanied by his draughtsman, Jossigny.

[107] ADS, Commerson's file, Commerson to Bernard de Jussieu, 6 February 1770. BCMNHN, MS 887, doss. III, piece 11, 'Cocos', n.d.

[108] BCMNHN, MS 277, I/ 3, 'P.Sc.', n.d.

(su bulborum arboreorum)), has never been published, it is stored in the archives of the Parisian Natural History Museum.[109] It was Sonnerat, then, who included the descriptions and drawings of the 'cocos de mer' in his *Voyage à la Nouvelle-Guinée*. Although the engravings resemble those done by Jossigny, Sonnerat does not make any reference to either him or Commerson. Finally, Sonnerat produced the 'Description du Cocos de l'Île de Praslin, vulgairement appelé Cocos de Mer' as part of the SE series of the Parisian Academy of Science.[110] That account, however, published in 1776 with the aim of impressing the botanists of Paris, gives no hint of the complex communications that I have detailed between the various actors.

Finally, then, let's look at how the practical know-how was described.

Documenting Know-How: The Example of Global Rice Cultivation

This section reinterprets Poivre's *Voyages d'un Philosophe* (1768), in which he documented the observations he made in the 1740s and 1750s, when he had been sent by CIO to East Asia in the hope of setting up commercial relations with Cochinchina. As highlighted in the previous section, there is a need to turn to unpublished material to trace local knowledge and the ways in which it was encountered, interpreted, and written down, as in most published accounts the original source of the knowledge was erased. Throwing a twist into this practice, Poivre had the accounts of his travels, *Voyages d'un philosophe* and *Suite des Observations sur l'État de l'Agriculture Chez différentes nations de l'Afrique et de l'Asie*, published at a point when he needed proof of and authentication for his wide-ranging agricultural knowledge and larger plan to turn Mauritius into an agricultural mini-state. In that sense, I argue that he used local and Indigenous knowledge as a political instrument to justify his efforts and publicise his expertise, praising local agricultural practices and plant knowledge, as I elaborate below.

Poivre's account was very much after the fashion of the eighteenth-century French agrarian thinkers for whom global systems of agriculture were sources of comparison for the new systems that were sometimes more successful and sometimes less so. Agronomical observations were

[109] BCMNHN, MS 279, file 3. MS 279, 280, and 281, contain all in all about 585 engravings, most of them by Jossigny, but some by Sonnerat, Morlaix, and Pecquet. The Cocos series contains forty-eight engravings. On Madagascar's palm trees, see also Allorge-Boiteau and Allorge, *Faune et flore de Madagascar*, 124–25.

[110] Sonnerat, *Voyage à la Nouvelle Guinée*, plates 3–6; Sonnerat, 'Description du Cocos de l'Île de Praslin', 263–6.

rooted in 'useful' philosophical, political, and economic notions. Physiocratic ideas, therefore, must be understood in debates about 'techniques', which would always be connected to a philosophical demonstration.[111] On the basis of this understanding, Poivre claimed that during his travels in Asia he was spurred by a most interesting question: how did China produce the huge quantities of food required to feed its huge population? Did the Chinese possess any secrets relating to the propagation of their cereals and foodstuffs? To investigate these questions, he began to interact with Chinese labourers, and soon he thought he had found the answer:

> Their secret consists simply of maturing their fields judiciously, ploughing them to a considerable depth, sowing them in the proper season, turning to advantage every inch of ground which can produce [even] the most inconsiderable crop, and preferring grain to every other species of culture as by far the most important.[112]

In eighteenth-century France, Chinese agriculture, with its 'exotic' technologies and cultivation methods, had fascinated agronomic thinkers long before the French Minister of Finance, Henri-Léonard-Jean-Baptiste Bertin (1720–92) began to encourage scientific enquiries into the Chinese methods of making porcelain. The *voyageurs* of the time, missionaries in particular, had begun to document the agricultural methods used in China.[113] Thus, Poivre's agricultural and economic observations in China emerged at a point when French agronomy was rapidly improving.

The Chinese agricultural methods were a model for Poivre's visions of an agricultural island state. As in the quote above, according to him the reasons for China's fertile soils were intensive work, profound labour, and a wide variety of fertilisers. At the same time, he observed the social framework around this type of agriculture: simple and regular habits, the equality of individuals, the paternal government of the emperor, and the protection and honouring of agriculture.[114]

For his comparative study, Poivre focused on agronomy, which he investigated in several parts of the world: the Cape of Good Hope, Madagascar, South Asia, South East Asia, and to a great extent in China.[115] In so doing,

[111] Bourde, *Agronomie et agronomes en France*, 20; on peasantry and peasants in enlightenment France, see also Roche, *La France des Lumières*, 99–126.

[112] Poivre, *Travels of a Philosopher*, 143. In order to avoid my own translations, I quote from the original 1770 English translation.

[113] For a discussion on Chinese agriculture and its place in France, see Bourde, *Agronomie et agronomes*, 440–4.

[114] Ibid., 443.

[115] On his travels in Asia and political economy, see also Maverick, 'Pierre Poivre', 165–77; Maverick, 'Chinese Influences upon the Physiocrats'.

he overcame the language barrier, since, as he argued, 'agriculture, in every climate, is the universal art of all mankind'.[116] Through observation, he traced the tacit knowledge and know-how acquired through agricultural experience.[117] He observed, analysed, interpreted, and documented what he saw through the lens of his own background and knowledge. He used practices as a common language, which made it possible for him to access knowledge. He observed the various types of rice and their cultivation in Cochinchina, and made clear references to Cochinchinese knowledge regarding the usage of certain plants.[118] In the speech at the Académie des Sciences, Belles-Lettres et Arts in Lyon, which he gave on 1 May 1759, on his return to France after his long travels in the Indo-Pacific, he declared:

> I would be happy if in the number of observations I have made in India and China, there is someone who can contribute to the perfection of our arts. The memory of my travels will become dearer to me if the knowledge, however imperfect, which I have brought back can help enlightened citizens to extend the light of our nation on trade.[119]

Reading Poivre's account from my cross-cultural perspective, I am able to gain insights into the transmission of knowledge during his travels, and how practices related to plant knowledge were explored in other parts of the world. His unpublished notes reveal the strategies he used to document knowledge. He stated, for instance, that he had engaged with local knowledge, relating to the seeds of the *diou*', red flower: 'I informed myself about the manner of sowing this seed and how to cultivate the plant. This is what the expert I consulted told me.'[120] This statement was followed by a detailed description of the requirements of the seeds depending on the season, the soil, and water. Like his unpublished notes, his *Voyages d'un philosophe* is full of his detailed observations about knowledge practices and how materials of everyday value were processed.

He was clearly well aware of the value of embodied knowledge; even in the 1750s as the CIO agent, he had proposed recruiting Chinese workers to go to Mauritius on a voluntary basis, to use their skills and transmit them. He successfully recruited a dozen men, among them bamboo and rattan craftsmen, silk cultivators, carpenters, a potter, and a cotton

[116] Poivre, *Travels of a Philosopher*, 9.
[117] On knowledge practice, see Klein and Spary, *Materials and Expertise*; Smith, *Entangled Itineraries*; Smith and Schmidt, *Making Knowledge*.
[118] For a global history of rice, see Bray et al., *Rice*.
[119] Poivre, 'Discours de Pierre Poivre à l'Académie de Lyon', 1 May 1759, in Dürr, 'Quatre inédits de Pierre Poivre'. The manuscript is kept in the archives of the Académie in Lyons, MS 187.
[120] ANOM C/1/2, ff. 161v–215r, 'Journal d'un voyage à la Cochinchine'.

worker.[121] While, however, he was making the final arrangements, he learned that the CIO prohibited Chinese people from being carried on its ships.[122] His splendid idea was scuppered.

Next, he compared several methods of cereal and rice cultivation in Asia and Africa, in order to find those most suitable to enrich European cultivation. Starting with Madagascar, he observed that Malagasies cultivated rice only during the rainy season. They used pickaxes to work the fields, starting with weeding: then five or six men would line up and, moving forward in a row, dig a series of small holes. Women and children followed, popping a grain of rice into each hole and using their feet to scuff a little earth over it; a swift and efficient method.[123] With regard to wheat, because the soils were so fertile it would self-propagate annually, growing higgledy-piggledy with a range of native herbs. Back to rice; in Siam (present-day Thailand), Poivre observed that the seeds were sown by putting them deep into a little hole in well-watered soil without covering them with too much earth.[124] Then once the plants had reached a height of 5 to 6 inches, the Siamese would pull them up by the roots and transplant them, in clumps of three to four sprigs about 4 inches apart, into small parcels of land, in deep clay soil which had been well ploughed. Poivre explained that transplanting the rice seedlings from their original soil to clay led to much better results than leaving them in the former. He thought that these methods would have come to Siam with the Chinese and Cochinchinese migrants who had worked the fields in the capital[125] and its surroundings, indicating once again the importance of embodied knowledge. Because these workers contributed so much to the improvement of rice cultivation and commerce in general, they enjoyed the government's protection from oppression.[126] Poivre, having observed the healthy relationship between legal protection of social or ethnic groups and their productivity, emulated it with the enslaved people in Mauritius, as Chapter 4 will reveal.

In Cochinchina, Poivre observed the cultivation of six different kinds of rice. The first three, the 'little rice' (the most delicate, whose grain was small, oblong, and transparent), the 'great long rice' (with a round grain), and the 'red rice' (whose name derives from its reddish husk), were produced in great abundance in paddy fields.[127] The more curious types of rice Poivre observed were two varieties that grew in dry soil and required only rainwater: these formed a considerable part of the Chinese trade. Dry rice could only be cultivated in the mountains and high

[121] ANOM, C/1/3, Poivre to the Secret Committee, Canton, 31 December 1750.
[122] Ibid. [123] Poivre, *Mémoires d'un botaniste et explorateur*, 53–4. [124] Ibid., 67.
[125] Possibly the city of Phra Nakhon Si Ayutthaya.
[126] Poivre, *Mémoires d'un botaniste et explorateur*, 68. [127] Ibid., 86.

ground, where the soils were worked with spades. Its seeds were sown 'as [Europeans] do wheat' at the end of December or the beginning of January, when the rainy season ended, giving cold and dry weather.[128] After only three months, the crop was plentiful. In the cultivation of common rice, the Cochinchinese agricultural practices were similar to those on the Coromandel coast.

Poivre described a tool which had been developed to water the rice fields.[129] Then he described the 'Malabar' tools used for working the ground; they resembled 'the *araire* and the *souchée* (plough) used in the south of France' and they required oxen (but more usually buffaloes) to pull them.[130] In Cochinchina, the ground was ploughed twice, then the rice was sown in a small field, covered with shallow water, and carefully tended.[131] This example shows that it was a common strategy to document knowledge and describe tools by drawing comparisons with those familiar to a European readership; not only did he compare the agricultural techniques used in Asia, he also made reference to the agricultural methods that would have been familiar to a French reader.

In Cochinchina, once the rice plants were about 5 to 6 inches high, the fields were harrowed and flooded before the plants were pulled up along with their seed plots; they were then transplanted into parcels of four or five stalks 6 inches apart. It was usually women and children who did this. Unlike in India, Poivre observed, there was no need of machines or other devices to flood the ground, because in these Cochinchinese fields a redirection of the many springs and small rivers rising in the mountains could be made to flood the land naturally.[132]

In addition to rice cultivation, according to Poivre, the most important cultivational activity in Cochinchina was that of sugar cane; he described this in detail as well. Two types were cultivated: one that contained little water and lots of salt, and another one (preferable) which was very large and high and contained lots of sap but little salt (hence used mainly as fodder). Most importantly, this plant should be eaten, by both animals and humans, while it was still green because this was when it contained the most nutrition.[133] The cultivation of sugar cane required the soil to be worked then the cuttings planted three by three in a line (similar to the way wine was cultivated in some provinces of France) and set about 18 inches into the ground about 6 feet apart (looking something like a chessboard).[134] This had to be done during the rainy season, because the cuttings needed plenty of water to sprout. If good care of the cane was taken during its first six months, the harvest could be reaped in twelve to

[128] Ibid., 87. [129] Ibid., 61. [130] Ibid. [131] Ibid., 88. [132] Ibid., 88–9.
[133] Ibid., 89. [134] Ibid.

fourteen months. Given that it was possible to grow sugar cane from cuttings and it was easy to experiment with cultivation and sap production, this plant was quite attractive to colonies (as had already been recognised in the Atlantic).[135]

In addition, Poivre closely examined the cultivation of the mulberry tree – the only source of food for silkworms – in Cochinchina. The prime requirement was sandy soil close to the sea and/or rivers . On the eleventh moon at the end of the monsoon season, the last year's growth was pruned, and the prunings were divided into small plants of 5–6 inches and placed them diagonally in lightly mixed soil. These were transplanted in the same way as lettuce: two months later all the plants would be covered with leaves, looking like a new vineyard.[136] On the 'second moon' (presumably, that is, after two months), the greenery was given to the silkworms. Every year, the process was started anew.

Poivre's observations on the cultivation of crops are extremely detailed: few other sources, sadly, give such rich information about local practices of cultivation. Some of them do, however, provide glimpses into the plants' use as remedies or foodstuffs and, more generally, how Europeans tried to understand them. Poivre, meanwhile, not only described the use of materials, but also tried to make them more accessible to a European understanding by comparing them with the uses and processes with which he, and possibly his readership, was familiar. In that way, he not only described local agricultural skills, but also the ways in which many other local populations used plants and prepared them for consumption.

Conclusions

Chapter 2 has examined the ways in which French actors obtained information about plant cultivation from informants in, mainly, East Asia and Madagascar. Its argument runs counter to the prevailing historical myth that eighteenth-century naturalists were interested in classification alone. In fact, the search for and collection of useful plants in the service of Mauritius was conducted very much in accordance with the eighteenth-century efforts to use the resources of the world as a wealth-creating enterprise and a form of natural management. Contemporary economic theories most certainly had an impact on the classification of plants, and botanists searching for wealth.

Rather than insisting on classification, this chapter has argued for the significance of utilitarian botany; it was the plants' usefulness which

[135] Cordier, 'Voyage de Pierre Poivre en Cochinchine', 81–121.
[136] Poivre, *Mémoires d'un botaniste et explorateur*, 89.

proved their status. Thus, the focus of the chapter has been on utility and especially on the ways in which useful knowledge might be communicated to French naturalists. At the same time, drawing on evidence from the Muséum's archives and from Malay vocabularies, this chapter has demonstrated how botanical knowledge was fundamentally, if not primarily, shaped by people other than trained naturalists.

Poivre, ardent self-promoter that he was, stressed in the treatise he wrote for his successor that during his five years in office he had looked for 'all useful products, may it be for commerce, ordinary usage of everyday life, or even only pleasure, in all four parts of the world, and particularly Asia'.[137] This chapter has traced these attempts, looking at the exploration, identification, communication, and documentation of the plants and knowledge collected in specific Indo-Pacific localities, ranging from Madagascar to China. It has drawn attention to local engagements and negotiation practices in several very different cultural contexts, including languages, knowledge practices, and customs. Local mediators facilitated the collection of plants and practice-based knowledge, involving several languages and naming systems as a means of identifying natural goods, their cultivation, and their usage.

French actors were curious to learn about local practices, because those practices could prove a plant to be of value for nourishment and everyday life, including hygiene, body care, construction materials, and artefacts – and even pleasure. Madagascar and China in particular were sources of knowledge and material for agriculture and utilitarian botany. Other places, such as Cochinchina and India, were of importance when naturalists made enquiries into their crops together with the related cultivational methods and preparation techniques.

In sum, this chapter has two main conclusions in relation to the documentation of plants: first, it was important for a plant's native/local name to be known to Europeans, in order to make it possible for them on returning to the area, to access that plant again. Second, European naturalists followed a twofold strategy to document and interpret knowledge. On the one hand, they relied on familiar European knowledge and their own experience and expertise, and one the other, they documented local practices in order to learn, reproduce, and apply them.

The pursuit of natural enquiry was a matter of local engagements with the natural world and the people who made use of it. Utilitarian botany was constructed at the intersections of French agrarian thinking and Indigenous knowledge about the plants' names, features, cultivational

[137] AD Loire, FGB, 15 J 8, 'Etat dans lequel j'ai remis la colonie de l'Isle de France à mon successeur', 23 August 1772.

methods, harvesting, and eventual utility. The Europeans accumulated practice knowledge along with the plants themselves, by interacting with and learning from local populations who had employed these plants for hundreds, if not thousands, of years. In other words, practice-based knowledge was crucial to the French agronomists unfamiliar with the methods employed in tropical climates.

Then, rather than categorising plants according to a system, the collection and examination of new plants and materials was done in order to study their climate, cultivational behaviour, and nutritional needs. Only then could their transfer, cultivation, and employment in a different environment become feasible, as will be addressed in the next chapter.

Here and elsewhere in this book, I have made the point that the analysis of unpublished material was more likely to reveal Indigenous knowledge, because in published material that knowledge tended to be removed prior to publication. So, it is unpublished material then that permits the tracing of knowledge in Mauritius and elsewhere. But it is not only Poivre's published descriptions of practices (tacit knowledge) and unpublished documentation more widely, that will be tested in Chapters 3 and 5, in the analysis of how transferred knowledge was applied in agricultural practice. There, I will explore to what extent knowledge transfer relied on the movement of people (embodied knowledge) as well as to what extent the actors in question were able to comprehend and put into practice descriptions written by other people.

3 Agriculture and Everyday Knowledge

> Who taught [man] agriculture, this art so simple, of which the most stupid man is capable, and so sublime so that the most intelligent animals cannot conduct it?
>
> Jacques-Henri Bernardin de Saint-Pierre.[1]

Let us consider Mauritius's role as an agricultural laboratory for the broader French Empire. While the previous chapter was primarily concerned with the acquisition of plant knowledge outside Mauritius, this chapter focuses on what happened on the island. I look at the plants that were imported (primarily foodstuffs and fodder, but also some industrial materials and a few ornamental plants) and the ways in which knowledge was applied. In addition, I trace how people on the island attempted to cultivate those newly arrived plants; because of the general lack of experience in tropical agriculture and little to no understanding of the plants' needs, the cultivation techniques were disparate and random. The creolisation of knowledge was an ongoing process.

Poivre engaged increasingly with the cultivation of staple crops, both for the provision of the colonists and the enslaved people, but his visions clashed with colonial reality: frustrated colonists; enslaved people who, starving, were dying in great numbers, and the climate. The island became a crucible of overpopulation and famine. Poivre set up his own garden to provide food for the planters and enslaved people working in it.[2] Apart from that, while its primary function was as a nursery for spices, he conducted experiments in it with different types of grains. At the same time, he encouraged the inhabitants of Mauritius to grow foodstuffs rather than cash crops such as coffee, cinnamon, and cotton.[3] Hence, the produce of the island must be seen as two rival ecologies: crops grown for profit, and local agriculture

[1] Bernardin , 'Etudes de la nature', 254.

[2] See AD Loire, FGB, 15 J 3, 'Instruction du roi pour le sieur régisseur de l'habitation du roi à Monplaisir', 1772, and MA, OA 127, the official document that attests to the sale of Poivre's Monplaisir to the Crown, 12 October 1772.

[3] ANOM, C/4/25, ff. 9r–10v, Poivre to the Duc de Praslin, 13 January 1769.

along environmental lines.[4] Historian Peter Jones emphasised that throughout the eighteenth century agricultural progress and growth was essential to European countries. There, 'agriculture ... was concentrated on the issue of the food supply and the risk of social disorder should grain stocks run low'.[5] But that was not the case during the Enlightenment alone; texts by ancient Romans such as Marcus Terentius Varro make it clear that for many centuries huge importance has been accorded to agriculture.[6] Jones stated, too, that 'diffusion, whether of information (useful knowledge), "know-how" (skill), or tools and technology' were central in the Enlightenment era.[7]

The acclimatisation of useful plants and the global transfer of natural materials more generally provides 'important evidence how human and natural powers interacted'.[8] In support of arguments brought forward by STS environmental historians, I will show in this chapter that the interactions on the island between the natural world and human agency were determined by its climate, the environment, and the sociocultural impact of the very people who cultivated the plants – the enslaved gardeners – and the planters, as well as the European colonists.[9] In light of this human–natural interaction, agricultural methods were adopted and modified according to both environmental conditions and the human possibilities and limitations in Mauritius. The island's climate in particular had to be taken into consideration: given that specific crops required specific seasons and cultivation know-how, the climate had a strong overall influence on the acclimatisation of plants and the cultivation activities.

Looking at the interplay of the numerous factors relating to agriculture, this chapter with nine sections proceeds as follows. First, it describes Enlightenment agronomy, and, second, the role of botanical gardens. Then it extends these elements into the island context, where it elaborates on the natural economy and the pressing question of how to feed the island's population. The purpose of these two sections is to clarify the meaning of the natural economy in the context of eighteenth-century Mauritius, with a particular focus on the dynamics between its overpopulation of enslaved people and the introduction of food crops. This is followed by an analysis of acclimatised crops from all over the world, for

[4] On the quarrel between two rival ecologies of commerce, see Jonsson, *Enlightenment's Frontier*, 122–6.

[5] Jones, *Agricultural Enlightenment*, 14. [6] Varro, *Varro on farming*.

[7] Jones, *Agricultural Enlightenment*, 11.

[8] Spary put forward this argument in relation to tropical fruits. Spary, 'Fruits of Paradise?'.

[9] Stockland, 'Policing the Oeconomy of Nature'; Reuss and Cutcliffe, *Illusory Boundary*; Carse, 'Nature as Infrastructure'.

the purpose of food, fodder, construction, and everyday materials divided into three sections. Lastly, I explore in two sections the cultivation activities and environmental concerns (in relation to climate, soil, and plagues) on the island after 1767, embedded in the relationship between feeding the population, balancing interest groups, and cultivating staple crops. In that section, I provide examples of the cultivators' goals – and their miscalculations.

Agronomy, Physiocracy, and the Role of Botanical Gardens

Historian of science Simon Schaffer once characterised agronomy as a mix of scientific and political work.[10] Similarly, André Bourde, in his important study on French agronomists in the eighteenth century, pointed out that in a scientific sense 'agronomy is not a systematic or exclusive doctrine'; likewise, it is not 'an art' but 'the vivid process of a 'science', which cannot be defined as 'rigorous, exact, demonstrative, abstract'. Bourde suggests instead that agronomy was an authentic 'new science' attached to 'good cultivation', with improved and new agricultural techniques to achieve the optimal result.[11]

In Enlightenment France, agriculture became a key element for physiocracy. Although this term literally means 'rule of nature', for French physiocrats nature was an order 'within [which] man was reinstated to his superiority according to a pre-established hierarchy'.[12] The physiocrats saw the earth as the only true 'producer', and felt that the economy had to be based on agriculture rather than vice versa. These physiocratic ideas, however, never resulted in any kind of practical reforms to improve production; instead, the movement busied itself with moral philosophy based on an ideal society shaped according to the image of Mother Earth.[13]

Histoire naturelle in Enlightenment France should not be understood solely, however, as merely a set of theoretical debates; to its practitioners it was inseparable from its social implications and its practical uses – from collection and classification to cultivation.[14] Then from the mid-eighteenth century onwards the academic study of plants became increasingly

[10] Schaffer, 'Enlightenment Brought down to Earth', 257–68. On utility and nature in the *Ancien Régime*, see also Spary, 'Peaches which the Patriarchs Lacked', 15–16.

[11] Bourde, *Agronomie et agronomes en France*, 12, 19; see also Béaur, *Histoire agraire de la France.*

[12] Herlitz, 'Art and Nature', 163.

[13] Gabriel-Robert, 'Bernardin de Saint-Pierre et la Physiocratie', 36.

[14] Spary, *Utopia's Garden*, 5. On a literary approach to French natural history, see also Garrod and Smith, *Natural History in Early Modern France.*

associated with horticultural knowledge related to gardens. Botany thus became a popular science at a point in time when commercial interest in plants was expanding, in a development more commonly known as economic botany.[15]

For commercial–scientific purposes, the botanical networks that had been created increased significantly in the Enlightenment era.[16] Particularly after the Treaty of Paris, which ended the Seven Years War in 1763, the question of acclimatisation went hand in hand with the question of how to feed France's colonial inhabitants, especially those of Mauritius. At that point, when the rest of the French empire was on the verge of breaking apart, the task of feeding the people in the colonies was seen as more important than their cultivation of cash crops.[17] Yet historians, instead of enquiring about agriculture in Mauritius, have focused on the French attempts to acclimatise spices there.[18] The frictions between the introduction of plants and Mauritius's identities became especially evident during the period under review, when the types of plants cultivated reflected how the island and its role under imperial strategies were perceived, and how they changed. Cash crop projects were inflicted on an island whose settlers wanted to cultivate food for their very survival rather than spices that were difficult to acclimatise in the first place. Yet, during the searches for plants of both everyday and economic value, the two types of project coexisted and sometimes even went hand in hand. Both enterprises required a sensitive balance to be maintained and intensive management that coexisted and sometimes clashed both for the collection of plants outside the island but also for their cultivation on Mauritius.[19]

In the spirit of the eighteenth century:

> cultivators and improvers claimed to be bringing nature to perfection by using human ingenuity to accomplish the project of making the world more habitable and pleasant for humans.[20]

Poivre sought to do more or less the same in the case of Mauritius. His dual function as intendant and agronomist in touch with physiocratic circles would certainly have played a key role in this enterprise. After his first retirement, he returned to Lyons, where he was attached to its Academy: this allowed him to develop his ideas regarding his engagement

[15] Easterby-Smith, *Cultivating Commerce.* I thank Sarah Easterby-Smith for her inspiration to my work and our fruitful discussions.
[16] Spary, 'Botanical Networks'; Dauser, *Wissen im Netz*; Drayton, *Nature's Government*; Jardine et al., *Cultures of Natural History*.
[17] Lacour, *La République naturaliste*, 506. See also Spary, *Eating the Enlightenment* and *Feeding France*.
[18] Ly-Tio-Fane, *Mauritius and the Spice Trade*, 2 vols.
[19] Roberts, 'Le Centre de Toutes Choses'. [20] Spary, 'Fruits of Paradise?'

with nature. He shared his physiocratic ideas, and his *Voyages d'un Philosophe* became well known within these intellectual circles.

It is important to understand the role of Poivre's trajectory in the context of the complex colonial politics, his patronage by high officials, and his earlier problems with the CIO and rival botanists. The fact that he promoted himself to the CIO and later to the court indicates that the Crown remained a powerful yet theoretical authority whose support he gained only after a long time. He published the *Voyages d'un Philosophe* when he needed justification for his further attempts and plan(t)s for Mauritius.

The Gardens of Mauritius

The gardens of Mauritius were used as nurseries for economic and medicinal experiments with the newly introduced plants. Governor La Bourdonnais founded Monplaisir (see Figure 3.2), the island's first garden, in 1735, and this was followed by the CIO's garden, Le Réduit (see Figure 3.1), in 1748. In order to maximise the chances of the successful acclimatisation of plants, more colonial gardens were encouraged. So, in 1761 the CIO set up an acclimatisation garden in Réunion, and with the shift of power after the Seven Years War, it became a royal garden under the administrator Jean-Baptiste-François de Lanux.[21]

Governor La Bourdonnais started cultivating the garden of Monplaisir in the 1730s, but he soon abandoned it. When his home was transformed into a hospital with twelve rooms for officers and soldiers, his lovely garden was abandoned, and in 1754 the Chevalier de Godeheu disparagingly commented that all that could now be seen of his beautiful residence was a hill of bricks and rotten wood.[22] Godeheu also criticised 'the craze of mankind for building something', referring to the fact that Governor David did not continue the work on Monplaisir but had instead built a villa in the Le Réduit region in 1746.[23]

The garden at Le Réduit did in fact provide medicinal plants, in particular when Fusée-Aublet worked on the island between 1752 and 1759.[24] Contemporary observers describe him as having:

> the character of a learned & curious botanist & corresponds with some of the French virtues of the greatest quality but he appears to be impound [*sic*] in his

[21] ADS, DB Lanux, twelve letters to Réaumur in the 1750s, see also Lougnon, *Correspondance du Conseil supérieur* and Allain, *Une histoire des jardins botaniques*, 61. See also BSG MS 1085 which contains letters of both Lanux and his son.

[22] Godeheu, 'Extraits du Journal de Godeheu', 152, entry from 2 June 1754. [23] Ibid.

[24] Fusée-Aublet, *Histoire des plantes de la Guiane françoise*. For Fusée-Aublet's accounts during his stay in Mauritius, see particularly BCMHN, MS 452, 453, and 579.

judgment very often and half mad, dwelling days and nights in the woods & fields in search of curious herbs and plants, & in making experiments on their virtues & effects, which had lately [nearly] cost him his life, smelling and tasting everything of the vegetable kind that comes in his way.[25]

Although he was primarily a pharmacist, he surveyed the entire island in terms of its topography, soil, minerals, animals, and plants.[26] To begin with, he cultivated medicinal plants, which were useful for the island's hospital and for the crews of French ships landing there. He also planted flowers, fruit, and other consumable products from 'all four parts of the world'. When the astronomer Pingré arrived at Le Réduit in summer 1761, he was welcomed by an avenue of rose bushes and an alley of flowering orange trees, and he observed that recently planted cherry, plum, apricot, chestnut, walnut, and even oak trees were in flower, and were promising a decent harvest. He exclaimed: 'I thought I was in a magnificent garden somewhere around Paris.'[27]

Pingré noted other European fruits and vegetables on the island: strawberries, peaches, and grapes. He continued, 'our peas, our broad beans and our artichokes were also quite good'.[28] According to Pingré, however, the principal fruits known in Mauritius were pineapples, bananas, papayas, and guava but also potatoes and other roots – that is, non-European foodstuffs. It was only in the late 1780s that food crops from Europe were increasingly acclimatised to the island.[29]

In contrast to the CIO's Le Réduit, the garden of Monplaisir became Poivre's secret spot, where he could act independently. While waiting for Galloy's shipments and his own return from East Asia in 1767, Poivre had bought it from the CIO, planning to acclimatise new arrivals there.[30] For instance, in February 1769, he received six containers of tree plants from China with almost every plant 'in good condition'.[31] They were to be conserved and naturalised in Monplaisir before their distribution – in the correct season of course – to the different areas of the island. But, as he complained in a letter to Minister Praslin, Mauritius's inhabitants lacked knowledge of tropical plants. So, in an attempt to improve his public image, he declared that he was taking care of the plants himself in his

[25] BL, Add. MS 33765, f. 12r, 'Geographical Collections of Alexander Dalrymple'. Dalrymple refers to Fusée-Aublet as 'Monsieur Oblette'.
[26] Fusée-Aublet, *Histoire des plantes de la Guiane françoise*, esp. vol. 2, 129–59.
[27] Pingré, *Voyage à Rodrigue*, 217. [28] Ibid.
[29] BCMNHN, MS 47, 'Mission de Joseph Martin à l'Isle de France, État des végétaux utiles qui se trouvent en France et qui manquent à nos colonies des Isle de France et de Bourbon', 1788.
[30] ANOM, C/4/18, f. 105r, Poivre to Praslin, 30 November 1767.
[31] Poivre to Galloys, 16 February 1769, letter printed in Laissus, 'Note sur les manuscrits de Pierre Poivre', 45.

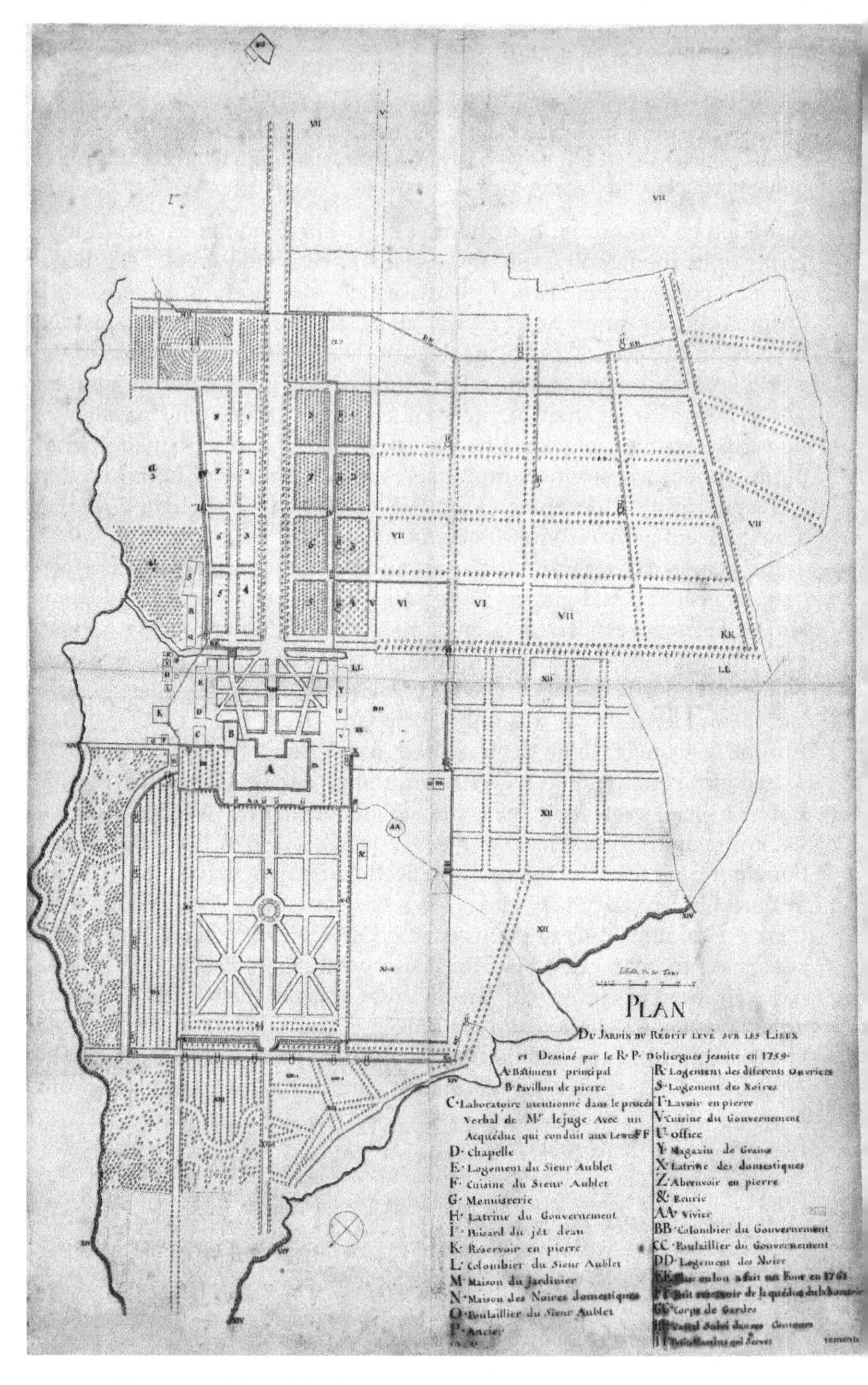

Figure 3.1 Map of the Garden of Le Réduit, by R. P. D'Oliergues, Jesuit, 1759, Bibliothèque nationale de France, https://gallica.bnf.fr/ark:/12148/btv1b55012946f/f1.item

garden, applying methods to assure the acclimatisation of, for example, Chinese plants 'as in their natural soil'.[32] In his garden, Poivre explained in another letter to Praslin, he also acclimatised a range of teas, the star anise, the oil tree, the camphor tree, almost every type of Chinese fruit tree, a great number of trees native to different parts of Asia and Africa, and grains and fruits from the Americas and Tahiti.[33]

Poivre argued that Monplaisir was the richest garden in the universe because of its immense collection of plants, which he – or rather his collaborators – had gathered from the four corners of the world.[34] He had wanted to keep these plants in his garden alone, but after his departure in 1772 the administrative structure changed and his role was divided between the intendant Maillart-Dumesle and the director of the garden.[35] At that point the garden as a whole was divided into the residence and the garden itself, which then became a colonial *jardin du roi*, also called the Garden of Pamplemousses. Provost, who had gathered spices for Poivre in the Indo-Pacific, became its '*commissaire*' ('inspector'), and he reported back to France on the garden's condition.[36]

The tasks of the garden's administrator were set out in writing, a system probably set up by Poivre himself in the hope of ensuring the continuation of his work, ready for the appointment of the young gardener Jean-Nicolas Céré (1737–1810) as director of the garden.[37] The instructions detail the needs and tasks of the garden's enslaved people, who took care of its plants, trees, and vegetables. It was asserted that the administrator's attention should be focused on the conservation of spices and trees of interest.[38] It was three years after Poivre's departure, in 1775, that Céré was appointed as director. Ten years after his appointment, he wrote to the royal inspector, M. le Brasseur, that since its formation by Poivre, the Garden of Pamplemousses had become the 'tributary by its rich peonies' of Pondicherry, Goa, the Seychelles, Bourbon, Madagascar, and the Cape of Good Hope and that it sent specimens and plants to other colonies in both the Indian Ocean and the Atlantic.[39]

[32] ANOM, C/4/18, f. 105r, Poivre to Praslin, 30 November 1767.
[33] ANOM, C/4/25, f. 49r, Poivre to Praslin, 10 August 1769.
[34] ANOM, C/4/29, ff. 13v–14r, Poivre to the minister, 2 April 1771.
[35] BCMNHN Ms 3378, 'Journal de culture du Jardin du Roi, 1785'.
[36] Rouillard, *Le Jardin des Pamplemousses*, 26.
[37] See AD Loire, FGB, 15 J 3, 'Instruction du roi pour le sieur régisseur de l'habitation du roi à Monplaisir', n.d. [1772–5 ?]. See also BCMNHN, MS 47, doss. 5, 'Notice sur la vie de Mr. de Céré'.
[38] AD Loire, FGB, 15 J 15, 'Inventaire des épices cultivées dans le jardin du roi de Monplaisir, effectué par le commissaire de la Marine', 1774.
[39] BCMNHN, MS 303 'Recensement de tout ce que renferme le Jardin du Roi, le Monplaisir, Isle de France', June 1785.

The garden gained international fame, and Céré was accepted into the scientific circles of the Parisian Académie des Sciences, the Jardin du Roi, and the Royal Botanical Garden in Kew, whose eminent director, Joseph Banks, cited the Garden of Pamplemousses (formerly Monplaisir, see Figure 3.2) as a model for West Indian planters. The garden administrators even exchanged knowledge with their counterparts in the Austrian Akademie der Wissenschaften (Academy of Sciences) and at the magnificent gardens of Schönbrunn, in Vienna.[40] At the same time, the colonial garden of Réunion began to blossom under the botanist Joseph Hubert. What La Bourdonnais had begun in the 1730s was augmented by Poivre during his directorship, and progressed well under Céré, who sought Poivre's expertise and advice in his retirement.[41]

Natural Economy and the Question of How to Feed the Island

Let us briefly reconsider the historical, geopolitical, and economic context, as that will allow us to explore the role of the natural economy. After the 1763 Treaty of Paris, Minister Choiseul planned to set up a naval base for French troops on their way to India. Whereas the Crown, through Choiseul and his successor (and cousin) Praslin, aimed at establishing an entrepôt for merchants and troops in the Indian Ocean, the local government of Mauritius, Poivre in particular, followed instead their vision of a self-sustaining island colony yet encouraging the project of acclimatising spices. Poivre's ideas and plans in turn were often challenged by the French troops who, having landed on the island, had to be fed and looked after. As mentioned in Chapter 1, the Anglo-French rivalry should not be underestimated; the presence of the French forces had a serious impact on the island's provisions and on the arrival of those vessels Mauritius struggled even more.

Poivre's ongoing problems with the imbalances of the island are vividly presented in his correspondence with Praslin, as are his criticisms of the Mauritian merchants, whose aim was, unsurprisingly, to make as much profit as fast as possible. These were ideas that the celebrated physiocrat Quesnay addressed in *Philosophie rurale* (1763) when he criticised the isolation of the colonial system. He studied several aims in the European colonies from their beginnings, pointing out that a main

[40] Ly-Tio-Fane, 'Contacts between Schönbrunn and the Jardin du Roi', 86. On Céré, see also Deleuze, 'Notice sur M. De Céré', 329–37; Ly-Tio-Fane, 'Botanic Gardens', 7–14.

[41] The success of the gardens in the 1780s, the exchange between Céré's and Hubert's garden as well as the gardens in the Atlantic can be traced vividly in BCMNHN, MS 3378, 'Journal de culture du Jardin du Roi', 1785.

Figure 3.2 Map of the Garden of Pamplemousses, n.d., Bibliothèque nationale de France, https://gallica.bnf.fr/ark:/12148/btv1b53105290p

concern had always been good ports, as the colonies were intended to serve France, and indeed feed it. While Quesnay agreed that the colonies should serve the mother country, he also claimed that they should be considered as provinces and integrated into the system as a whole.[42] In physiocratic thought, then, the role of the colonies was ambivalent: on the one hand claims were made that the cultivation of the colonies was nothing more than the result of the kingdom's efforts; on the other was a pretence that it was the factories of the kingdom that established and maintained the colonies.[43] Further, the CIO's monopoly on trade was seen as unfair, because it ignored the needs and the interests of both merchants and consumers.

Poivre and Governor Desroches explained in a *mémoire* to Praslin that although the CIO had indeed intended to increase agriculture on the island, the project had not succeeded because of bad management, particularly in the face of the rising numbers of inhabitants and vessels stopping over, without securing the stable production of basic necessities.[44] Poivre, using the CIO as a scapegoat, sought to use these circumstances for his own colonial visions when criticising the CIO's ignorance of the deforestation and, indeed, the destruction of the island. He argued that the Company had left a mess of abandoned land without any trees to offer protection against wind and sun.[45]

Echoing her husband's claims, Françoise Poivre[46] explained in 1769 in a letter to Bernardin that:

> the project to make the uncultivated lands profitable is very good. But the government did not do anything except establish colonies. If agriculture was dear, honoured, and protected, all these lands would have been weeded bit by bit.[47]

In order to create a better lifestyle for its inhabitants, the island needed a boost. Agriculture, with its multiple production and active commerce, was to become the key aspect for the domestic establishment of the island.[48] In a report which Poivre must have written shortly before his departure in 1773,

[42] See Røge, 'A Natural Order of Empire', esp. 33–5. On the physiocratic movement in France in general, see Weulersse, *Le mouvement physiocratique en France*.
[43] See also Turgot's criticism on this ('Lettre aux auteurs du Journal de l'Agriculture', 32–40).
[44] Poivre and Desroches to Praslin, n.d. [1771?], 'Extrait d'un mémoire intitulé, Iles de France et de Bourbon', BnF, NAF 9347, f. 196r.
[45] 'Discours prononcé à son arrivée à l'Isle de France aux habitants de la colonie', in Poivre, *Mémoires d'un botaniste et explorateur*, 119–38.
[46] More about Françoise Poivre in the next chapter.
[47] EE BSP_0143, Françoise Poivre to Bernardin, n.d. [1769?].
[48] On agricultural reform in early modern France, see Stockland, 'La Guerre aux Insectes'.

he observed that the island's community could not be established through the cultivation of sugar cane or indigo, as these (unlike in France's Atlantic colonies), were totally unsuitable for the small island. He argued that if the production of sugar, indigo, or coffee (as during the CIO's tenure), was prioritised, the lands and fields would yet again be completely denuded of trees – trees which were indispensable for the successful cultivation of food crops in that climate.[49] So, as those crops required a lot of land – land that was unavailable – the colony would always be tightly bound to mainland France, and its expenses would be overwhelming. In times of war, the island could not feed the troops stationed on it, and it would always be vulnerable to the enemy causing a famine by cutting off the seaborne food supplies. The fear of hunger, particularly in wartime, caused significant and ongoing uncertainty throughout the island.

Cereals had been the first crop, cultivated by the Dutch in the seventeenth century. Maize, native to the Americas, had been introduced to Mauritius under Governor Wreeden in 1670, but locusts destroyed the crops.[50] The problem of locusts was significant and remained a problem throughout the French rule of the island. Due to factors both human and non-human, the local administration was struggling to manage the island, in that while there were mouths to feed, the cultivation of spices was intended to generate a financial boost. One might well sympathise with the island's settlers, who were not enthusiastic about spice plants that could not feed them – and were, to boot, labour-intensive and tricky to acclimatise. The governor and the intendant had to balance the interests of different groups by increasing agriculture in the more fertile parts of the island. Each of the eight regions of the island (Pamplemousses, Montagne longue, Rivière du rempart, Flacq, Port Bourbon, Moka, Plaine Wilhem, and Port Louis) produced a certain number of crops such as maize and wheat. The astronomer Le Gentil observed during his stay in Mauritius that Pamplemousses and Flacq (a northern area whose name derives from the Dutch word '*Vlakte*', 'plain') were the most fertile areas of the island: the latter provided good rice and some wheat, while the former provided the best wheat, maize, and manioc.[51] The area around Port Louis, however, did not produce enough (only a little wheat, rice, and maize), and basic necessities had to be imported, as Governor Desroches reported to Praslin in July 1770.[52]

The cultivation of cereals was to be encouraged because, as outlined in an article, dated 17 February 1773, in Mauritius's gazette (weekly paper),

[49] AD Loire, FGB, 15 J 17, 'Observations de monsieur Poivre sur l'introduction de la culture des épiceries aux îles de France, de Bourbon et de Seychelles', 12 September 1773.

[50] Rouillard and Guého, *Les plantes et leur histoire à l'Ile Maurice*, 622.

[51] Le Gentil de La Galaisière, *Voyage dans les mers de l'Inde*, vol. 2, 630.

[52] ANOM, C/4/27, Desroches to Praslin, 20 July 1770, f. 149r.

'the necessity of attaching ourselves to those cereals, which, while assuring the life of the defenders of our homes, can alone protect us from the enterprises of a watchful enemy'.[53]

The article argues that the island was too remote and vulnerable, particularly in comparison to France's Atlantic colonies. Even in peacetime, the island could not build up stores and would thus remain an expensive enterprise because basic necessities still had to be imported from elsewhere. Poivre's argument in relation to deforestation in particular went hand in hand with his plea for the cultivation of fine spices, which would not require a lot of space; this was consequently, he felt, the only possible solution for the island. The demand for spices was high in Asia, he argued, whereas the European demand for coffee, indigo, and sugar was causing prices to rise in the Asian market, which could never be beaten by the production in Mauritius.[54] So staple grains were to be the principal object of the colony rather than those cash crops. Yet, in a letter to the minister of 23 February 1769, Poivre and Steinauer (Dumas's short-term successor before the arrival of Desroches as governor) pointed out the bad condition of the islands' storage depots, and asked for financial support for improvement works.[55] They argued that the island could only develop its full potential with financial aid from mainland France.

The argument was constantly used in the correspondence with Versailles; only with financial aid from the Crown could agriculture become the key to a flourishing island and a healthy population. In the hope of persuading the French ministries to cooperate, it was argued that the spice trade would bring a fortune to the island. Desroches envisaged that 'the Isle de France will be the centre of our commerce in Asia'. He continued:

> This view seems to have been shared by those who wished the Isle de France to be the warehouse of the nation's trade in the Indies, but they did not know about the correct application required by the knowledge of details of the trade with the Indies and the current state of the colony or because they have not thought about it – All that the Isle de France will not produce commodities of exchange, all that its population will be numerous and that consequently the matelots will be lacking and that the labour will be expensive there, as long as the food will be rare . . . and

[53] 'Mémoire concernant des observations sur différents points essentiels de la culture de l'Isle de France, avec un précis de la méthode générale du haut du Quartier de la Montagne Longue', in *Annonces, affiches et avis divers pour le colonies des Isles de France et de Bourbon*, 17 February 1773, 4.

[54] AD Loire, FGB, MS 15 J 17, 'Observations de monsieur Poivre sur l'introduction de la culture des épiceries aux îles de France, de Bourbon et de Seychelles', 12 September 1773.

[55] ANOM, C/4/25, f. 30, Poivre and Steinauer to the minister, *Demandes à la Compagnie pour les besoins de la Colonie*, 23 February 1769.

to say in passing that as its port will not be secure against storms the negociants of Europe will not use it as a stopover even when seeking to trade with Asia.[56]

In short, according to Desroches, in 1771 Mauritius produced enough for only three months: without the resources imported from Madagascar, India, and continental Africa, famine on the island would have been unavoidable.[57] The battle to feed the people of Mauritius was unremitting throughout the eighteenth century.

The financial support from France was insufficient, and the establishment of the island was therefore a matter of local management, drawing on local resources and knowledge. But at the end of 1771, after almost four years of improvements in agriculture, Poivre was still struggling to supply enough food for Mauritius. Writing angrily to Réunion's *ordonnateur*, Crémont, who was about to go off on a sightseeing trip to the island's active volcano, Piton de la Fornaise, instead of carrying out his tasks in the service of the island, Poivre exclaimed:

This is not the time, my dear Ordonnateur, to leave your post and to make a journey to the volcano while having ten thousand men to feed, we are lacking bread here and you are wasting the only two months of the year in which you can make us part of your abundance by stocking us up.[58]

Turning Mauritius into a self-sustaining island was such a difficult task that it was never achieved. Poivre complained in a letter to Praslin when he had to ask for provisions from Europe in 1769:

I arrived on an island which had been in continuous debt for ten years and which had experienced the horror of famine. I had found its cultivation abandoned and almost down to zero.[59]

But even though agriculture had been encouraged on the island, the harvests were still uncertain; Poivre tried his best to convince the colonists that the fruits of their labours would be for themselves and not for the Crown's stores (i.e. to supply the troops). He argued in his letters that he could not order the inhabitants of a colony facing near-starvation to sow and harvest for the Crown: they had to do so for their own needs, and then, once the cultivators were more at ease with the situation, the harvest could be shared.[60] Hence, the agriculture was to be increased to provide

[56] BnF, NAF 9347, f. 196r, Poivre and Desroches to Praslin, n.d. 'Extrait d'un mémoire intitulé, Iles de France et de Bourbon'.
[57] ANOM C/4/30, f. 260, Poivre to de Boynes, 7 February 1772.
[58] ADR, PR 12C, Poivre to Crémont, 12 November 1771.
[59] ANOM, C/4/25, f. 55r, Poivre to Praslin, 12 August 1769. [60] Ibid.

the troops, the navy, and the enslaved people of the island at the expense of the cultivation of cotton, coffee, and cinnamon.[61]

From his arrival in 1767, Poivre had instructed the colonists to focus on the cultivation of the staple grains, even though these had been neglected and badly maintained, that grew best on the island. Every year, between June and October, the harvests consisted of wheat, barley, oat, and beans and, between November and May, rice and maize. But the fields were never left to recover, nor were they worked thoroughly.[62] Poivre argued that these circumstances spoke in favour of the island's fertile soil, which was however seriously threatened by insects and other 'scourges'. Despite all the problems Poivre had to face in terms of agriculture, he was convinced that the land could produce a considerable amount of wheat, white rice, maize, and different kinds of beans: in its current state, however, the island could not utilise half of its productive potential except in terms of manioc, with an annual production of millions of roots.[63] Manioc could be planted from late August to late January because it was robust – a big advantage in terms of the island's difficult climates, and one which I will discuss later, in the next section, on food crops and slavery.[64]

So, the governor and the intendant instructed the settlers to focus on food crops. They were to sustain themselves by cultivating their gardens, as illustrated by the case of a private residence with fields in the Pamplemousses district. In the late 1770s, the occupants of a *habitation*, a private house, should produce livestock and bread, to be sold in the market, plus timber and fruits, including the mango, the *annona squamosa* (*attier*) with its sugar-apple (*pomme-canelle*), the French fig, the *annona reticulata* with its custard apple, also called the bullock's heart (*cœur de bœuf*), and the *jamrosa* (a type of guava).[65] But not all planters followed these instructions, instead deciding to make a bit of money on the side from coffee as a cash crop.

Nonetheless, Poivre reported at the end of 1769 that they had had a breakthrough, with a superbly rich harvest, to a degree that had never been seen before on the island.[66] Then in July 1771, Governor Desroches documented that the agricultural activities had increased considerably as a result of asking the population to plant for their personal advantage rather

[61] ANOM, C/4/25, ff. 9r–10v, Poivre to Praslin, 13 January 1769.
[62] Brest, DM, MS 89, no. 41, 'Agriculture', Poivre to Praslin, 30 November 1767.
[63] Ibid.
[64] 'Suite du Mémoire contenant des Observations sur la Culture de l'Isle de France', in *Annonces, Affiches et Avis divers pour le Colonies des Isles de France et de Bourbon*, 9 June 1773, 2.
[65] *Annonces, Affiches et Avis divers pour le Colonies des Isles de France et de Bourbon*, 15 January 1773.
[66] ANOM, C/4/24, f. 281r ff, Poivre and Steinauer to Praslin, report for the year 1769.

than forcing them to plant for the king.[67] With this relaxation, Desroches envisaged, the island would become less and less dependent on external supplies. But this plan did not work out; it was still impossible to feed everyone on the island adequately, let alone build up a stock of supplies.

The clash in Poivre's visions with those of Governor Dumas and his successor, Desroches, led to confusingly contradictory regulations being distributed to the cultivators. In a letter to Minister de Boynes in February 1772, Poivre expostulated that Desroches had discouraged the cultivators by reducing the price of goods whose price Poivre had fixed back in 1769.[68] In response, they had grown other crops, which they saw would be more profitable for them. Due to new internal administrative reforms, the regional wardens, who did not have much knowledge of agriculture, Poivre claimed, were under the governor's supervision, and Poivre, as intendant, had little influence over them. Those '*harlequins*' ('clowns') as Poivre described them in a private letter to an unknown recipient, had been brought into the island's administration, with duties and power over its various regional systems, including the police, the roads, the construction of churches, and parts of the justice system. Poivre commented ironically 'soon they will say Mass'.[69]

To cut a long story short, Poivre followed the strategy of satisfying the colonists' needs by allowing them to keep their harvests, and only then extending their cultivation efforts to fill up the storage. Building up a stock of locally grown produce was a necessity for the island's community but not for France, which was more interested in its vessels being supplied. The island was so frequently on the verge of famine, so its government was increasingly engaged with the search for basic necessities while desperately trying to grow food crops.

Food Crops and the High Number of Enslaved People

This section is concerned with the question of how the huge number of enslaved people in Mauritius shaped the cultivation of the food crops. Famines occurred more frequently with the increasing number of inhabitants, most of them the enslaved people brought in against their will. There is a clear relation between the slave population and the cultivation of certain crops. Even back in 1756, Alexander Dalrymple had confirmed that under Governor La Bourdonnais:

[67] ANOM, C/4/27, f. 159r, Desroches, 21 July 1771.
[68] ANOM, C/4/30, f. 260r, Poivre to the minister, 7 February 1772.
[69] ANOM, C/4/27, f. 53v, Poivre to unknown, 4 November 1770.

the inhabitants increased daily and tho' they feared a want of provisions, yet it was not without the force that Monsr. de la B[ourdonnais] could get them to till the ground, and plant manioc or cassava root, to prevent that disaster. The great number of slaves they had bought from Madagascar and Bourbon were almost starved.[70]

He stated that although manioc was also used in the West Indies to feed enslaved people and although La Bourdonnais had instructed every inhabitant to plant 500 square feet of manioc for every slave in his possession, the instructions were not followed or – worse still – the planters secretly poured boiling water over the seeds at night, to destroy them.[71] Commerson's writings allow us to understand more about this incident: the colonists were so suspicious of the manioc La Bourdonnais had introduced from Rio de Janeiro in 1739 or 1740 that they did this to demonstrate that the plants could not acclimatise in Mauritius.[72]

This example confirms that the introduction of certain staple crops – in relation to the increasing slave population in particular – had clearly begun under La Bourdonnais's tenure, when he and his successor had hoped to create an infrastructure on the island that echoed the formation of the island's identity.[73] Manioc, in fact, would only slowly prove itself useful as a foodstuff in the Mascarenes because the roots were toxic when raw, whereas in West Africa, once its preparation was mastered it became self-sustaining.[74] Nevertheless, in Mauritius the cultivation of manioc was encouraged from 1767 onwards because it was intended to be the main food source for the enslaved people and for cattle. It was particularly suited to Mauritius, because it grew in very dry soil.[75] The case of manioc thus combines three main factors of Mauritius's consistent challenges: the high number of enslaved people, the lack of agricultural know-how, and the challenging environment.

The 'Mémoire contenant des Observations sur différens points essentiels de la Culture de l'Isle de France', printed in the island's gazette, makes the point that manioc was not only a food source for enslaved people but was also used to make bread.[76] To supply the enslaved people,

[70] BL, Add. MS 33765, f. 2v, 'Geographical Collections of Alexander Dalrymple'. On the enslaved people's supply in Bourbon under the administration of the CIO, see ADR, C° 1476; 1477; 1478; 1479; 1480; 1481; 1482.

[71] BCMNHN MS 302, f. 11r, Commerson's journal.

[72] Ibid., f. 11r, Commerson's observation on the manioc in Mauritius.

[73] Roberts, 'Le Centre de Toutes Choses', 321, 322–7.

[74] Carney and Rosomoff, *Shadow of Slavery*, 54.

[75] Brest DM, MS 89, no. 41, 'Agriculture', Poivre to Praslin, 30 November 1767.

[76] See *Annonces, Affiches et Avis divers pour le Colonies des Isles de France et de Bourbon*, 15 January 1773. See also the editions of 17 and 24 March, in which an anonymous reader's recipe for bread-making, sweet crêpes, and pastry was printed.

it was recommended that 'the smartest and best Black person' of the *habitation* should be ordered to check the state of the manioc on a daily basis by testing its roots and picking the ripe ones to provide the other enslaved people with their daily rations.[77] Later, manioc was used as a substitute for wheat; the inhabitants were encouraged to make flour from it because it could be grown easily and in great quantities on the island. The article's author concluded:

> Among the objects of substance which attract the care of our colonists, manioc is unquestionably of the first rank by the importance of its function to feed the Black people, the resource of which it has been in times of scarcity.[78]

An anonymous letter to the gazette championed the advantages of manioc, in response to which the editor printed a recipe to make flour out of the cassava root.[79] The recipe not only referred to dough for bread but also suggested that manioc flour could be used for biscuits if eggs and butter were added. In the previously cited *mémoire*, however, the author referred to two main ways of using manioc to feed enslaved people: either by baking the root whole, or preparing it as a galette (a thin pancake) – less rich, though, than the baked chunks. Although the galette was regularly prepared for at least one-third of all the enslaved people on the island, this was, according to the authors, an absolute waste in light of the other method, which required less time to prepare and was more substantial.[80] In contrast, bread made from potato flour did not work well, so it was replaced by the *songo* root (a type of taro still harvested in the Mascarenes today for its corms, leaf stems, and leaves), yam, and other starchy crops mixed with wheat flour or cornflour.[81] In addition, an article in the island's gazette argued that bread made of the latter ingredients would be much better for the enslaved people because it would surely contribute to their health and strength.[82]

In times of famine, it was of course the enslaved people who were the first to be affected by the reduced maize harvests, not only in Mauritius

[77] Anon., 'Manière œnonomique d'exploiter une piece de manioc', in *Annonces, Affiches et Avis divers pour le Colonies des Isles de France et de Bourbon*, 10 August 1774, 2.

[78] Anon., 'Suite du Mémoire contenant des Observations sur différens points essentiels de la Culture de l'Isle de France', in *Annonces, Affiches et Avis divers pour le Colonies des Isles de France et de Bourbon*, 10 March 1773, 3.

[79] *Annonces, Affiches et Avis divers pour le Colonies des Isles de France et de Bourbon*, 3 March 1773, 2–3.

[80] Anon., 'Suite du Mémoire contenant des Observations sur différens points essentiels de la Culture de l'Isle de France', in *Annonces, Affiches et Avis divers pour le Colonies des Isles de France et de Bourbon*, 3 March 1773, 3.

[81] Anon., 'Moyen économique d'améliorer le pain de pommes de terre', in *Annonces, Affiches et Avis divers pour le Colonies des Isles de France et de Bourbon*, 8 September 1773, 2.

[82] Ibid.

but in Bourbon as well. Although Poivre had promised Réunion's director, Crémont, a large amount of maize for Réunion, Poivre could not keep his promise, due to shortages in the harvest in 1771.[83] He explained to Crémont that the maize provision for enslaved people was not secure and that he had had to send the Crown's enslaved people to the island's inhabitants to seek food (rather than giving them their rice and maize on their work sites).[84] Many planters, however, had been replacing their cultivation of cereal, wheat, and beans with coffee, so their grain stores were empty. As mentioned earlier, the plan in 1767 had been to discourage the colonists from cultivating coffee to focus on grain instead, but these regulations had not been followed, because the desperate settlers had hoped to make a little money from cash crops.[85]

The problem of supplying the enslaved people was not a new one. As early as 1754, the CIO had even made them subsist on *songo* roots.[86] *Songo* is similar to the *chou-caraïbe*, both belonging to the arum family.[87] The *chou-caraïbe* had arrived in Réunion via Mauritius in 1778 (maybe even before), when Céré and Cossigny had sent seeds to the gardener Joseph Hubert.[88] Its flower was said to have a warming effect: according to a printed *mémoire*, Hubert conducted experiments by employing a blind woman (his mother) to confirm the source of heat, which encouraged Hubert to use Réaumur's thermometer.[89] Even though the printed text states otherwise, Hubert includes in a handwritten note in the margins that *chou-caraïbe* was not native to the Mascarenes, but it came from Madagascar – like the *songo*, which grew wild along the river banks in Madagascar.[90]

Early in the 1750s, in light of the struggle with escapees rebelling and hiding in the forests, it was suggested that all the *songo* plants on the island should be destroyed, because the rebels were using them.[91] At the same time, Fusée-Aublet stated that *songo* could be a very useful food source in times of famine, and indeed it proved its value over the course of the

[83] Poivre to Crémont , 3 August 1771, printed in Ly-Tio-Fane, 'Problèmes d'approvisionnement de l'Ile de France', 112.
[84] Ibid.
[85] Brest DM, MS 89, no. 41, '*Agriculture*', Poivre to Praslin, 30 November 1767.
[86] ANOM, C/4/27, 'Statistiques pour l'année 1770'.
[87] Charpentier de Cossigny , Lettre à M. Sonnerat , 33.
[88] ADR 4 J 61, entry 78, 'Nottes des Graines que j'ay eu de l'Isle de France tant du Jardin du Roy que de Palma', August 1778.
[89] ADR, MS 4 J 72, 'Mémoire du Citoyen Joseph HUBERT, habitant de l'île de la Réunion … sur la chaleur naturelle des fleurs d'une espèce d'Arum indigène, à l'île de la Réunion'. This is a printed account of an extract of certain *Mémoires des sociétés savantes*. Hubert made handwritten side remarks to the printed text, stating that he said it was his mother and that he never said that the arum was native but that it came from Madagascar.
[90] Ibid. [91] Fusée-Aublet, *Histoire des plantes de la Guiane françoise*, vol. 1.

eighteenth century. Reports from 1769 reveal, however, the dangerous side effects of this root vegetable, which were said to reach epidemic levels among the enslaved people. But these descriptions may not have been true. *Songo* has a slight numbing effect, which made the Europeans fear it, and it offered enslaved people a potential opportunity to trick their masters by faking the symptoms. In this sense, the case of *songo* clarifies the relationship between plants and resistance.[92] The value of *songo* should therefore be seen as somewhere between foodstuff and a powerful weapon against oppression. In the later 1770s, Cossigny de Palma asserted that the *songo* root was not poisonous at all, and that he himself had eaten it during a famine on the island in the 1770s. In his writings, Cossigny de Palma used the Malagasy pronunciation of the word, *sonzes*, and said that the enslaved people sometimes ate the roots of two different kinds, a white and a red one, which were similar in taste.[93] Indeed, *songo* became an everyday foodstuff in the French colony, as shown by the bread recipe above.[94]

As explained earlier, the lack of food concerned not only the enslaved people but the island's population in general. The increase in agriculture and the rising number of enslaved people that came with that increase had a negative effect on each other, creating a vicious circle in the lack of provisions. Three new storage depots were established, one in Port Louis, one on the Great River, and a third in the Flacq area, in the hope that they might one day be filled.[95] In Flacq, manioc and a little cotton were planted: a sugar refinery and a forge made the area particularly attractive, even though it lacked useful timber.[96] For the agricultural development of the Pamplemousses area, Poivre required an additional 2,000 enslaved people. The Flacq area was said to be the most fertile part of the island, as several rivers ran through it; as it produced a lot of rice, coffee, and timber, Poivre planned to increase its slave population by 4,000.[97] In the other areas, he also planned to increase the slave population greatly, to work the 'provisions plantations' and the fields. In order to turn the island into an island of provisions, as envisaged by Minister Praslin, Poivre decided to send 5,000–6,000 enslaved people out to the poorest colonists engaged in agriculture, in the hope that their production would

92 See, in particular, Schiebinger, *Plants and Empire*.
93 Charpentier de Cossigny, Lettre à M. Sonnerat, 33; on the role of the Malagay dialects in the Mascarenes serving as a contact language, see also Larson, 'Enslaved Malagasy'.
94 Anon., 'Moyen économique d'améliorer le pain de pommes de terre', in *Annonces, Affiches et Avis divers*, 8 September 1773, 2.
95 Ly-Tio-Fane, 'Problèmes d'approvisionnement de l'Ile de France au temps de l'intendant Poivre', 106.
96 ANOM, C/4/18, f. 381r, Poivre to Praslin, 30 November 1767. 97 Ibid.

increase.[98] He planned to source 'those arms which agriculture lack[ed] on this island' from Madagascar and from Mozambique.[99]

As mentioned earlier, Poivre tried to solve the problem of supply by introducing more and more enslaved people. It was slavery that shaped the food crisis. The vicious circle between the enormous number of enslaved people and the lack of food is clearly illustrated by Commerson's expedition to Réunion's volcano. The group consisted of five masters and thirty-two enslaved people, who carried the food, Commerson stating that 'it is their [the slaves] number which made us hungry'.[100] The enslaved people outnumbered the French explorers by over six to one, and those enslaved people also had to eat. But the food they carried was not enough. In other words, just as in Mauritius in general, by trying to solve the problem of supply through introducing a high number of enslaved people, the French had created a new, and insoluble, problem.[101] Poivre's long-term plans could not succeed in just five years, especially through, in trying to solve a problem, creating a new one. This can be illustrated, too, by Poivre's strategy to fight locusts by introducing an insect-eating bird – which became a pest in turn because it bred too successfully.[102] This said, I obviously do not intend to compare enslaved people to birds, but simply point to the pattern in Poivre's strategic thinking in which he followed a clear path while ignoring possible consequences.

Fodder and Farm Animals

The problems of supply and the environmental challenges in Mauritius also applied to livestock; Mauritius lacked the animals needed to provide enough meat, milk, and clothes for the humans. So the search for useful plants aimed also at finding suitable fodder for animals, which was why grasses such as the *fatak* were introduced. Poivre explained:

> In many quarters there are large tracts of tilled ground, covered with grass of an extraordinary size, which grows to the height of five or six feet; the natives [the Malagasies] call it *Fatak*; it is excellent for nourishing and fattening their horned cattle.[103]

[98] Brest DM, MS 89, no. 41, 'Agriculture', Poivre to Praslin, 30 November 1767.
[99] Ibid.
[100] Commerson to Curé Beau, 16 February 1772, printed in Montessus de Ballore, *Martyrologe et biographie de Commerson*, 155.
[101] See also Mandelblatt's work on the relationship between food provision and slavery in the French Atlantic ('How Feeding Slaves Shaped the French Atlantic').
[102] See Stockland, 'Policing the Oeconomy of Nature'.
[103] Poivre, *Travels of a Philosopher*, 29. *Voyages d'un Philosophe* was published in 1768 and translated into English in 1770 as *Travels of a Philosopher*. Here, I quote directly from the English translation.

The problem, though, was that *fatak* started growing at the beginning of the rainy season, when the colonisers fed their cattle. But the grass soon died down, leaving only dry ground, with the attendant risk of wildfires during the dry season.[104]

Instead of importing more and more farm animals from Madagascar, Poivre's plan was to breed those on the island by instructing the inhabitants how to do so. He wrote to Praslin that the colonists understood that keeping herds was 'a most solid thing to do'.[105] In addition, the local administration was convinced that trading enslaved people with Madagascar instead of cattle was much more profitable, because the animals could be bred locally. So they planned to introduce other animal species from beyond Madagascar, in order to breed them; when Galloys was sent to China in 1767, he was instructed to collect not only specimens of useful plants but also animals, which could secure the meat supply on the island.[106] His haul included two Chinese rams and four ewes, several breeds of pheasants, hens, roosters, and ducks of Tibetan, Indian, and Chinese breeds, and birds which eat harmful insects, particularly the Chinese *hwamei* (*hoa-mi*).[107] Then a new species of fish was introduced to Mauritius. Native to the Moluccas and coming to Mauritius via Batavia in 1760 or 1761, the *Gourami* was imported. This fish, as Poivre explained to Bertin in a letter dated 26 June 1777, was far superior to all other kinds, requiring little space and water while breeding prolifically and fattening easily.[108]

In 1767, his first year in office, Poivre calculated that 12,000 livestock animals should be sufficient to multiply *ad infinitum*, resulting in a sufficient supply for the future.[109] Mauritius's pastures were excellent, he pointed out, calculating that they could feed more than 30,000 horned cattle without any reduction in the cultivation of the staple grains. These, however, turned out to be serious miscalculations, especially in light of the climate. Meat consumption was too high, Poivre then claimed, and the 'miserable butcher' had to produce meat in high quantities since he was instructed not to refuse meat to anyone, meaning that 160 to 180 bullocks were slaughtered between late summer 1769 and December 1770.[110] So the cattle-breeding plan did not work out, and Mauritius was obliged to continue its cattle trade with Madagascar.

[104] Poivre, *Mémoires d'un botaniste et explorateur*, 58.
[105] Brest DM, MS 89, no. 41, 'Agriculture', Poivre to Praslin, 30 November 1767.
[106] ANOM, C/4/18, ff. 16r–21r, 'Instructions pour M. L'abbé Galloys', n.d. [1766?].
[107] Ibid., f. 20v.
[108] Poivre to Bertin, 26 June 1777, printed in Cordier, 'Les Correspondants de Bertin, Secrétaire d'État au XVIIIe Siècle, IV. Pierre Poivre'.
[109] Brest DM, MS 89, no. 41, *'Agriculture'*, Poivre to Praslin, 30 November 1767.
[110] ANOM, C/4/25, f. 55r, Poivre to Praslin, 12 August 1767.

Mauritius remained dependent on its trading networks in the South-West Indian Ocean. For instance, in 1770, Poivre had to import meat from Réunion, writing to Réunion's administrator, Crémont, that 'we are lacking in many things, but above all meat'.[111] In addition, the administrators in Madagascar had not been able to send the required amount of '*viandes salées*' ('salt meat'), and this forced Poivre to acquire meat direct from France. Preserving meat in that tropical climate was far more difficult than in Europe, and in fact only possible at all in the cool season of May: at any other time of year meat would rot in transit from Madagascar to Mauritius.[112]

In an attempt to secure the stock of beef, two important *ordonnances* were declared on 5 May 1771: meat was reserved for the king's troops, the crews of royal vessels, and the sick in hospital.[113] In an earlier order, dated 24 October 1767, the local government had hoped to protect deer in the woods, brown goats, and game birds ('*le gibier à plume*'), particularly during the breeding season.[114] In the same *ordonnance*, fish were also to be protected, because they were an important food resource to supplement beef. Further regulations dated 3 July 1769 fixed the prices for poultry and other birds, plus milk, wheat, beans, peas, eggs, manioc, barley, and oats.[115]

These regulations bring us back to the beginning of the chapter: the question of how to feed every hungry mouth on the island. Although it was constantly and energetically tackled it was never solved. The island was overpopulated. The visiting troops added woes to the island's plight. In the planning of the acclimatisation and breeding of both plants and animals, the environmental factors hampered the administration's attempts. But climate was not the sole reason for the difficulties in introducing new livestock and fodder; the problems had arisen due to the lack of understanding, creating mismanagement, that I reveal throughout this book.

Trees, Materials, and Objects Required for Everyday Life

Mauritius was in need of plants to provide the materials required not only for agriculture and food production, but also for ship and house construction (timber, fat, wax, and gum), plus baskets and other everyday objects.

[111] Poivre to Crémont, 6 December 1770, letter printed in Ly-Tio-Fane, 'Problèmes d'approvisionnement de l'Ile de France au Temps de L'intendant Poivre', 108–9.
[112] ANOM, C/4/25, f. 55r, Poivre to Praslin, 12 August 1769.
[113] *Ordonnance* signed by Desroches and Poivre, 5 May 1771, in Delaleu, *Code des Isles de France et de Bourbon*, 226–7.
[114] *Ordonnance* signed by Dumas and Poivre, 24 October 1767, ibid., 214–16.
[115] *Règlement* signed by Desroches and Poivre, 3 July 1769, ibid., 220–1.

Commerson's notes relate not only to practically useful plants but also those grown for pleasure, such as the *hortensia* or 'Japanese rose' and the Malagasy *nymphaea*, a species of water lily.[116] Trees were of course indispensable, for shade and water; Poivre tried his best to protect the forests, and planted trees along the streets in Port Louis and along rivers.[117] Similarly, the shade of banana trees, for instance, was required for the cultivation of Ceylonese cinnamon (which needed a lot of watering, as well).[118] In 1769 Commerson examined the trees important for everyday subsistence, such as the ebony.[119]

It was in fact ebony that was perhaps the most useful Mauritian tree; its presence there had been one of the main reasons for the Dutch to take possession of the island. During his earlier travels in Cochinchina, Poivre had identified the *chao chagne*, a type of teak or hardwood, as the most suitable timber for construction. It was said to be 'more solid than that in India'.[120] Indeed, the teak (*tecq*) had been introduced to Mauritius from India; when Commerson examined it, probably shortly after 1770, he adjudged it an excellent tree for shipbuilding.[121] As Fusée-Aublet also referred to teak in his notes relating to the late 1750s and early 1760s, one species at least must have been introduced before or during his time on the island. The teak tree, however, was very vulnerable to storms, so did not grow successfully on the island, with its frequent cyclones.[122]

The environmental conditions continued to create problems. For instance, although Cossigny de Palma claimed in 1767 that he introduced the *benzoin* from Bengal, by 1779 there was just one little tree left in the Jardin de Pamplemousses.[123] Then the Chinese mulberry tree had been introduced under La Bourdonnais, for the production of silk on the island, but a report by the gardener Céré dated 1796 states that they were unsuccessfully cultivated due to insuperable physical causes.[124] It was replaced by the '*bois-noir*' ('blackwood'), from India, which was good for ship construction.

[116] BCMNHN, MS 1904, file IX, 'Labor botanicus in gallicana insula denno susceptus anno 1772'.
[117] Brest, MS 89, no. 57, Poivre to Praslin, 30 November 1767.
[118] ANOM, C/4/22, no. 66, Poivre to Praslin on 16 June 1768, secret letter.
[119] BCMNHN, MS 277, file. 2, piece 1, 'Ebenus (vera) declarata'; BCMNHN, MS 1904, file 9, 'Flore de Isle de France', 1772.
[120] ANOM, C/1/2, ff. 161–215, 'Journal d'un voyage à la Cochinchine', Poivre's journal, entry from 28 January 1750.
[121] BCMNHN MS 1904, file 10, 'Le Tecq'. On the teak tree, see also MS 887, III, and MS 1904.
[122] Rouillard and Guého, *Les plantes et leur histoire à l'Ile Maurice*, 414.
[123] Ibid., 340–1.
[124] MA, OA 127, Céré, 'Mémoire sur le Jardin National de l'Ile de France, adressé au Général Decarn le 1rt. Nivose an IV', 1796.

More successful examples were the ravensara and the ravenala, which acclimatised well in Mauritius. Incidentally, it was not, according to legend, the French who imported the ravenala but the Malagasy princess Bétia, when she emigrated from Madagascar to Mauritius in the 1750s.[125] The ravensara was apparently of such great interest to the people in Mauritius that they were unwilling to send it to Paris in 1768 because the seeds could only be planted when fresh: they were very rare and precious because only they contained the 'true taste and scent'.[126] The ravenala, meanwhile, remained an object of interest throughout the eighteenth century, and its seedlings and specimens were even sent to Paris. In August 1778, Céré sent specimens of the creole 'species of the Malagasy palm tree' – that is, the ravenala – to the botanist Le Monnier, together with a brief explanation of the Malagasy ways of using the leaves, which were waterproof, as plates and roofs.[127] Poivre had made similar observations in 1756. As it was easy to propagate, the ravenala was one of the island's most useful trees. Its leaves being very strong, a roof made of them would last for six or seven years. As the leaves were big, when they were supported by bamboo, they could even be used as walls and to make boxes. The trunk was used to build floors, while the flowers were used to add flavour to meals.[128]

Similarly, other trees were used in order to produce materials for food and ordinary objects. In his list of *Plantes cultivées au Réduit en 1759*, for instance, Fusée-Aublet included certain plants which he had of heard from the enslaved people.[129] These included the '*sauvurus anmus folie tiliaceo*', a sort of betel which the enslaved people called '*beaume*', and a certain '*tabernae montana*', which might have been the toad tree, native to eastern Africa and which Fusée-Aublet described as 'the timber with which the Black people make their spoons'.[130]

The coconut palm was said to be the most useful tree, because of the many sorts of things that could be made out its fruits, leaves, and timber. Poivre even exclaimed that in India 'the most useful of trees in their orchards is indisputably the coconut palm', especially because of its

[125] Rouillard and Guého, *Les plantes et leur histoire à l'Ile Maurice*, 504. More on Bétia in Chapter 4.

[126] BCMNHN, MS 277, III, 'Envoie de Graines de l'Isle de France en 1768 avec des observations y relativent', to Turgot.

[127] 'Envoi a M. Le Monnier', 30 August 1778, printed in *La Revue rétrospective de l'Ile Maurice 4* (1953), 299–300.

[128] BCMNHN, MS 1265, 'Le Raven-ala', n.d.

[129] BCMNHN, MS 452, 'Plantes cultivées au Réduit en 1759'.

[130] Ibid. As I encountered difficulties in reading and interpreting this manuscript, the spelling of some names might differ. See also the entry on '*melopepo*' (some sort of squash?) or '*bonduc*' and their uses by enslaved people.

nuts.[131] In the late 1750s he had researched the 'Malabar' agricultural methods for growing the coconut palm. He explained that trees were sown 25–30 feet from each other, and that they flourished best in mixed soil or sand. These trees were worth the effort: although they took about ten to twelve years to bear fruit, they would last for more than fifty.[132] The coconut palm tree was generally praised for its fruit by contemporary observers.[133] In contrast, however, Bernardin observed in Mauritius that the coconut tree was 'one of the most useful trees for commerce in the Indies, even though it hardly served except for its bad oil and bad ropes'.[134]

Many different kinds of timber were introduced to Mauritius to be used for construction and everyday objects. Sometimes, trees would be introduced without any of the (embodied) knowledge that came with it. Then, as several instances show, it was often the enslaved people who would inform the French settlers about ways of using the various parts of a tree.

Indian Jackfruit, South American Tomatoes, and European Strawberries

Relying on Poivre's description of the useful plants he introduced during his office, I observed that he enriched the island with fruit trees, like the breadfruit tree, and those with oily seeds.[135] The *rima*, wild breadfruit tree, had been introduced to Mauritius in 1755, when Poivre sent it from the îles Mariannes.[136] The 'true' breadfruit tree, however, discovered by James Cook (1728–79) in Tahiti in 1768–70, came to Mauritius only in 1797. According to a report entitled 'General State of Cultivated Trees Native to Several Parts of the Indies, Reunited on Isle de France', dated 24 May 1768 and written up by one De Reine, Mauritius was already quite well 'equipped' in terms of its natural variety.[137] By then, its useful flora

[131] Poivre, *Mémoires d'un botaniste et explorateur*, 63. Note that '*le cocotier*' is translated as 'cocoa-tree' in the English translation of the *Travels of a Philosopher*, 54; the translator probably confused it with the '*cacaotier*' or '*cacaoyer*'.

[132] Poivre, *Mémoires d'un botaniste et explorateur*, 64.

[133] Bernardin, OC, vol. 2, 98; vol. 1, 172, 178, 180–3. Though Bernardin was still convinced that its fibre, liquor, and nut could not compete with French linen, wine, and hazelnut.

[134] Bernardin, 'Voyage à l'Isle de France', 66–7.

[135] AD Loire, FGB 15 J 8, 'Etat dans lequel j'ai remis la colonie de l'Isle de France à mon successeur', 23 August 1772.

[136] Rouillard and Guého, *Les plantes et leur histoire à l'Ile Maurice*, 476; on breadfruit, see also Smith, 'Give Us Our Daily Breadfruit', 53–75.

[137] BCMNHN, MS 357, 'État général de arbres cultivés de divers parties des indes, Reunis a l'isle de france, et qui sont aux ordres de sa Majesté, pour l'ornement de [ces] terres chaude[s], et pour enricher ses colonies qui sont en deça de la ligne offert a Sa Majesté'.

ranged from 400 mango trees of different species, cinnamon from Ceylon and Manila, 64 coconut palm trees, 190 orange trees from China, Manila, Madagascar, Réunion, and others, date palms, banana trees from different parts of the East Indies covering 1,101 square feet of field, plus different species from China, including the litchi, mulberry, and two 'trees whose name is unknown', fruits and trees from Senegal, species from the Americas such as avocados, and some additional species from Madagascar.[138] In short, as stated in the statistics on cultivation activities for 1770: 'A taste for the plantation of all kinds of trees took hold in the colony.'[139]

Let us explore in greater detail the useful plants that were acclimatised on the island. During his voyages in the 1750s Poivre had already started sending plants for foodstuffs. For instance, he sent mango live plants from Java and the litchi and tea from China, exclaiming that he had no doubt that these plants could acclimatise perfectly in Mauritius, where they would be 'a sweet treat' for the colonists and the sailors.[140] Then in about 1750 Governor David introduced the jackfruit, up to 70 cm long and covered with rough coarse hair.[141] Other edible fruits which were successfully acclimatised in Mauritius were the gingko (also for remedies?), the jujube (Chinese red date), the date, the mangosteen, and the litchi. Indeed, they were growing long before Poivre's incumbency in Le Réduit in the 1750s and 1760s.[142] With the help of Cossigny, Poivre introduced the mabolo ('*diospyros philippensis*') from the Philippines, a tree which bears reddish-brown fruits with soft sweet flesh, which is why it is also called the 'velvet apple' nowadays.[143]

In the 1750s, Fusée-Aublet domesticated a number of other fruits and vegetables in Le Réduit, such as the 'margose' or 'bitter melon' ('*momordica charantia*'), a vegetable with a slightly bitter taste 'which the Indians in Mauritius eat in their curry'. Fusée-Aublet also cultivated two types of the luffa (*pipangaye* in French), a vegetable of the cucumber family ('*cucurbitaceae*') native to India. Fusée-Aublet identified it as '*Petola bengalensis, Cucumis actuangulus, papongajo*' ('*luffa acutangula*'), which required little cultivation and grew in almost every *habitation* on the island. It did so alongside the '*pipangaye unie*' ('*luffa cylindrica*'), which was identified as '*Petola, Momordica luffa*'.[144] Fusée-Aublet also acclimatised the '*poivron*',

[138] Ibid. [139] ANOM, C/4/27, f. 279r, 'Plantations', 1770.
[140] ANOM, C/1/3, Poivre to the secret committee, 31 December 1750.
[141] Fusée-Aublet, *Histoire des plantes de la Guiane françoise*, vol. 2, 139, 160.
[142] BCMNHN MS 452, Extract from Poivre's journal entry on the current state of Le Réduit, August 1767.
[143] Rouillard and Guého, *Les plantes et leur histoire à l'Ile Maurice*, 341.
[144] Ibid., 275. Fusée-Aublet used Rumphius's work as a reference point to identify these domesticable plants. Rumphius, *The Ambonese Herbal.* On Rumphius, see, for instance, Cook, *Matters of Exchange*, 329–38.

the *capsicum annum* or pepper, ranging from bell pepper to chilli pepper, which had been growing in Mauritius since the mid-eighteenth century.[145] In the late 1760s, Bernardin observed two types of '*petit piment*' ('small chilli pepper'), one of which was known in Europe and the other native to the island, whose fruits were very small and shiny like 'a grain of coral' on a 'nice green leaf'.[146] He also observed that the 'Creoles' used the latter for their stews, and that it was so spicy that it burned the lips like 'a caustic', which is why it was called '*piment enragé*' ('enraged chilli pepper'). These descriptions of the bitter melon and chilli pepper allow us to understand not only that foreign food crops were introduced but also *how* they were employed by the population in Mauritius, and above all how people hoped to remake their lives in a land foreign to them by employing ingredients 'from home' in their food.

The acclimatisation of plants included mainly vegetables native to the Indian Ocean world. These included the aubergine (or '*béringelle*'). By the late 1760s there were at least two species of this plant, a round yellow one native to Madagascar, and a fig-shaped violet one. Bernardin observed that the latter was 'not bad' when 'well-seasoned and grilled'.[147] Later, Céré cultivated the '*béringelle gigantesque*' in the Garden of Pamplemousses.[148] By the mid-eighteenth century, several species of '*brèdes*' ('greens') existed on the island, namely the '*brède martin* (*solanum nigrum*)', which Fusée-Aublet indicated and Bernardin identified as '*l'espèce de morelle*' and was eaten like spinach, and was also distributed to enslaved people. There was also the '*brède malgache*' or '*brette de Madagascar*'.[149] Fusée-Aublet identified other types of greens, including the '*brède malabar à piquants* (*amaranthus spinosus*)', which he named '*blitum spinosum*', and Céré cultivated the similar '*brède malabar*' and the edible amaranth considered as '*brède tricolor* (*amaranthus tricolor*)' in the Garden of Pamplemousses in the mid-1780s.[150] In addition, the betel, a plant native to Malaysia which had probably found its way to Mauritius via Madagascar, was introduced not later than the point when Fusée-Aublet worked on the island.[151] In addition, they introduced Chinese food crops to the island, such as the so-called '*blittum*', a vegetable widely consumed in China, eaten like spinach.[152] A different vegetable introduced to Mauritius was the so-called '*pè-tsai*', or pak choi, which Poivre compared to the European leek, explaining that, '*pè*' was Chinese

[145] Rouillard and Guého, *Les plantes et leur histoire à l'Ile Maurice*, 371. [146] Ibid., 372.
[147] Bernardin, 'Voyage à l'Isle de France', 212.
[148] Rouillard and Guého, *Les plantes et leur histoire à l'Ile Maurice*, 379. [149] Ibid., 380.
[150] Ibid., 431.
[151] Ibid., 440; see also Fusée-Aublet, *Histoire des plantes de la Guiane françoise*, vol. 2, 156.
[152] This is what Poivre explains in a letter to Bertin. Poivre to Bertin, 23 October 1778, printed in Cordier, 'Les Correspondants de Bertin, Secrétaire d'État au XVIIIe Siècle, IV. Pierre Poivre', 327.

for 'white' and '*tsai*', 'vegetable', and that it was of a better quality than ordinary leek.[153]

The species mentioned above reveal that the main sources of the food crops cultivated in Mauritius were Madagascar, India, and China. Yet, looking at the useful crops introduced to Mauritius shows the Atlantic–Indian Ocean connection as well, since several crops from the Americas were acclimatised; for example, the pineapple, native to the Americas, was introduced to Mauritius as early as 1753. The Abbé de la Caille observed it during his stay on the island, and it was later cultivated in the Garden of Pamplemousses under Céré.[154] Further, tomatoes, also known as '*pommes d'amour*' ('apples of love'), arrived from the Americas only in 1785, much later than other species.[155] It is difficult to ascertain precisely when the potato, and indeed precisely which species, was introduced to Mauritius. Bernardin, for instance, refers to the '*pomme de terre, S. americanum*', as being no bigger than a walnut.[156] In general, potatoes were easily confused with sweet potatoes, yams, and cambers/cambars/cambarres (a type of yam with a purple flesh). Fusée-Aublet observed the yam in Mauritius in the 1750s, naming it '*ubium vulga*' or '*ubium digitatum*', and Céré cultivated the '*cambare ordinaire*' in the Garden of Pamplemousses.[157] Belonging to the same botanical family, the air potato, native to India, was also imported to the island and continued to be grown as '*cambare indigène*' ('native camber') under Céré in the late eighteenth century.[158] A different kind of yam, the '*cambare Bety*' was apparently introduced by the Malagasy princess Bétia, in the 1750s. It became known as '*Igname de Mozambique*' under Fusée-Aublet and Céré, the latter describing it as better than all the cambers already known.[159]

This westwards cross-oceanic connection also reveals the flora native to Europe, which Fusée-Aublet domesticated in Mauritius in the late 1750s and early 1760s.[160] Later, Poivre conducted experiments with French strawberries, using seeds sent from France by the naturalist and gardener Duchesne in Versailles.[161] François-Etienne Le Juge (who had been

[153] 'Observations sur le mémoire envoyé de Chine, au sujet du PÈ-TSAI', n.d., in Cordier, 'Les Correspondants de Bertin, Secrétaire d'État au XVIIIe Siècle, IV. Pierre Poivre'.
[154] La Caille, *Journal historique du voyage fait au cap de Bonne-Espérance*, 237.
[155] Rouillard and Guého, *Les plantes et leur histoire à l'Ile Maurice*, 375. [156] Ibid., 381.
[157] Fusée-Aublet, *Histoire des plantes de la Guiane françoise*, vol. 2, p. 146.
[158] Rouillard and Guého, *Les plantes et leur histoire à l'Ile Maurice*, 541, 904. The name itself is a paradox, of course.
[159] Ibid.
[160] BCMNHN, MS 452, Extract from Poivre's journal, entry on the current state of Le Réduit, August 1767. As Fusée-Aublet's opponent, Poivre, pointed out, these had not been planted properly, and were not protected against the sun.
[161] BM Nantes, MS 2212, f. 2059, in his letter, Poivre thanks Duchesne for grains, 21 August 1769, a transcription is available at www.pierre-poivre.fr/doc-69-8-27.pdf (last retrieved 29 April 2016).

a commander in Gorée, then came to Mauritius as *conseilleur supérieur*, and died in 1766) also tried to acclimatise European fruit trees alongside Asian ones.[162] In an anonymous treatise, probably dated 1769, the author asserts that out of all the European fruit trees, only the peach bore fruit in the Mascarenes, but that trees from Africa and Asia fared much better.[163] Bernardin, however, said that European fruits and vegetables introduced to Mauritius failed not in terms of their ability to prosper but because of their taste. He explained that by sampling all fruits in Mauritius, he ran the risk of poisoning himself because of their low quality. He claimed that the guava tasted of bugs, the *atta* or sweet-sop was boring, the rose apple had a slightly sugary, insipid taste, and the papaya seeds tasted of cress. The mango alone never did any harm, despite its 'odour of turpentine', and could compete with European fruits.[164] These examples show not only that unknown fruit might have tasted strangely to a historical actor (and still do today), but rather illustrate the difficulties of acclimatising a European plant in the Mauritian climate, so different from that of France. Climate and environmental conditions such as rain, sunshine, and soil – as today – had an impact not just on the success or failure of the actual growth of a plant but also on the quality of its fruit.

Despite all the efforts made, documents related to the late 1780s make it clear that only few trees acclimatised well and continued to grow in the Garden of Pamplemousses (only four guava trees, four mango trees, four jackfruit trees, four avocado trees, four sweet orange trees, four litchi trees, four date palms, two breadfruit trees, two Malay rose apple trees, four mulberry trees, one coffee bush, and some seeds of Chinese plums).[165] Different documents from the same year show that live plants were made ready to be sent to Saint-Domingue, indicating that other plants and trees had been acclimatised successfully. These included several species of nutmeg, cloves, cardamom, and cinnamon, fruits such as the raspberry, the betel, the jackfruit, the Filipino velvet apple, the kaffir lime, the sweet orange, the mango, the litchi, the sago palm, the coconut palm, the ravensara, the ravenala, the raffia, plus '*vonampy*', '*voukoa*', '*bilimbi*', and '*biboa*', which I was unable to identify.[166]

As this section has shown, plants for foodstuffs came from all over the world: Europe, Africa, the Americas, and Asia. It shows, too, not only that

[162] Malleret, 'Pierre Poivre, L'abbé Galloys et l'introduction d'espèces botaniques et d'oiseaux de Chine à l'Ile Maurice', 124.
[163] Malleret refers to this anonymous memoir but does not provide a source, ibid., 128.
[164] Bernardin, OC, vol. 1, 172, 178, 180–3.
[165] MA, OA 127, 'Copie de la notte des plantes', 10 May 1788.
[166] MA, OA 116, 'Envoi des plantes à St. Domingue sur le navire particulier l'Alexandre', 10 March 1788.

a certain crop was introduced but *how* it was used and processed, sometimes revealing Indigenous preparation techniques which came to Mauritius, separately from the actual plant, as embodied knowledge. But even if plants did naturalise and bear fruits, the quality of the edible ones differed from that of those plants grown in their native climate, if we are to believe Bernardin. Acclimatisation was a delicate process, particularly if their native environment was so different from that of Mauritius. More about this in the next section.

Climate and Cultivation Techniques

How were plants cultivated in an environment similar to that of their origin but not in the same geographical setting? How did the transplantation process take place, and what were the factors that determined the successful naturalisation of a plant? In answer to these questions, this section follows the cultivation of several species and practices in relation to selected plants. I explore what happened to those crops once they arrived, and how they underwent their transition into the new environment.

When new trees and plants were acclimatised to Mauritius and multiplied there, the French gardeners needed tropical agricultural knowledge and skills. Drawing on various examples ranging from the cultivation of rice to cultivation techniques, I show that cultivation in Mauritius was a complex creolisation of expertise deriving from the local populations of the plants' native countries (including the knowledge of those who might have been enslaved), which meshed with the horticultural knowledge in the European colonists and the African and Asian enslaved people on the island.

Poivre had argued in the 1750s that although the most skilled settlers in Mauritius were those with knowledge about *European* agriculture, their knowledge was not suitable for the tropical island. He claimed that there was not a single one who knew how to treat the seedlings. Further, he affirmed that settlers lacked patience, diligence, and dedication in cultivating the foreign plants which could be of interest to the French kingdom but did not promise immediate profit, such as spices that were difficult to acclimatise.[167] And this was the only thing sought in Mauritius, as Poivre complained in 1754. He remarked – in so doing, promoting himself – that Mauritius was in need of a man who had travelled in the Indies or in tropical climates similar to that of the island, someone who was familiar with the agricultural methods practised by the people who lived in such climates.[168] Certainly, he argued, the island had no need of a person who,

[167] ANOM, C/4/8, letter 20, Poivre to Michau de Montaran, 10 January 1754.
[168] Ibid.

knowing only about European agriculture, expected to cultivate nutmeg in Mauritius 'like chestnut in France'.[169]

Even though one must consider Poivre's self-fashioning, his arguments in relation to know-how, cultivation, and climate seem logical. The European plants transmitted to tropical climates also required different treatment as a result. According to Poivre, his rival botanist Fusée-Aublet did not take into account the special needs of European plants in tropical climates. Poivre's frustration with 'poor cultivators' in their lack of tropical agricultural skills remained one of his biggest problems.[170] He reported to Praslin in November 1767 that the colonists had started to increase the cultivation of cinnamon and pepper, but not in the proper manner; he concluded that he had not seen a single pepper plant cultivated in the correct way.[171]

Indeed, handling plants and getting them to grow well required a considerable degree of natural knowledge. We shall start with rice. Governor La Bourdonnais had already introduced several species of rice and maize into Mauritius, cultivating fields in the south east of the island.[172] When Poivre travelled in East Asia shortly after La Bourdonnais's tenure, he found out about the dry rice, a type that required little water, warmth, and humidity to grow. He had great hope for dry rice when he shipped it to Mauritius because, so he argued, it was finer in taste, produced a greater crop than any other species, and was easier to cultivate because the fields did not need to be flooded.[173] Once back on the island, he decided to employ the cultivation skills he had documented during his travels. As he observed in *Voyages d'un Philosophe*, Mauritian cultivation of wheat, rice, and maize was conducted as in Madagascar, and these methods were effective; although cereal cultivation was somewhat neglected and not properly understood, or so Poivre claimed, Mauritius did manage to produce a fair quantity of those cereals.[174]

Dry rice, however, required different treatment. Following the Malagasy method, seed was sown at the beginning of the rainy season. Although some of it failed, rain was sufficient for most of the dry rice to

[169] Ibid.

[170] For instance Poivre to Galloys, 16 February 1769, printed in Laissus, 'Note sur les manuscrits de Pierre Poivre', 45.

[171] Brest DM, MS 89, no. 41, 'Agriculture', Poivre to Praslin, 30 November 1767.

[172] Rouillard and Guého, *Les plantes et leur histoire à l'Ile Maurice*, 611.

[173] Poivre, *Mémoires d'un botaniste et explorateur*, 87.

[174] Maize is indicated as '*le blé de Turquie*'. Commerson elaborates on the name and points to the fact that 'maize' was not known in France but rather '*gros bled*' (Commerson to Cossigny, 1770, printed in Cap, *Philibert Commerson naturaliste voyageur*, letter 10. Poivre, *Mémoires d'un botaniste et explorateur*, 57–8).

germinate, grow, and ripen.[175] The crops, however, suffered when rain and humidity lacked. Not only that, but while dry rice ripened fifteen to twenty days earlier than the other kinds, so could be harvested before the cyclone season, the colonists did not appreciate 'nature's gift', as Poivre called it. Worse, the enslaved people, badly instructed, had failed to sow the different types of rice in separate plots, then closely observe the dry rice in order to harvest it when ripe, had mixed the different types of rice together in the fields. So, slowly but surely, the dry rice was lost.[176] When crops unknown to the island's population arrived, the agricultural techniques (such as for rice) were experiential, and traditional or people's own methods of cultivation could not be used because dry rice had to be handled differently. But the Malagasy methods of rice cultivation were the dominant form in Mauritius – they had come, of course, with the large number of Malagasy enslaved people.

Poivre claimed that even though food crops and other useful plants, such as coffee, cotton, indigo, sugar cane, pepper, cinnamon, and the mulberry tree, had already been introduced under La Bourdonnais, they had not been handled properly.[177] He argued that the experiments conducted in only a superficial manner had not been successful. If a simple plan had been followed, Poivre argued, to 'secure bread' on the island, Mauritius would have been flourishing before his arrival in 1767. Thus, upon his arrival as intendant, the cultivation of fruits and vegetables was to be increased around the port by encouraging 'skilled cultivators in the gardening parts' and stocking the depots with wheat and rice from all garden owners.[178] Furthermore, Poivre decided to distribute acclimatised specimens from his garden to all 'skilled cultivators', hoping that their successful propagation would one day form the wealth of the colony.[179] In 1770, cultivators began to plant fruit trees: even the people who lived in the port, who had previously ignored the need for horticultural activities, were as ordered by law to plant trees along the roads and in their houses.[180] Even Governor Desroches, who was not much of an agronomist, started a small garden at the foot of one of the mountains in order to encourage the colonists to do the same.[181]

[175] Dupont de Nemours, *Notice sur la vie de M. Poivre*, 62. On the tools employed in Monplaisir, see MA, OA 127, 'Etat des outiles et essentiels d'agriculture dépendant de l'habitation de Monplaisir'.

[176] Poivre, *Mémoires d'un botaniste et explorateur*, 87.

[177] Poivre, Travels of a Philosopher, 39–40. See also ANOM, C/4/27, f. 257r, 'Graines et approvisionnement du cru de l'Isle', 1770.

[178] ANOM, C/4/27, f. 257r, 'Fruits & légumes', 1770.

[179] ANOM, C/4/27, f. 279r, 'Plantations', 1770. [180] Ibid. [181] Ibid.

To boost the spread of information around the island, the first printing house, the Imprimérie Royale of Mauritius, was founded in 1768 in the rue de l'Hôpital in Port Louis.[182] A printing house in Réunion followed, in 1777. The idea was to inform the island's inhabitants of national news, agricultural know-how, and topics important to the colonists, such as slavery and foreigners. The printing house grew in importance when, for instance, instructions on the planting of certain plants and spices were distributed to the inhabitants, as in 1772.[183] The need for precise instructions probably has a prehistory. When seeds and grains were imported in 1770 and handed to the inhabitants for distribution, they could not fulfil Poivre's expectations. He complained in 1771:

> Our colonists are not quite farmers [enough] to have given the plants the precise care which they require, and I would have done much more to the advantage of the colony if I had charged myself with the cultivation of all plants and seeds.[184]

In 1770, the Parisian printer Pierre Nicolas Lambert became director of Mauritius's Imprimérie, where he founded the gazette *Annonces, affiches, et avis divers pour les colonies des Îles de France et de Bourbon*, which came out on 13 January 1773.[185] The gazette, which was printed and distributed to the colonists every Wednesday between 1773 and 1790, was exactly what Poivre would have needed in order to keep his settlers informed and to distribute instructions more easily.[186] But it was only a year after his tenure, that newspapers were regularly distributed, informing the colonists and inhabitants about the state of agriculture, cultivation methods, recipes, and general news concerning island life and Europe. For instance, an anonymous *mémoire* printed in the gazette illustrates the difficulties of cultivation with regard to finding the right season and the influence of climate. Apparently, the inhabitants claimed that the seasons for cultivation differed from one region of the island to another, but the administration asserted that the season for cultivation was the same in all its regions:

> Despite what some cultivators say about it [the time favourable for the planting of each species], the differences of seasons, from one quarter of the Isle to another,

[182] Toussaint, *Early Printing in the Mascarene Islands*, 29.

[183] AN, MAR G/101, file 4, 'Instructions sur la manière de planter et cultiver avec succès les plants et graines de gérofliers et muscadiers. A l'usage de M. M. Les Habitants des Isles de France & de Bourbon', 1772.

[184] ANOM, C/4/29, f. 13r, Poivre to Praslin, 2 April 1771.

[185] *Annonces, affiches, et avis divers pour les colonies des Îles de France et de Bourbon*, no. 1, 13 January 1773, 3.

[186] See http://dictionnaire-journaux.gazettes18e.fr/journal/0030-affiches-des-iles-de-france-et-de-bourbon (last accessed 15 April 2016). Feyel, *L'annonce et la nouvelle*; Labrosse and Rétat, *L'Instrument périodique*; Toussaint, 'Les débuts de l'imprimerie aux Iles Mascareignes', 7.

are not very noticeable, and it may be said that the vegetation in the colony is subject to the influences of the same climate. The rains, more or less frequent from one part to another, come from foreign causes, which should not prevent the observation, throughout the island, of the same weather conditions for plantations of the same kind.[187]

The rainy season was a very important factor. For instance, the season for planting maize, several species of peas (*'pois du Cap'*, *'pois d'Achery'*), cucumbers, and rice started at the end of November and went on until late January, the start of the rainy season.[188] As, though, there was a risk of losing the crops because of downpours in October, it was recommended not to try to kill two birds with one stone by preparing the same fields too early for the cultivation of wheat (whose season started in late May and lasted until mid-July). Doing so could result in the young and delicate plants being washed away by floods.[189] The season for planting oats and most vegetables, potatoes in particular, was March; meanwhile, the transplanting of fruit trees and young trees for timber could be done between mid-June and mid-August.[190] These examples illustrate the difficulty of ascertaining the right season for the cultivation and transplanting of seeds and plants, and the general ongoing mingling, rupture, and rejection of horticultural techniques. They also make it clear that climate was not the only reason why it was difficult to introduce new livestock and fodder; this was also due to a lack of understanding, resulting in mismanagement.

Soil, Cyclones, and Plagues

Mauritius was, and is, a relatively small island in the middle of the Indian Ocean. Because such a small ecosystem can be easily unbalanced, European settlement led to rapid environmental changes. The Europeans were indeed facing an island that was unfamiliar to them not only climatically but also in its landscapes and soils. As Richard Grove has explored in great detail, the ecological ramifications of human influence over the course of the eighteenth century, deforestation in particular, were devastating.[191] Mauritius was often buffeted by wild storms, destroying not only its crops but also the cargoes of ships in harbour. Another factor was the quality of the soil, which played an important part in the cultivation of plants. Then the human-generated factors, such as soil

[187] 'Suite du Mémoire contenant des Observations sur la Culture de l'Isle de France', in *Annonces, Affiches et Avis divers pour le Colonies des Isles de France et de Bourbon*, 9 June 1773, 2.
[188] Ibid. [189] Ibid. [190] Ibid.
[191] Grove, *Green Imperialism*. See also Brouard, *Woods and Forests*.

erosion and deforestation, added to the mix. Despite the efforts the French settlers made to mitigate all those, it was the environmental conditions and the plant material itself that determined the outcome of the Mauritius project. Only slowly did the French colonists become aware of novel climatic features which meshed with the required adaption of horticultural knowledge in a tropical climate.

Successful cultivation was (and still is) determined by factors both non-human and human, the former including soil, sun, cyclones, plagues, and the plant material itself; the latter including cultivation know-how, its correct application, tools, and an understanding of the plants' needs. Poivre stressed the high quality of the Mauritian environment, which was, according to him, generally more fertile than that of Europe.[192] In terms of the soil, he asserted that it was one of the best to be found in the world.[193] Yet, finding the right season for planting any specific crop was only one factor of the many determining success. In addition to plagues of insects, storms limited the spread of the island's crops;[194] in March and April 1772, cyclones hit the island so hard that not only were the crops destroyed, but also the costs of the reparations for the buildings, and the vessels sheltering in Port Louis, were higher than any before.[195] The storm devastated both Mauritius and Réunion; there is evidence of forty dead and much of the plantation destroyed.[196]

Soil was another important factor. Some sources, such as the account by Dalrymple, state that rather than Mauritius, it was Réunion that provided natural resources of higher quality, such as pure air and superior soil:

> This island of Bourbon . . . was preferred to [Mauritius] which had nothing but its port to recommend it, on account of the purity of its air, and fertility of its soil, in which it far exceeded the other.

And:

> Mauritius being of a barren soil, much inferior to that of Bourbon, it was difficult to improve it sufficiently to furnish provisions necessary for the inhabitants and the ships that touched there.[197]

[192] ANOM, C/4/18, f. 398r, 'Forges de l'Isle'.
[193] ANOM, Col C/4/27, f. 252r, 'Mémoire sur l'Isle de France'.
[194] On cyclones in nineteenth-century Mauritius, see Mahony, 'Genie of the Storm'.
[195] Poivre, 'Mémoire concernant le précis de l'examen qui a été fait par arrêt du Conseil, de l'administration du S. Poivre, intendant de la Marine aux isles de France et de Bourbon', 15 March 1774, printed in Laissus, 'Note sur les manuscrits de Pierre Poivre', 54–6.
[196] BSG, MS 2551, f. 86r, Lanux junior to Pingré, 30 July 1772. Lanux even refers to three storms (29 February, 2 March, and 14 April 1772).
[197] BL, Add. MS 33765, ff. 2r–3r, 'Geographical Collections of Alexander Dalrymple'.

Other sources, such as an account by the Briton Richard Smith in 1763, tell us that Réunion's soil was:

> in general fertile tho the mould is not above 2 or 2 ½ feet deep under which is a [layer] of rock, and produces plentifully with very little wheat, oats and more other Europe[ean] grains, roots and pulse and yields two crops a year.[198]

In addition to the problems relating to its soil, Réunion was 'much infested by caterpillars, locusts and other insects and by rats and ... birds' which damaged the crops, sometimes excessively.[199] Similar problems occurred in Mauritius: plagues of locusts would destroy the crops. According to Poivre's ecological plan, these were to be fought with the help of the *oiseau martin*, a bird used to regulate the locust population.[200] These problems with insects, rats, and monkeys affected the island badly as early as the 1750s.[201] Natural conditions – climate, plagues, and storms in particular – had a strong influence on not only the degree of successful acclimatisation of crops, but also their harvest, and throughout the period under review nature remained a salient factor.

Conclusions

The attention paid in this chapter, together with Chapters 4 and 6 in particular, was to focus on knowledge-making in Mauritius itself, rather than knowledge transfer from the colony to France. I have presented the extraordinary variety of plants and animals that, introduced to the island over the course of the eighteenth century, were used for the population's nourishment, subsistence, and everyday life. I have explored not only the fact that plants of different kinds and different purposes *were* introduced but more importantly *how* they were meant to be cultivated and used by focusing on local forms of knowledge. Furthermore, I have located agricultural augmentation and cultivation activities on the island's challenging location where it stands, small and remote, in the middle of the Indian Ocean.

Closely linked to the question of feeding the island's population was the huge number of enslaved people forced to migrate to it over the course of the eighteenth century. In comparison to the cost of foodstuffs, enslaved people were cheap, and they were employed for the cultivation of domestically produced stock. I have not only elaborated on the relationship between the high number of enslaved people and the need for agriculture,

[198] BL, MSS Eur/Orme OV 4, f. 69v, Smith, 'Account of the Island of Bourbon in 1763'.
[199] Ibid., f. 72v.
[200] See on this particular matter, Stockland, 'Policing the Oeconomy of Nature'.
[201] See BL, Add. MS 33765, f. 11r, 'Geographical Collections of Alexander Dalrymple'.

but also given examples of the introduction of certain food crops in relation to slavery. Plants such as manioc, maize, and *songo* were cultivated with the specific purpose of feeding the island's forced labourers. It was slavery that created and shaped the island's crisis of provision.

The local knowledge far exceeded European knowledge and its knowledge practitioners, as well as the European plants. Here, I have argued for the complex interplay of human-generated factors, environmental conditions, and the plants' needs when acclimatising them on the island. I have elaborated on the diverse factors influencing the cultivation of the crops, which would often fail due to the impact of humans on nature. In an island context – and surely also in Europe at the time – systematic inquiries and methods of increasing supplies faced immense problems and challenges, ranging from the actual search for useful crops to their acclimatisation and harvest. The attempts at agricultural and economic improvement made by the local administration were based on many uncertain factors, both human and natural.

Closely connected to this were the environmental and administrative obstacles which the local administration had to overcome, and I have shown how the governor and intendant tried to manage and balance the various interest groups. The chapter also touched on the spice trade, and how it must be understood from the perspectives of both the local administration and the colonists. While the island's administration was trying to bring the uncertain situation under control, even though the cultivation of consumable plants was more important than the cultivation of spices, the instructions given by the authorities to the islanders were often contradictory, creating confusion and a lack of concrete results.

4 Enslaved People as Knowledge Carriers

When Pierre Poivre moved to Mauritius, his wife Françoise, née Robin (1748–1841; almost thirty years his junior), went out there as well.[1] From 1768 to 1770 she carried on a lively correspondence with Bernardin while he was in Mauritius. In 1769, she wrote to him about his manuscript on the enslaved people's situation on the island, later published as part of his *Voyage à L'Isle de France* (1773). In her letter, Françoise commented on his manuscript, which she felt drew a dark picture of her island home. She asserted:

> If this country was cultivated by free men it would be a very happy place. A climate which does not allow an always green countryside, a land which produces two harvests a year without ever resting, very beautiful woods, many rivers, not very pleasant, it is true, but which always fertilises. You cannot picture this island after a long stay at the port. The port does not resemble at all the rest of the country.[2]

Françoise Poivre's ideas echoed her husband's speech on his arrival as intendant of the Mascarene Islands in 1767; he had announced that an agricultural island colony such as this one, which was precious to the French colonial attempts in the East Indies, must be cultivated by 'free hands'.[3] This was an ambitious vision, and this chapter highlights the practical realities of Poivre's acclimatisation garden Monplaisir (see Figure 3.2), and in Mauritius more widely in the second half of the eighteenth century. What could 'free hands' mean in practice on that island, populated as it was by thousands of enslaved people? As Suzanne Miers put it: 'No definition of slavery can be separated from the definition of its antithesis – freedom.'[4] Likewise, it is important to understand what 'Black' meant. In *Creating the Creole Island*, Megan Vaughan has explored slavery and bonded labour with regard to Mauritius, concluding that the

[1] When Poivre died, she remarried the economist Dupont.
[2] EE BSP_0140, Françoise Poivre to Bernardin, 1769.
[3] 'Discours Prononcé par P. Poivre, à son arrivée à l'Isle de France, aux habitants de la Colonie assemblés au Gouvernement', in Poivre, *Mémoires d'un botaniste et explorateur*, 130.
[4] Miers, 'Slavery', 2.

term '*Noir*' ('Black') was a very slippery one. It was used to denote not only someone of African ancestry, the African mainland or Madagascar (and some Malagasies also must have had Indonesian heritage because of an earlier forced migration of Indonesians), but also those of Asian origin.[5] 'Noir' was by no means a single, stable term: indeed, the sources rarely, define whether 'Noir' meant 'enslaved' or referred to skin colour; nor do they offer information about where 'Noir' originated. In brief, the term 'Noir' and the boundaries between free and unfree were complex and messy, determined by race and ethnicity.[6]

It is against the background of slavery, ethnic prejudice, economic motives, and anti-slavery sentiments that this chapter analyses the contribution of the enslaved people to the plants used for subsistence, healing, and commerce in Mauritius. It puts a particular focus on cultivation activities, plant knowledge, and the tasks that enslaved people carried out in the gardens of the island.[7] In so doing, it draws attention to food, practices, and cultural identity, yet also to the human agency embedded in the social structures and – in a wider sense – the power relations that shaped life in Mauritius. I explore the relationship between the colonisers who employed the enslaved people and the free people of colour from Africa, Madagascar, and South Asia, who worked on botanical expeditions and in the colonial gardens.[8]

After a general discussion of enslaved people in the historiography of early modern science, my first step is to elaborate on forms of slavery and bonded labour in the Indian Ocean world. Second, I explore the links between slavery itself, the treatment of enslaved people, and political economy. It is important to understand slavery in Mauritius in the contexts of both Enlightenment thought and that of bonded labour in the Indian Ocean more generally. Then in my third step I probe the relationship between caste, race, and hierarchy in relation to the tasks undertaken by the enslaved people. Finally, by looking at the daily garden management, I examine the cases of the enslaved gardeners Hilaire and Charles Rama, focusing on the impact of both skills and interpersonal relations in the context of slavery.

[5] Vaughan, *Creating the Creole Island*; Larson, *Ocean of Letters*; Marsh, 'Territorial Loss and the Construction of French Colonial Identities', 1–13.

[6] On the legal aspects and slavery in the French colonies, see Boulle and Peabody, *Le droit des noirs en France au temps de l'esclavage*. For a discussion of law and slavery in the Atlantic world, see Peabody and Grinberg, *Free Soil in the Atlantic World*.

[7] On eighteenth-century anti-colonialism, anti-slavery, and free/unfree labour, see Bénot, *Les Lumières, l'esclavage, la colonisation*; Oudin-Bastide and Steiner, *Calcul et morale*; Duchet, *Anthropologie et histoire au siècle des Lumières*; Duchet, *Le partage des savoirs*.

[8] On the rise of the free population of colour in Mauritius, see Allen, 'Economic Marginality', 126–50.

The Place of Enslaved People in Natural History

Even though not rooted in the history of science as such, in the West Indies context, scholars such as Virginia Bernhard have long argued that in the establishment of Bermuda enslaved people were introduced not merely for their labour but in many cases for their expertise as pearl divers or as cultivators.[9] In the history of science some important studies have been written, recognising and appreciating the knowledge of those who were enslaved. The perception of enslaved people as agents of natural history makes a significant contribution to the points that scholars began to raise on the connections between knowledge production and the slave trade.[10] The significance of an informal network of enslaved people as carriers of knowledge can be seen in, for instance, the case of Henry Smeathman in Sierra Leone.[11] Similarly, Anna Winterbottom showed that in the settlements of the East India Company, where enslaved people formed considerable percentages of the populations, they were valued not only for their physical work but also as a source of knowledge.[12] Similarly, Kathleen Murphy concludes that the famous British apothecary and natural historical collector James Petiver 'transformed the routes of the British slave trade into a means of collecting collectors'.[13] Also, James Delbourgo's most recent research on the enslaved people in the world of the British physician and collector Sir Hans Sloane emphasises the importance of enslaved people in the collection of materials and the circulation of the goods that came with the slave trade.[14] As a corollary to that, as Judith Carney and Richard Rosomoff explain, the oral traditions of Black descendants of runaway enslaved people in north-eastern South America relate that enslaved African women would, prior to being taken aboard the slave ship, conceal rice seeds in their hair or in that of

9 Bernhard, *Slaves and Slaveholders in Bermuda*; Bernhard, *Tale of Two Colonies*; Carney and Rosomoff, *In the Shadow of Slavery*; Carney, *Black Rice*; see further on the question of slavery, skills, and artisanship Newton and Lewis, *Other Slaves*.

10 Delbourgo, 'Gardens of Life and Death', 114; Sweet, *Domingos Álvares*; Gómez, *Experiential Caribbean*; Carney and Rosomoff, *In the Shadow of Slavery*; Carney, *Black Rice*; Murphy, 'Translating the Vernacular'; Scott Parrish, 'Diasporic African Sources of Enlightenment Knowledge'; Schiebinger, *Secret Cures of Slaves*; Delbourgo, *Collecting the World*. Although Brown is not primarily concerned with natural knowledge, his study about slavery and death in Jamaica provides significant insights (*Reaper's Garden*). For examples of plants and herbs used by Africans in the Atlantic World, see, for instance, Grimé, *Ethno-Botany of the Black Americans*.

11 Starr, 'Making of Scientific Knowledge in an Age of Slavery'.

12 Winterbottom, *Hybrid Knowledge*, 164.

13 Murphy, 'Collecting Slave Traders', p. 659.

14 Delbourgo, *Collecting the World*; Delbourgo, 'Sir Hans Sloane's Milk Chocolate'.

their children.[15] Most likely, other natural goods would have travelled similarly via unofficial – often untraceable – channels.

Most of these studies focus on the British and Iberian Atlantic, analysing the contribution of enslaved people in that area to European science. In relation to the knowledge traditions of those who had been enslaved, Kathleen Murphy observed that in terms of African knowledge in the eighteenth-century British Atlantic, the distinction between 'know-how' and 'scientific knowledge' was inextricably connected to issues of race and ethnicity.[16] She asserted that although the colonials would appropriate enslaved people's knowledge and expertise, it was nevertheless regarded as simple 'know-how' rather than genuine knowledge. Similarly, Susan Scott Parrish has posited that diasporic African knowledge was a source of Enlightenment knowledge; Europeans overseas admitted their 'reliance on colonial correspondents and, beyond that, on the Africans who did the collecting for them. Indeed, there were some facts that, unless they originated with Africans, had no credibility.'[17] However, as Parrish argues, this reliance was very complex because of the political conditions and the plantation culture, and would often go hand in hand with the demonisation of diasporic African traditions.[18]

Atlantic studies have achieved a great deal by looking into the ways in which local communities remade their lives. For instance, in her groundbreaking study on the origins of rice cultivation, Judith Carney explored the Americas.[19] She corrected the fallacy that it had been Europeans who had introduced rice to West Africa, whence, together with the relevant agricultural know-how, it travelled to the Americas. She further suggested that distinctive agricultural knowledge 'transmitted through practices and technologies to make nature yield ... followed different paths of development'.[20]

Many studies have now been written on Black medicinal knowledge. Londa Schiebinger's latest study on the French and British Atlantic worlds has a similar theme, examining the complexity of the circulation of human agents, sickness, and medicinal plants along with the knowledge of them. For cures and healing, Amerindian, African, and European knowledge competed on West Indies plantations – yet not all knowledge

[15] Carney and Rosomoff, *In the Shadow of Slavery*, 88, 88–94.
[16] Murphy, 'Translating the Vernacular'.
[17] Parrish, 'Diasporic African Sources of Enlightenment Knowledge', 283. On the Dutch Cape garden as a middle ground, see also Fleischer, '(Ex)changing Knowledge and Nature at the Cape of Good Hope'; Augusto, 'Knowledge Free and "Unfree"', 136–82.
[18] Scott Parrish, 'Diasporic African Sources of Enlightenment Knowledge', 283.
[19] Carney, *Black Rice*. [20] Ibid., 5–6; Carney and Rosomoff, *In the Shadow of Slavery*.

traditions were seen as equal and, for a variety of reasons, the African therapies were not even tested.[21] In the history of medicinal knowledge in particular, scholars have endeavoured to understand the knowledge production outside the frameworks of European science and medicine, as in, for example James Sweet's work on an African healer, Domingos Álvares.[22] The study by Pablo Gómez on seventeenth-century Black and free Caribbean communities shines a light on how the leading figures of these communities made use of medicinal plants for their own purposes. He argues that the testing of remedies, disease origins, and cures made colonial settlers and Caribbean people view 'experiential knowledge' as powerful and competitive.[23] Since then, Kalle Kananoja, basing his study on West Central Africa, has published a ground-breaking study of medicine in Africa and the wider Atlantic world, arguing that African knowledge was central in the shaping of responses to illness.[24]

Building upon the important works cited above, I consider enslaved people in Mauritius as signal agents and active carriers of plant knowledge, from the agricultural to the medicinal, acting both as collectors and as skilled cultivators in various plant-related contexts. I argue that enslaved people's plant knowledge was inevitably a source of knowledge for the Europeans there (in Mauritius), and embraced by them in order for them to accumulate knowledge and to conduct experiments with plants.

In Mauritius, however, this expertise was not necessarily denigrated as simple 'know-how'; on looking at the actual practice of knowledge, it becomes clear that the embodied plant-related knowledge was crucial, and that the French actors were aware of that. On the island, the enslaved people were important, and seen to be so, not only because of their physical work but also because of their plant knowledge and skills, which sometimes allowed them to hold more important posts in the gardens, as will be shown in the last part of this chapter.

Forms of Slavery and Bonded Labour in the Indian Ocean World

It is important to understand the meanings of slavery in Mauritius, and of slavery and bonded labour in the Indian Ocean more generally. This is where the complications start; Gwyn Campbell has observed that there is 'no scholarly consensus as to the meaning of slavery in the Indian Ocean World', partly because, except for the plantations in the Mascarene Islands, Madagascar, and the East African coast, exact numbers and

[21] Schiebinger, *Secret Cures of Slaves.* [22] Sweet, *Domingos Álvares.*
[23] Gómez, *Experiential Caribbean.* [24] Kananoja, *Healing Knowledge in Atlantic Africa.*

relations with the movements between the Indian Ocean and the Atlantic rarely existed.[25] The history of the Indian Ocean world is a highly complex one, as was the slave trade within it, since it was multidirectional, with destinations that shifted. It involved both female and male enslaved people, who were obliged to fulfil roles ranging from field hand to domestic servant, to soldier, among others.

Forced labour in the Indian Ocean world must be understood as something very different from the slavery in the ancient world or indeed the slavery in North America.[26] 'Free' and 'unfree' were very vague terms, encompassing as they did complex nuances and various degrees of what it meant to be enslaved.[27] It is important to realise that 'free labour' at that time cannot be understood as it is today, especially as back then it included debt bondage. Alessandro Stanziani distinguishes between two basic systems of Indian Ocean slavery. On the one hand, he notes the 'open system' in the commercialised and cosmopolitan cities of South East Asia, where slavery and other forms of bondage were indistinct and fluid. On the other, there were the 'closed systems' of South and East Asia, in which 'the stigma of slavery made it inconceivable for a slave to be accepted into the kinship systems of their owners as long as they remained enslaved people; instead they were maintained as separate ethnic groups'.[28]

The bondage system in and around the Indian Ocean was strictly hierarchic, where any individuals, or sometimes a group or caste, ranked as inferior, had to obey their superiors, who in turn had obligations to a superior caste and so on. In relation to the slavery system in South Asia, Richard Eaton concluded that 'the only common denominator was the slaves' total dependency on some powerful person or institution'.[29] In the majority of cases in that area, people became enslaved through debt. This differed from debt bondage: whereas the latter was usually set up in advance in a voluntary arrangement as a credit security strategy, the former resulted from actual debt.[30] Some people were forced into migration through being prisoners whose sentence had been commuted from death to penal servitude for life. When the conditions of new colonies were such that they did not attract free labour, such convicts were shipped overseas to them, to stand in.

[25] Campbell, *Structure of Slavery*, x–xi. [26] Stanziani, *Bondage*, 11.

[27] On the question of the shades of 'free' and 'forced' labour, see Stanziani, 'Free Labor – Unfree Labor', pp. 27–52; Augusto, 'Knowledge Free and "Unfree"'.

[28] Stanziani, *Bondage*, 176. [29] Eaton, 'Introduction', 3.

[30] Campbell, 'Slavery and the Trans-Indian Ocean World Slave Trade', 291; Campbell, *Structure of Slavery*.

In the French colonies another common form of labour was a contract of *engagement* (indentured labour); this was a seventeenth-century system whereby white people, especially from Normandy and Brittany, would agree to go to the West Indies.[31] In the eighteenth century, however, the *engagés* (indentured labourers) could also be of colour, and their existence on the Mascarene Islands was one of the dominant forms of dependence, alongside the plantation economy and forms of slavery similar to those of North America.[32] The line between *engagé* and slave, seriously indistinct, brought legal difficulties with it, especially in nineteenth-century Réunion, where no legal or factual line between the two forms of labour existed.[33] Meanwhile, in the Indian Ocean area most of the forms of slavery had been in existence long before the arrival of Europeans, albeit the system affected women and children for the most part, and was not linked to any demand for plantation labour.

In this chapter, I look at two basic forms of labour in Mauritius: first, that conducted by the enslaved people, and, second, that conducted by indentured Black labourers from Africa, Asia, and those born in Mauritius ('Creoles'). As I shall show, even if an enslaved person's status allowed them to enjoy a certain degree of latitude and some privileges, they were still enslaved. Ultimately, even when an enslaved person was emancipated, we know little about how they attempted to rebuild their life.

Slavery, Enlightenment, and Political Economy

In eighteenth-century France, French case law 'made no clear distinction between hiring a person for services and "hiring" a thing'.[34] For much of the eighteenth century in France, the perception of slavery in the Enlightenment philosophers and economists remained ambiguous, their attitudes influenced by economic and political considerations. Only in the 1780s were these attitudes radicalised, in connection with the first slave revolts in the Antilles.[35]

In the mid-eighteenth century, the enslaved people in Mauritius enjoyed a certain latitude as a result of underlying economic motives. But was the sympathy with the enslaved people sentimental, or was it the result of a new awareness of a humanistic or, more precisely, social problem? Michèle Duchet argues that in physiocratic thought 'human-ism' cannot be understood in the same way as in the works of

[31] Debien, *Les engagés pour les Antilles*; Dechêne, *Habitants et marchands de Montréal au XVIIe siècle.*

[32] Campbell, *Structure of Slavery*. [33] Stanziani, *Bondage*, 184. [34] Ibid., 160.

[35] Ibid., 3.

Enlightenment thinkers such as Montesquieu or Diderot; instead, it should be understood in its context and with its goals.[36] Following this assumption and to add weight to Richard Grove's points, I read Poivre's writings against slavery and the announcements he made in his *Discours* upon his arrival on Mauritius in 1767.[37] In it, he set out the reforms to be made to slavery on the island.

At this point, let us recall the botanical goals that Poivre set out for Mauritius. His economic argument, that enslaved people were less productive than labourers, was by no means a common one. Although physiocratic thinkers had mooted the idea, it was not radical – but it was highly complex and it was ambiguous. Moral concerns became mixed with economic motives and, above all, the idea was fundamentally racialised, as shown in the next section.[38]

Why did Poivre use an economic argument, and what role did Enlightenment thought play in the criticism of slavery in the colonies in a wider sense?[39] Despite critical physiocratic voices raised against slavery, the physiocratic theory of slavery and the 'good treatment' of enslaved people were, in fact, related to profit and commerce: a 'lazy slave' was counterproductive to the economy. Poivre's own writings reflect this theory between behaviour and productivity. His ideas, as set out above, were neither abolitionist nor radical, but were based on physiocratic paradoxes and must be understood in the context of the time.[40]

Under Poivre's tenure, the agricultural activities in Mauritius were closely connected to his idea that it was the extent of the population's freedom that shaped the state of agriculture.[41] He believed that agriculture in Africa was lost because its people were enslaved: otherwise it could have been, he felt the 'storehouse of the universe'.[42] He illustrated his ideas about 'free hands' in his publication *Voyages d'un Philosophe* (1768), a high point in the liberal treatment of plantation workers producing sugar. He claimed that in Cochinchina it was 'free hands' that cultivated sugar cane and worked in the sugar refineries, and that in comparison to

[36] See particularly on this question, Duchet, *Anthropologie et histoire au siècle des Lumières*, 119–49; on the Enlightenment, science, and slavery in the English Atlantic, see Otremba, 'Enlightened Institutions'; for slavery, race, and anatomy in the eighteenth century, see Curran, *Anatomy of Blackness*.

[37] Grove, *Green Imperialism*, 205–6; Dobie, *Trading Places*.

[38] See on this ambiguity in particular, Bénot, *Les Lumières, l'esclavage, la colonisation*; Oudin-Bastide and Steiner, *Calcul et morale*; Duchet, *Anthropologie et histoire au siècle des Lumières*.

[39] See in particular on this matter, Røge, 'Question of Slavery in Physiocratic Political Economy', 149–69; Røge, 'Natural Order of Empire', 32–52; Cheney, 'Aufklärung und die Politische Ökonomie des Kolonialismus', 207–28.

[40] See also Rönnbäck, 'Enlightenment, Scientific Exploration and Abolitionism'.

[41] Poivre, *Mémoires d'un botaniste et explorateur*, 112. [42] Ibid., 113.

the Americas, where it was enslaved people who worked the sugar plantations, the production achieved by the Cochinchinese non-forced labourers was at least double. He concluded that it should be the law of the land that determined agricultural productivity, and that he did not see productivity flourishing when people were not free.[43] His criticisms of slavery went hand in hand with the concepts of profitability and unfree labour.[44]

The bad treatment of enslaved people, Poivre argued, did not mean that they were more efficient. On the contrary; in purely economic terms, badly treated enslaved people were not profitable. His attitude was more than ambivalent and was not necessarily a *moral* disagreement.[45] His anti-slavery sentiments were closely connected to – maybe even driven by – economic reasons: as free hands produced more than non-free, he claimed, Mauritius should be cultivated only by free people because only then could the island be more productive. In light of his plan to turn Mauritius into a self-sufficient island rather than a colony dependent on outside help, I conclude that his anti-slavery sentiments must be seen as a means to an end. His arguments were influenced by economic reasoning, and to him this was not necessarily a paradox in relation to his moral concerns.

In theory, he aimed at introducing a 'good master–good slave' policy, as he had declared in his *Discours*. Enslaved people who were treated well would, he argued, always serve their masters well, whether in war or in peace; they would neither flee to the woods nor desert and join the enemy.[46] He suggested, therefore, that the people should be left in peace, particularly with regard to religion. Then, he claimed, the enslaved people would serve their masters with joy and fidelity because they could consider themselves 'free and happy', even when enslaved.[47]

In political philosophy as a whole the dilemma regarding slavery and the status of labour was fundamental: the concept of slavery was seen as distinct from that of the slave trade. While Poivre's overall ideas related to the good treatment of the enslaved people, he apparently saw no problem with the slave trade itself. Back in May 1755, Poivre, then a CIO agent, set

[43] Ibid., 93.

[44] On the physiocratic movement and slavery, see Seeber, *Anti-Slavery Opinion in France*; Røge, 'Question of Slavery in Physiocratic Political Economy'. On slavery and public opinion in eighteenth-century France, see also Ehrard, *Lumières et esclavage*.

[45] See also Harvey, 'Slavery on the Balance Sheet', 83.

[46] 'Discours Prononcé par P. Poivre, à son arrivée à l'Isle de France, aux habitants de la Colonie assemblés au Gouvernement', in Poivre, *Mémoires d'un botaniste et explorateur*, 133.

[47] 'Discours prononcé à la première assemblée publique du noveau Conseil Supérieur de l'Isle de France, le 3 août 1767, par P. Poivre, commissaire pour sa majesté aux Isles de France et de Borubon, et président des Conseils Supérieurs qui y sont établis', ibid., 149.

up a contract for slave trading with the Portuguese on the island of Timor. Even later, he remained very much aware of the 'good market' of the slave trade. In his private journal from 1767, he explained that twenty-three Creole men and four women had been sent to Bourbon in exchange for horses at the rate of two enslaved people for one horse – which was, according to him, 'a very lucrative exchange'.[48]

The Cape of Good Hope and Madagascar played a significant role in the Mauritian slave trade.[49] Between 1670 and 1769, Madagascar remained the biggest supplier, contributing 70 per cent of the enslaved people arriving in the Mascarenes. Nineteen per cent came from the Swahili Coast and Mozambique, 9 per cent from South Asia, and 2 per cent from West Africa. Then between 1770 and 1810, East Africa – Mozambique and the Swahili Coast – became the most important supplier of enslaved people, providing 60 per cent of all imports.[50] Richard Allen suggests that the high numbers of enslaved people were required because the mortality rate on the island was so high.[51] In addition, on looking more closely into the sources I realised that during the *époque royale* the many enslaved people were needed in Mauritius in order to realise Poivre's idea of turning it into a proper island community with substantial infrastructure and housing and to urbanise parts of the island, Port Louis in particular.

On Madagascar, the local elites were deeply involved in the business of slavery. Take the case of the Malagasy Princess Bétia, known to Europeans under her French name Marie Elisabeth Sabbabadie Betty or her nickname 'Betti' or 'Béti'. Her father had been King Ratsimilaho, the former chief of Foulepointe (on the east coast of Madagascar), so she was the aunt of his grandson Yavi, the new chief. In the 1750s, Bétia had moved to Mauritius, where she bought land and people.[52] She lived in Mauritius from the early 1750s, the owner of plantations, enslaved people, and herds.[53] On 15 January 1773, the weekly newspaper *Annonces, Affiches et Avis divers pour le Colonies des Isles de France et de Bourbon*, included a list of imprisoned enslaved people from June 1771 to

[48] ANOM, C/4/19, ff. 167v–185r, Extract of Poivre's journal (7 July to 24 November 1767), entry from 20 and 21 August.

[49] On the Franco-Dutch trade between the Mascarenes and the Cape of Good Hope, see Thiébaut, 'Informal Franco-Dutch Alliance'.

[50] Allen, 'Mascarene Slave-Trade and Labour Migration', 36–37.

[51] Ibid., 37. Meghan Vaughan makes a similar claim in *Creating the Creole Island*.

[52] On Bétia, see Allen, *Slaves, Freedmen and Indentured Laborers in Colonial Mauritius*. On free women of colour and their status in Mauritius, see especially Allen, 'Free Women of Colour and Socio-Economic Marginality in Mauritius'. For free women in the revolutionary Atlantic, see Clark, *Strange History of the American Quadroon*.

[53] ANOM, E 184, f. 198r, Poivre to de Boynes, 12 February 1772.

August 1772, which includes a certain Denis belonging to Bétia, but without any further clarification.[54] So there is evidence that her name was widely known on Mauritius.

In 1772, after her father's death, planning to claim more enslaved people and cattle as part of her inheritance, she returned to Madagascar. There, however, she apparently started a coup against her nephew, encouraged by the French corporal, Jean Filet Olésime, known as La Bigorne, who had been stationed in Madagascar since the 1750s and who later married Bétia. He then started playing games with Mauritius's administration by trying to manipulate Bétia for his own interests in Madagascar; he apparently hoped to create an alliance with the Ancoves people inland, to join Bétia's forces against Yavi's in Foulepointe.

In 1771, Foulepointe had been one of the major suppliers of white rice and cattle for Mauritius. If the coup had been successful – it failed in any case because La Bigorne dropped dead during the attempt – Poivre argued that the Malagasies would have lost their trust in the French, and hidden in the mountains, abandoning the ports, so the French vessels there would no longer have been secure.[55] It becomes evident that the French were very dependent on Foulepointe and on their commerce with the Malagasies. French agents could not claim humans as captives in Madagascar and export them as enslaved people to Mauritius because this ran the risk that the local elites would stop supplying corn, rice, and cattle. So, the French integrated their activities into the existing African slavery systems. The Malagasy elite contributed to the French slave trade and entered into commercial relations with the French not only for basic necessities but also for labour.

Despite the critical views of Poivre and others on the treatment of enslaved people, the slave trade was a lucrative business. But while to Poivre slavery and slave trade were two distinct concepts, hence not contradictory, Françoise Poivre was apparently a little more radical in her views; in 1770 she declared that she would no longer buy enslaved people. When Bernardin presented a slave – possibly Côte (more about him below) – to Françoise before his return to France, she refused to accept his '*petit noir*' ('little Black person'), because she no longer bought enslaved people and likewise could not accept them as gifts.[56]

[54] 'Denis, belonging to Béti'. All the other enslaved individuals are indicated as '[name] belonging to [name]' or '*au roi*'. Béti, however, did not need any further explanation, *Annonces, Affiches et Avis divers pour le Colonies des Isles de France et de Bourbon*, 15 January 1773.

[55] Ibid. [56] EE BSP_013, Françoise Poivre to Bernardin, 1770.

But while Françoise Poivre might have rejected the idea of buying enslaved people for her house, her husband, as intendant, nevertheless continued the slave trade in his pursuit of the establishment of the island. Looking at his garden of Monplaisir from this perspective, his double standards become even clearer. While his wife (and perhaps Poivre too) had stopped buying new enslaved people for their *habitation*, the slave trade for the Mascarenes continued, and the number of enslaved people introduced increased significantly.

French Naturalists, Regulations, and the Treatment of Enslaved People

The annual report for 1769 provides significant insights into the numbers and treatment of enslaved people employed in the king's service. The report lists 19,000–20,000 Black people, of whom 6,000 were responsible for the cultivation of the fields – a number that the local administration wished to increase by another 2,000. Slavery and free/unfree labour with all their different shades of meaning were by no means reserved for males; as we have seen earlier, women and children were also involved.[57] Poivre and Desroches tallied 1,617 enslaved people in the service of the Crown, of whom 1,000 were men, 358 women, and 259 children; in 1770 the government planned to increase the total to 3,000.[58] The enslaved people were to receive their stipend in the form of rice instead of a monthly wage. During the CIO's tenure they had received a monthly 'salary', and the new royal government, according to Poivre and Desroches, did not want to make them believe that they were treated worse than under the CIO. So at the outset the new administration also paid them in cash – but then, they argued, with that money the enslaved people would buy '*boissons pernicieuses*' – liquor – which in the eyes of the slave owners was counter-productive. Since the monthly food allowance of 60 pounds of corn per head was not sufficient, enslaved people fell ill from malnutrition, the provision of rice instead was seen as a doubly successful strategy for the French, in that it was more economical for them and it could be modified in case of scarcity. Again, an economic argument was presented to justify the change: Poivre and Dumas asserted that with this improvement in their well-being, the enslaved people could work harder.[59]

Although Poivre apparently did not realise it, the high number of enslaved people clearly conflicted with his idealism and his argument

[57] See also the list of jail sentences announced in *Annonces, Affiches et Avis divers pour le Colonies des Isles de France et de Bourbon* (1773–5), which includes children and women besides men, ranging from Bengali to Malagasy origin.

[58] ANOM, C/4/27, f. 274r, 'Statistiques pour l'année 1770', 'Noirs'. [59] Ibid.

that the cultivation of spices required less forced labour. He actively supported the slave trade for the Mascarenes, and he instructed colonial officials to supervise the treatment of enslaved people in the colonies. For instance, he wrote to Réunion's director Crémont in July 1767 that he should ensure government protection for the enslaved people and see that their masters treated them well.[60] This meant that the enslaved people in the Mascarenes could negotiate their working and living conditions to at least some degree, but they definately did not have the status of a freeborn.

Indeed, in Mauritius the enslaved people working the gardens were given a certain degree of latitude, which can be illustrated through various examples. As I indicated at the beginning of the chapter, Bernardin sent Françoise Poivre a manuscript of his *Voyages à l'Isle de France* in which he described the inhuman treatment of the enslaved people in the French colony, and she commented in a letter to Bernardin that he should consider the fact that in the Americas the enslaved people were treated worse.[61] Bernardin had assumed that slavery did not exist in the British colonies, and Françoise Poivre corrected him.[62] In addition, his criticism of slavery in Mauritius derived from his first-hand observations, whereas Françoise Poivre might not even have seen the situation of the enslaved people inland: she was probably only familiar with them at Monplaisir. This indicates that enslaved people were treated much less harshly in the Poivre household than Bernardin observed in other parts of the island.

That enslaved people experienced harsh treatment in parts of the island other than the gardens is clear from manuscripts, private accounts, and printed matter. The pharmacist Fusée-Aublet, who worked in Mauritius between 1753 and 1761, was shocked by the treatment of enslaved people, in particular their poor nutrition. He claimed that families would be torn apart, sometimes leading to suicide.[63] While from 1767 the royal administration tried to improve the enslaved people's situation, murder and death were not unusual: within the first month of Poivre's tenure he reported eight dead bodies on the island, one of them a Black woman who had been hanged (or had hanged herself?) in the Moka area.[64] An account dated 1756 notes that it was possible to tell when enslaved people had been punished because of the burn marks on their skin. Fusée-Aublet was outraged when observing that the enslaved people were shackled and 'doomed to the most painful labour & drudgery during

[60] ANOM, E 99, Poivre to Crémont, 26 July 1767.
[61] EE BSP_0121, Françoise Poivre to Bernardin, n.d. [1769?]; and EE BSP_0140.
[62] EE BSP_0143, Françoise Poivre to Bernardin, n.d. [1769?].
[63] BCMNHN, MS 452, Fusée-Aublet, 'Observations sur le traitement des nègres'.
[64] ANOM, C/4/19, ff. 167v–185r, Extract of Poivre's journal (7 July to 24 November 1767), entry from 22 August.

their unhappy lives'.[65] He also claimed that the CIO's regime had permitted the colonists to abuse enslaved people, which was why Fusée-Aublet fulminated against its regulations.[66]

Those regulations, the *Code noir* ('Black code') – originally passed in 1685 under Louis XIV – defining the conditions of slavery in the French colonial empire, attracted criticism from the local actors in the colonies. Bernardin claims they were regulations only in theory, and that when it came to the treatment of enslaved people they were neither respected nor implemented. But the judges overseeing the tribunals were always the enslaved people's 'first tyrants'.[67] The *Code noir* did not try to improve the enslaved people's situation but only to regulate it.[68] Bernardin observed that although according to the *Code noir* enslaved people were not to work on Sundays, and were to receive meat every week and clothes throughout the year, these regulations were not implemented.[69] This was the first time he had been exposed to the reality of slavery, which was in his eyes an allegory of immorality and corruption.[70] He criticised the terrible working conditions, including the requirement to labour almost naked in the burning sun, and the poor nutrition (he claimed that the enslaved people were given only boiled corn), and insufficient water supply: in addition, these unfortunate people were forced to pray to the Christian God.[71]

Bernardin processed his observations on the treatment of enslaved people in his utopian novel *Paul et Virginie.* The story is of a desperate and abused Black enslaved woman who begs the main character, the girl Virginie, for food, exclaiming: 'While there are still good whites in this country, we must not yet die.'[72] Virginie and her friend (and love) Paul help the Black woman and ask her master for mercy. The young people's helpful actions are returned in kind; at night they get lost in the woods and a group of maroons who had observed the incident earlier carry them home.[73] Throughout the novel, enslaved people who have been freed help Paul and Virginie's families to work their fields and gardens.[74] For instance, a slave called Domingue has green fingers and turns the families'

[65] BL, Add. MS 33765, f. 10v, 'Geographical Collections of Alexander Dalrymple'.
[66] BCMNHN, MS 452, Fusée-Aublet, 'Observations sur le traitement des nègres'.
[67] Bernardin, 'Voyage à l'Isle de France', 61. See the most recent edition with commentary (Niort, *Code noir*).
[68] EE BSP_0143; see also Françoise Poivre to Bernardin [1769?].
[69] Bernardin, 'Voyage à l'Isle de France', 60.
[70] Ngendahimana, *Les idées politiques et sociales de Bernardin de Saint-Pierre*, 175.
[71] Bernardin, 'Voyage à l'Isle de France', 59. [72] Bernardin, Paul et Virginie, 43, 42–4.
[73] Bernardin, 'Voyage à l'Isle de France'.
[74] Bernardin, *Paul et Virginie*. On the relationship between literary accounts and their use in the discipline, see also Pimentel, *Testigos del mundo*, 299–328. See also Pacini, 'Environmental Concerns in Bernardin de Saint Pierre's "Paul et Virginie"'.

land into fertile plots, growing corn, rice, cucumbers, potatoes and sweet potatoes, sugar cane, coffee, banana trees, and some tobacco.[75]

Non-fictional sources provide more insights into the treatment of the enslaved people working in the gardens. There is evidence that those working in the acclimatisation gardens and habitations were treated with more latitude than those in other parts of the island, and that the former had certain privileges. While accounts of the general treatment of the enslaved people in Mauritius describe a 'most miserable people' who were treated in 'inhumane ways', who worked all day and who, underfed by their masters, had to look for food at night, there are some accounts that provide a different view. For example the engineer and naturalist Cossigny de Palma argued that such destructive treatment would clearly run counter to the interests of the masters, so the enslaved people in his garden were treated differently; as in Monplaisir, the better treatment of the enslaved people was due not to sentiment but for economic reasons.

We should not, however, be tempted to draw an idyllic picture of the enslaved people's situation in Mauritius, particularly in view of the efforts made by some of them to escape. Some useful insights are provided by reports written by British officers, such as that by Colpoys during his stay in Mauritius:

> As to the Black troops mentioned in Capt. Johnstone's remarks, he was certainly misinformed, for instead of their being assistant to the Island, they are at present its greatest enemy. I have been well assured that there are now in the woods between three and 4000 of those runaway slaves who have formed themselves into bodies, and have appointed officers for themselves; and they frequently make marauding parties, on which they not only plunder plantations, but often kill the white people and set the Black people at liberty. Everybody wonders that when the regiments were here, they were not sent out to destroy so alarming an enemy as these promise to be.[76]

These escaped enslaved people – maroons, as they were known – had, as Colpoys explained, united long before the French settlement in 1721, resisting being drafted into forced labour.[77] Due to the increasing number of escapees, a law was drawn up in 1730 prohibiting female and male enslaved people from living together. But the number of maroons increased nevertheless, and even organised raids. In 1732 the maroons raided Flacq, evicting all its French settlers, who fled to the Grand Port.[78] In March 1739 French soldiers retaliated; one Black man and two Black

[75] Bernardin, *Paul et Virginie*, 19–20, 21, 41–2, 165.
[76] BL, IOR/H/111, f. 135r, 'Mr. Colpoy's Remarks', 1772.
[77] Cheke and Hume, *Lost Land of the Dodo*, 91.
[78] Barnwell and Toussaint, *Short History of Mauritius*; Adolphe, *Les Archives démographiques de l'Ile Maurice*, 40.

women were shot. The following June the troops discovered a maroon encampment – where an Indian slave had also been hiding for four years – which they then 'depleted'.[79] According to a report by the *Conseil de l'Isle de France* to the CIO, dated 15 January 1740, the number of maroons was estimated at twenty-one men and eighteen women, who were suffering from hunger and disease in their hideouts.[80] In his journal on Mauritius from 1754, the astronomer Nicolas-Louis de la Caille described four soldiers who caught a Black woman and treated her like 'a beast'.[81] Maroons, however, continued to live in the forests and when they were caught were punished severely, even killed.

Hierarchies, Castes, and Prejudices

In the Mascarenes, enslaved people were usually identified as belonging to one of four broadly defined castes linked to their ethnical and geographical origin: Creole (born in the archipelago), Indian, Malagasy, and Mozambican (including the Swahili Coast).[82] Contemporary observers such as Bernardin suggested that the Asian enslaved people enjoyed a higher status than that of Africans or Malagasies, because the Asians were considered both more intelligent and less suited to hard physical labour.[83] It was the Malagasies who worked on the land in Mauritius; as Bernardin put it, 'it is in Madagascar where one fetches the Black people for the cultivation of land'.[84] In contrast, the Malabars (mostly Tamil speakers, but as the French did not make a clear distinction this term referred to people from anywhere in India), together with lascars (Indian merchant seamen/skilled sailors) and 'free people of colour' ('*noirs libres*'), constituted, with just 400 to 500 of them, a tiny minority of the island's Black population. Poivre sought to increase their number because, he argued in a letter to the minister in 1767, 'these people are industrious, hardworking and frugal. We drew great services from them.'[85] Bernardin confirmed that many Indians living in Mauritius came from Pondicherry, 'where they offer themselves for a number of years'. They were not the 'true Black people', and were treated more like Europeans.[86] These people were probably *engagés* from India, forming 40 per cent of the

[79] Ly Tio Fane-Pineo, *Île de France*, 319.
[80] Ibid. On marronage during the CIO administration, see ADR C 2859; ADR C 2863.
[81] La Caille, *Journal historique*, 223–6, 225.
[82] Allen, 'Mascarene Slave-Trade and Labour Migration', 35.
[83] Bernardin, 'Voyage à l'Isle de France', 120–2. On stereotypisation and ethnic prejudice in the French empire, see especially Lamotte, *Making Race*.
[84] Bernardin, 'Voyage à l'Isle de France', 59.
[85] ANOM, C/4/18, f. 381r, Poivre to Praslin, 30 November 1767.
[86] Bernardin, 'Voyage à l'Isle de France', 59.

free women and men of colour in Mauritius (and 15 per cent of the servile population).[87] At that point, La Bourdonnais (in office 1735–46), felt sure that he could find skilled and experienced workers in India.[88] According to him, lascars were both cheaper than French engineers and more willing to work.[89]

But even if lascars were appreciated for their sailing skills, and normally arrived on the island as free men, by the 1750s Mauritius had become notorious for its forced labour conditions. As Poivre explained in a letter to the CIO in 1755, on the passage from South East Asia to Mauritius, his ship had a complement of twenty-two Indian sailors, but on a stopover in Manila fifteen of them deserted because it was well known that Mauritius was a place of slavery.[90] But lascar sailors were needed in Mauritius, so Poivre asked the governor of Pondicherry in 1768 to send some there. He informed Minister Praslin about this order, explaining that the lascars were 'gentle' and one had to treat them with 'lots of respect'. He continued:

> they are Muslims and very superstitious, they have all the prejudices of their religion, they require lots of moderation and do not listen to mocking in that respect. They do not know our orders or the strict discipline in service provision. I highly doubt that the officers of the king's navy can force them [to work].[91]

After the Seven Years War lascars were also shipped across the Atlantic to Saint-Domingue, where they told the enslaved people there how to extract coconut oil and process it.[92]

In 1770, some of the Malabars and lascars owned houses in a specific quarter of Port Louis.[93] It had been planned in 1769 to separate the white population from free people of colour, and it was designed as their village, with fields to make small gardens for cultivation. These free labourers worked according to their abilities.[94] As Richard Allen remarked, however, Mauritius's free people of colour were relegated 'to the margins of colonial social and economic life'.[95] Bernardin confirmed that he never saw any Malabar working in the fields on the island.[96] Governor Dumas

[87] Stanziani, *Bondage*, 193.
[88] Hazareesingh, 'Religion and Culture of Indian Immigrants in Isle de France', 241.
[89] Vaughan, *Creating the Creole Island*, 40.
[90] This is what Poivre explained in a letter to the Secret Committee, ANOM, C/4/9, letter 21, Poivre to the Secret Committee, 15 November 1755.
[91] ANOM, C/4/22, f. 133r, Poivre to Praslin, 17 June 1768.
[92] Hodson, *Acadian Diaspora*, 110.
[93] 'Ordonnance de M. M. les Géneral & Intendant concernant la plantation des arbres dans les rues & les entourages des emplacemens & maisons du Port-Louis, Isle de France. Du dix-sept Juin 1769', in Delaleu, *Code des Isles de France et de Bourbon*, vol. 1, 314.
[94] ANOM, C/4/27, f. 274, 'Statistiques pour l'année 1770'.
[95] Allen, 'Economic Marginality', 126. [96] Ibid.

stated that Malabars were '*noirs libres*' ('free Black people'), who seemed to him 'useless' because they were 'shy and feeble', and could never serve as defenders or cultivators.[97] There is evidence that these free people of colour helped the maroons and hid them, as indicated by an announcement in Mauritius's weekly newspaper: Louis, a boy aged between 15 and 16 who had arrived on the island in March 1773 from Bengal, had escaped from his master, Senne le Jeune, who was looking for him. There were rumours that Louis was looking for wood in the priests' quarter in order to sell it to the Indians, who in return hid him in their houses.[98]

There are further indications that some South Asian labourers – probably not just those who were free – had privileges such as being allowed to practise their religious customs. A rare private letter, which a certain Jean Pierre Charles sent from somewhere in the Indian Ocean to his wife in France on 6 August 1778, bears out this hypothesis. While his letter mainly expressed his love and longing for his wife, he also enclosed a small captioned (see Figure 4.1).[99]

The drawing is of two men with a small house, borne on a pair of bars, on their shoulders. At the upper right is a sketch of a heart pierced by an arrow. Aside from the exceptional value of this private letter, the fact that the lovelorn Jean Pierre Charles was able to draw this is significant. He may also have observed the Hindu Maha Shivatri festival celebrating Lord Shiva or the Cavadee festival, a ceremony to worship the Tamil god of war. The latter is still celebrated in today's Mauritius by Tamil communities, and involves participants carrying large shrines with statues in them through the streets.[100] That Jean Pierre Charles might have been able to observe such a procession, as well as the house portage achieved in a similar way, suggests that the Tamil inhabitants were able to practise their customs. Contemporary travellers also observed that the island's Indians came together to celebrate the 'Yamsey' festival.[101]

The question of religious tolerance was of concern to Poivre's administration. The religious customs of the Malabar and lascar families were tolerated – but, according to the king's decree, were not to be exercised in

[97] AD Montauban, MS 20J-130, Dumas's files, Dumas to Praslin, 10 June 1768.

[98] *Annonces, Affiches et Avis divers pour le Colonies des Isles de France et de Bourbon*, the 'Avis divers' section, Wednesday 26 May 1773. See also ibid., 15 January 1773, which announces that the police arrested a young slave boy who could neither say his name nor that of his master.

[99] Kew, HCA 30.290, Jean Pierre Charles to his wife in France, 6 August 1778.

[100] On the Kavadi festival today, see for instance Geaves, 'Pilgrimage to Tourism'.

[101] Allen, 'Free Women of Colour and Socio-Economic Marginality in Mauritius', 186.

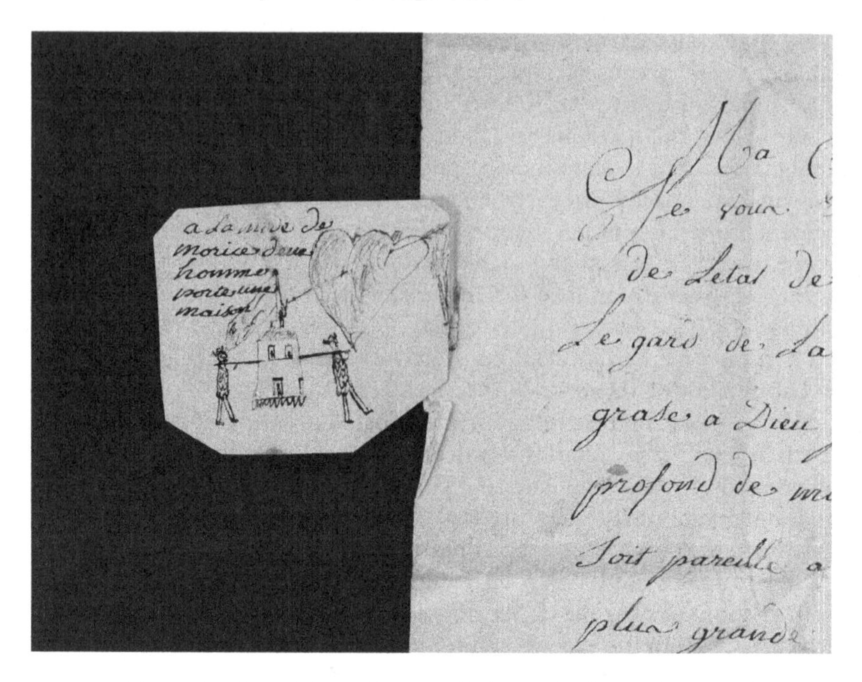

Figure 4.1 Jean Pierre Charles, 'A la mode de morice deux hommes porte[nt] une maison' ['According to the custom of Mauritius, two men carrying a house'], 1778, National Archives (UK), ref. HCA 30/290

public.[102] Hence, my observations are by no means arbitrary. Nevertheless, historians have shown that the caste system and the reasons for allowing South Asian labourers, and indeed some enslaved people, certain privileges was a much more complex matter. The main reason for this, some historians have argued, was linked to the fact that in Southern Asia domestic and/or urban slavery were the most common forms of forced labour and that Europeans adopted these patterns and practices when sending South Asians to the colonies.[103] Following this reasoning, this sensitivity to local practices might well have influenced the slavery system in the Mascarenes.

Ethnicity and race were, unsurprisingly, linked to social status and the tasks that came with it. Membership of a certain caste also meant the

[102] ADR, 12C, Praslin to Dumas and Poivre, 1 April 1769. For Dumas's letter raising the question of religious tolerance, see AD Montauban, 20J-130, Dumas to Praslin, 10 June 1768.

[103] Caplan, 'Power and Status', 170, 173, 182, 189.

allocation of certain tasks. Most of the enslaved people were not employed in the plantations or gardens but on building the island's infrastructure. In the 1750s, the colonists started building stone houses, and the enslaved people had to dig the foundations. Labourers and enslaved people worked side by side. Few European masons were employed by the CIO: those who were free demanded one to two 'Spanish dollars' (piastres) per day, which made the construction of a stone house very expensive.[104] In the town centre there was 'a large spot of ground surrounded by a strong and high stone wall, wherein are the Company's slave houses, stables etc'.[105]

Under Poivre, there was also a small group of *Noirs de Maréchaussée* (a 'Black police') under his direct orders.[106] They were treated better than others: the force was made up of eighty-nine people of colour, of whom only eighteen were free: many were married 'or like married, having wives and children', which increased their number to 155.[107] These men, together with their wives and children, had certain privileges. Though they were generally given 60 pounds of maize per month, they also received biscuits, salt meat, and some eau de vie.[108] Although the *Noirs de Maréchaussée* had certain privileges and the Crown paid for their housing and food, they were bound to the Crown by the requirement for tasks such as keeping law and order in their quarters and helping hunt down maroons.[109] In 1768, they were emancipated, and this was enacted in Mauritius on 12 February 1770.[110] That day, 116 enslaved people, mostly of Malagasy and Mozambique origin, but also some Creoles and some from Timor, India, Guinea, and Malaysia, were freed, as had been promised to them when they had joined the *Maréchaussée*.[111] Some other

[104] BL, Add. MS 33765, f. 7r, 'Geographical Collections of Alexander Dalrymple'. On currency in the Indian Ocean, see Allen, *European Slave Trading in the Indian Ocean*, xvii; Françoise, *L'empire de la monnaie dans les Mascareignes au XVIIIe siècle*. The Spanish dollar or piastre was widely used in the region. In the French possessions, the currency was the livre, and its exchange rate depended on time and place. For example, between 1760 to 1767 the piastre's value against the livre rose from 11 to 33, but from 1778 to 1786 it remained relatively stable at between 8 and 11 livres.

[105] BL, Add. MS 33765, f. 7r, 'Geographical Collections of Alexander Dalrymple', 1756.

[106] See Praslin's response to Poivre's demands, in which Praslin confirms that the *Maréchaussée* are directly ordered by Poivre, ADR, PR 22 C, Praslin to Poivre, 19 August 1768.

[107] ANOM, C/4/18, Poivre to Praslin, 30 November 1767. See also Paul, *Deux siècles d'histoire de la police à l'Île Maurice*, 24.

[108] ANOM, C/4/18, Poivre to Praslin, 30 November 1767. On the *noirs de Maréchaussée*, see also ANOM, C/4/18, f. 110r, Poivre to Praslin, 30 November 1767.

[109] ANOM, C/4/27, n 52, 'Noir', in 'Statistiques pour l'année 1770'.

[110] ADR, PR 22 C, Praslin to Poivre, 19 August 1768.

[111] MA, OA 75, ff. 11–13, 'Registre pour servir à l'enregistrement des actes de liberté accordée à des esclaves', 29 December 1768–5 February 1785, 12 February 1770.

enslaved people who had been given responsible work, for example, the Guinean Charlie and his Creole wife Magdeleine, who cared for the king's cattle, were declared free.[112]

During the transition, that is while the island's administration was passing from the CIO to the Crown (1767, the first year of Poivre's tenure), the CIO's enslaved people were sold to the Crown; Poivre bought 653 enslaved people, of whom 352 were employed in the two ports and on marine-related tasks, and when large numbers of ships arrived he tried to increase their number. Enslaved people were used in particular on the fortifications, for the construction of hospitals, granaries, and barracks, and for carrying water from the river to the port.[113] He considered importing 10,000 enslaved people for the work, because the birth rate among the enslaved people was very low.

The fact that he treated them well had been part of his plan; his idea was to encourage childbearing, hence increase the number of enslaved people without any financial effort on his part. But the enslaved people from Madagascar resisted the idea of bearing children into slavery. As Poivre observed, 8,000 new Malagasy enslaved people were introduced to Mauritius but these people did not 'multiply' because the women practised 'a terrible usage to destroy their fruit' – they induced abortions.[114] The enslaved women were valued for their reproductive capacity, but they could use this against their masters as form of resistance.[115] Because of the low birth rate of the Malagasy enslaved people, Poivre concluded that Malagasies were 'a very bad breed of slaves' and that 'Indians and Guineans multiply well, Mozambicans less, and Malagasies not at all'.[116] So he planned to introduce more enslaved people from the island of Gorée (in Senegal) because 'the Black people of this coast of Africa are stronger and more robust, more loyal than those of other countries; only they multiply here, marry, have a household, and build families. The Black people from Mozambique and Madagascar are libertines and deserters.'[117]

As this chapter shows, in the French case, it was stereotypical thinking that decided what types of task an enslaved person had to carry out.

[112] Ibid., 10 January 1772.
[113] Brest, Ms 89, no. 80, Poivre to Praslin, 30 November 1767.
[114] ANOM, C/4/18, f. 381r, Poivre to the minister, 30 November 1767.
[115] On enslaved women, see in particular Campbell et al., 'Women in Western Systems of Slavery'. On plants and female resistance, see Schiebinger, *Plants and Empire.*
[116] ANOM, C/4/18, f. 381r, Poivre to Praslin, 30 November 1767. On the question of race in early modern France, see Peabody, *There Are No Slaves in France*; Peabody and Stovall, *Color of Liberty*; Boulle, *Race et esclavage dans la France de l'Ancien Règime*; Sebastiani and Schaub, 'Genealogy and Physicality'.
[117] Brest, MS 89, no. 73, Poivre to Praslin, 30 November 1767.

People from Madagascar, for example, were praised – and feared – for their marine skills: although they could have been of great interest for work on French vessels, Governor Desroches wrote against the idea of employing them in the navy or the ports because, he said, they would do anything to return to their country.[118] In an account of Mauritius written by Charles Grant, Baron de Vaux, relying on his father's manuscripts, quotes Malagasies 'in [their] corrupt French':

Sa blanc là li beaucoup malin: li couri beaucoup dans la mer là-haut; mais Madagascar li là ['The White is very clever, he spends much time sailing up there in the ocean; but Madagascar is just over there'].[119]

Even though the Malagasies would not bear the children that Poivre and Desroches had hoped for, they considered them 'gentle and honest people having much more esprit than the other Black people, [who] are normally lazier and indolent'. Because of these characteristics, it was said, the Malagasies were not suitable for work on the land but better in the home.[120]

Enslaved People and Botanical Explorations in the Mascarenes

Despite his antipathy towards the slave trade, Bernardin took advantage of enslaved people. During his botanical expeditions *à pied* in Mauritius, Bernardin was accompanied by two enslaved men whom he named Côte and Duval; he had bought Duval shortly before the start of the trip on 26 August 1769, and had named him after his friend Louis David Duval, to whom he wrote letters in *Voyage à l'Isle de France*.[121] Bernardin described Duval as 'well-made but ... could not speak French'. Côte, however, spoke a little French, and was quite short (generating the name newly bestowed on him), very strong, and a brave 'man of fidelity'.[122] Before Bernardin returned to France, he had freed Duval, as he wrote in 1772 to his friend, Louis David Duval.[123] He had also considered freeing Côte – but, as mentioned above, there is evidence that he instead offered

[118] Quimper, 12 D (11/15), Desroches files, Desroches to Praslin, 6 February 1770.
[119] Grant, *History of Mauritius*, 297. See also on Charles Grant, Selvon, *Comprehensive History of Mauritius*, 145.
[120] ANOM, C/4/24, no. 36, Poivre and Desroches to Praslin, 1 September 1769.
[121] Bernardin, 'Voyage à l'Isle de France', 73. See also the letter in which Bernardin writes to his friend Duval directly, EE BSP_0215, Bernardin to Louis David Duval, 29 July 1772.
[122] Bernardin, 'Voyage à l'Isle de France', 73.
[123] EE BSP_0215, Bernardin to Louis David Duval, 29 July 1772.

him to Françoise Poivre as a gift prior to his departure.[124] There is evidence that Duval worked in Poivre's Monplaisir.

In a letter of 1771 to the minister, Poivre mentioned the good service of a skilled gardener, Duval, who was probably Bernardin's former slave. Duval had come to Mauritius in summer 1769 and was set to work in Le Réduit (see Figure 3.1), cultivating the medicinal plants needed in the hospitals: he worked with 'much competence', and this might well have been the reason why Bernardin engaged him for his botanical expedition.[125] Duval's name appears several times in the correspondence between the local administration and Versailles, and when 'the poor Duval' passed away shortly before April 1771, Poivre urgently sought a replacement.[126] This brief history provides evidence that Bernardin's slave Duval probably spent the rest of his life – free or unfree – in the service of Poivre.

Bernardin's account refers throughout to the help given by and to the enslaved people, and necessity for them in the exploration of the island. For instance, they helped each other when paths were narrow and steep, they helped prepare food and search for firewood and water, and they shared the fish they caught. When Duval could no longer carry his heavy load, they were allowed to take a break.[127] In the middle of the expedition, Duval cut his foot badly on a rock; Bernardin tended to his wound and gave him and Côte some eau de vie. Not only because of Duval's wound but also because of the heavy baggage, Bernardin planned at least two breaks for the enslaved people, concluding that because he treated them well they would follow him to 'the end of the world'.[128] Clearly, his account portrays himself in a flattering light, but, nonetheless, he and Duval seemingly developed an increasingly interpersonal relationship. But were the good relations between master and enslaved people as utopian as Bernardin described them in *Paul et Virginie*? In his account of this expedition he related that, just as in a scene in his novel, he got to know a young husband and wife who happily lived with their enslaved people in a house in the middle of the island.[129] When the young woman saw Duval's wound, she insisted on treating it herself, with a potion she had made from two different herbs, plus sugar, wine, and oil.[130]

The naturalists Commerson and Sonnerat, too, hired enslaved people for their expeditions, valuing them in that they were sometimes able to identify plants unknown to the Europeans. During his exploration of Réunion, Commerson found the '*bois tambour*' (*Tambourissa elliptica*

[124] EE BSP_013, Françoise Poivre to Bernardin, 1770.
[125] ANOM, C/4/25, Poivre to Praslin, August 1769(?).
[126] ANOM, C/4/29, Poivre to Praslin, 2 April 1771.
[127] Bernardin, 'Voyage à l'Isle de France', 74, 75. [128] Ibid., 75. [129] Ibid., 76.
[130] Ibid.

(Tul.) ADC) near the volcano. He climbed up, together with his draughtsman Jossigny, the intendant Crémont, the chevalier Bancks (a land surveyor in Réunion), Jean-Baptiste Lislet, who assisted Commerson as herbalist, a chevalier de St.-Lubin, M. Payet (their guide), and thirty-two enslaved people, who helped carry the equipment.[131] Enslaved people were not only important for carrying the equipment and food: the naturalists also relied on their plant knowledge. For instance, Commerson's enslaved people informed him that one could chew the '*bois tambour*' like betel.[132] Further, Commerson's notes illustrate the relationship between the French and the enslaved people in that that they ate together as a group. Rather than this expedition being reliant on human exploitation, Commerson's handwritten notes depict a jolly group having fun in the countryside.

When Sonnerat carried out his own explorations in India, he asked for the slave who had been of so much help to Commerson, as described by intendant Maillart, Poivre's successor, in a letter to the minister.[133] This was probably Joseph, whom Commerson mentions in letters to Cossigny de Palma.[134] At the Cape of Good Hope, when Sonnerat explored Table Mountain in mid-January 1773 with the Swedish botanist Thunberg, they hired two enslaved people who accompanied them during their expedition.[135] There are more examples of botanical explorations in which enslaved people helped. Yet they feature hardly at all in those accounts, or do so only – as Neil Safier suggested in the case of Joseph de Jussieu's explorations in the Americas – as part of an anecdote. This section confirms that enslaved people played an important role in the transmission of plant knowledge for identification and usage.[136] But in Sonnerat's published accounts Joseph's name does not appear: his expertise was not acknowledged in official writings nor was his very existence apparent.

Enslaved People, Garden-Related Tasks, and Healing Knowledge

For the work in the plantations, it was enslaved people from Madagascar who were mainly recruited; they were said to be dextrous and intelligent – and

[131] Geoffroy, 'Notice sur le Voyage de M. de Crémont', pp. 361–5. On the exact number of thirty-two (!) enslaved people, see Commerson's letter to Curé Beau, 16 February 1772, printed in Montessus de Ballore, *Martyrologe et biographie de Commerson*, 155.

[132] BCMNHN, MS 1904, 'Suite des descriptions [des plantes observées dans le voyage au volcan de Bourbon]', June 1769.

[133] ANOM, E 372, Maillart to the Minister of the Navy and the Colonies, 31 August 1775.

[134] See their correspondence printed in Cap, *Philibert Commerson*.

[135] Hansen, *Linnaeus Apostles*, vol. 6, 98.

[136] Safier, 'Fruitless Botany', 209.

easily offended in matters of honour.[137] Yet they were not the only nationality who worked out on the land. The inventories of private slave-owning colonists, probably from the late 1760s, list their *habitations*, their production of grain, rice, maize, coffee, and manioc, and the number of domestic stock (bullocks, sheep, pigs, horses, and poultry) and enslaved people. Their enslaved people are shown as either 'Black' or 'cultivator', and had different origins, shown as 'caste'. For the most part, they were Malagasies, but the others were 'Creole' or, more rarely, Indian or *caffre* (from the African mainland, East Africa in particular). They were listed as male, female, and sometimes children, with their (French) name and age.[138] Some of the women were, like the men, actively employed as cultivators, and some of them were responsible, as in Cossigny's Garden of Palma, for packing the grain to be distributed on the island.[139]

Sometimes only Creoles are listed, and Malagasies were an exception; for instance, a woman called Rose who was aged, according to the list, 58 and worked for a certain Jacques Boyes. He had seven 'Black people', six 'Black women', and four 'negrillons' ('Black children'), of whom the youngest was the little Benoît, aged 3.[140] All sixteen are shown as 'cultivators'. However, I infer that this does not necessarily mean that they all worked in the fields, but simply that the list follows a pattern: Boyes, for example, lists two '*noirs invalides*' ('handicapped Black people'), two Malagasies named Jouan (aged 75) and Philippe (30), in a separate column from the regular Black people, but they are all listed as 'cultivators'.[141] Another example is that of the Creole Antoinette Dsile, aged 17, and her 2-year-old son, Elie, together with a Creole enslaved woman, Marianne, aged 20. Both the little Elie and his mother, Antoinette Dsile, are indicated as 'cultivators' – which in light of her son's age, is doubtful.

Looking into the registers in Réunion from 1782, which are categorised in a similar way (name, caste, age, occupation), it is interesting to note that there the enslaved people are listed only rarely in relation to agriculture. In fact, the list does not show any '*cultivateurs*' at all, but lists

[137] ANOM, C/4/24, no. 36, Poivre and Desroches to Praslin, 1 September 1769. See also Bernardin 'Voyage à l'Isle de France', 59.

[138] BCMNHN, MS 280, vol. 1, Extract of this inventory, n.d. This is an extract of an alphabetical list, covering letters B to D. There are indications that it must be dated shortly after 1765.

[139] See letter from Cossigny to Céré, 1778, printed in *La Revue rétrospective de l'Ile Maurice 1953* (4), 103.

[140] BCMNHN, MS 280, Extract of inventory, n.d. On enslaved children, see also the edited volume by Campbell et al., *Children in Slavery through the Ages*; see also Campbell et al., 'Children in European Systems of Slavery', 163–82. On the Mascarenes, see Allen, 'Traffic Repugnant to Humanity', 219–36.

[141] BCMNHN, MS 280, Extract of inventory, n.d.

carpenters, sailors, domestic workers, blacksmiths, informers, keepers, cooks, wheelwrights, couriers, bakers, and artisans. Out of the approximately 800 enslaved people on that island only four are listed as gardeners in the king's service: the Malagasy Ferlaquel (aged 34), the Creole Jamb de Baccus (aged 48), the Malagasy Antoine S. Paul (aged 50), who is listed as handicapped because he was asthmatic, and the one-eyed Kaffer Bazica (aged 59).[142] Although this list shines a light on the ethnic diversity of gardeners (Malagasy, South African, and Creole, the latter a slippery term, of course), it does not reveal their precise tasks. In no list is there any reference to '*noir esclave*' ('enslaved Black person'). Instead, simply '*noir*' ('Black person') is used, making it difficult to ascertain what 'enslaved' really entailed. It is also worth noting that there were Creole women who ran households and even a *habitation*, such as a certain Françoise Lucy Magarde DesIsle, aged 32, and her three daughters, aged 10 to 14. As there is no record of a husband, DesIsle was probably a widow: on her land, she oversaw twenty-four male and nineteen female enslaved people as well as seven children.

Back in 1757, when the CIO's botanist and pharmacist Fusée-Aublet was the responsible gardener at Le Réduit, he also supervised and instructed the enslaved people who worked there.[143] In his manuscripts on his work in Mauritius in the 1750s and 1760s, he emphasised the important contribution of the enslaved people to the acclimatisation garden of Le Réduit; in an unpublished 'Eclaircissement particuliere' he declared: 'I have established a garden. It was done at great expense and the considerable assistance of slaves who served me.'[144] Similarly, in a letter to the minister from 1757, the then governor of Mauritius, Magon, specifically mentioned the enslaved people's responsibility for and importance to the cultivation of tropical plants.[145] Fusée-Aublet was struggling to establish Le Réduit with appropriate local plants: in the hospital, medicinal plants were almost out of stock, with only a few old and dry ones remaining. He thus aimed to look for alternatives but could do this only slowly. Instead, he focused on local 'grasses', particularly for

[142] ADR PR 10 C, entries 129, 177, 727, 733, 'Matricule général', 1782. See also *Annonces, Affiches et Avis divers pour le Colonies des Isles de France et de Bourbon*, 15 January 1773, in which a *habitation* is to be sold including its fifteen enslaved persons ranging from Malagasy and Mozambican carpenters to an Indian cook, to Malagasy and Creole women looking after the poultry or selling things at the bazar.

[143] ANOM, E 296, Magon to the minister, letter dated 20 March 1757. Magon divides the slave population into four groups, the first and largest part responsible for construction works, the second for the navy, the third for cultivation, and the fourth for working wood.

[144] BCMNHN, MS 452, Fusée-Aublet, 'Eclaircissement particulière'.

[145] ANOM, E 296, Magon to the minister, letter dated 20 March 1757.

cattle fodder, the supply of which was – he complained – one of his least important but most demanding missions.

After some unsuccessful attempts with several species, he focused on 'fatack', which may have been introduced under La Bourdonnais, as explained in Chapter 3.[146] Fatack, native to Madagascar, was eaten by the maroons in Mauritius.[147] The grass was an excellent source of fodder for cattle, too, and Fusée-Aublet was relieved that he had finally found a plant to feed the livestock in his garden rather than having to look for fodder in other parts of the island. This meant, he argued, that the enslaved women did not have to bring the fodder in, and as he no longer needed to feed grain to his animals it could be given to the enslaved people instead. Thus, he saw fatack as not only food for cattle but also something that could ease the enslaved people's lives, those of women in particular.[148]

Initially, two Creole boys[149] from Réunion were sent to Le Réduit in 1753, but since they were fed insufficiently, and became too weak to work, Fusée-Aublet sent them to the hospital for treatment.[150] But their health did not improve, so he sent them back to Réunion and engaged soldiers and enslaved people instead. He praised the enslaved people for their help in the CIO's garden, claiming that their assistance was indispensable and his duty was to treat them well. He acknowledged them for their help. Instead of keeping records and registers, he explained, he saw his task as taking good care of them.[151]

The enslaved people in Le Réduit, between forty and eighty men and women, carried out a range of tasks. Besides assisting in the cultivation of useful plants, they worked in the *habitation* (see Figure 3.1), helped the troops who came to Le Réduit, cut wood, carried stone and timber from the forests, and made lemon juice (whose purpose is not indicated).[152] They also, of course, cultivated the garden produce, particularly fatack, fed the cattle, and built the enslaved gardeners' housing.[153]

For the cultivation of medicinal plants, Fusée-Aublet insisted on enslaved people rather than French agents. In addition, enslaved people were sent out to destroy locusts in January, the breeding season, and to chase them away from the crops between June and August, when their attacks were the worst. Similarly, the enslaved people would have to hunt

[146] BCMNHN, MS 452, Fusée-Aublet, 'Prairies, paturages, a troupeau'.
[147] Ibid., Fusée-Aublet's mémoire. Poivre also makes a reference to fatak in Poivre, *Travels of a Philosopher*, 29; also 'fatak' or 'fattak'.
[148] BCMNHN, MS 452, Fusée-Aublet, 'Prairiers paturager a troupeau'.
[149] There is no indication as to whether they were free or unfree. From the context, possibly the latter.
[150] BCMNHN, MS 452, Fusée-Aublet, 'Pharmacie'. [151] Ibid.
[152] Ibid., Fusée-Aublet, 'Eclaircissements particuliers'. [153] Ibid.

down the rats and monkeys that threatened the crops.[154] Le Réduit's enslaved people were also detailed to transport grain and dried herbs to the ports and vessels, and to gather medicinal plants for the hospitals – where yet more enslaved people looked after the sick and injured ones Poivre employed thirty-six enslaved people in the hospital, particularly in the pharmacy,[155] and in 1772 this increased to 100.[156] This increase is evidence that enslaved people with medicinal knowledge were actively engaged in the hospitals and may even have augmented the medical knowledge with their own forms of healing. But above all, as these examples show, it was the enslaved people who were indispensable to the successful operation of the garden.

In Poivre's Monplaisir enslaved people carried out tasks similar to those in Le Réduit. Poivre advertised his garden as the richest the island could ever have received, and stated that after his departure at least seventy enslaved people would need to be employed in order keep the garden in order and productive. Otherwise, he claimed – in an attempt to outdo his rival – it would become like Le Réduit, just a 'pasture land' for 'all kinds of tropical curiosities'.[157] The enslaved people in Monplaisir originated from Guinea, India, Malaysia, Mozambique, Mali, and Madagascar. They lived and worked in the *habitation* (see Figure 3.2), where they were the cooks, domestic servants, carpenters, and gardeners. The enslaved people who worked in the house are listed as six families consisting of father, mother, children, and grandchildren.[158] In all cases the father would carry out a specific task, which was sometimes shared by the women and, more rarely, their children; for example, there was François, son of the Guinean cook Crispin and his wife, and the baker Mousagué, who also looked after the poultry.[159]

Monplaisir's twenty-two gardeners are listed individually; all of them came from Mozambique, except for the 'little Indian' Fohi. Apparently, it was only the men who worked as gardeners; women are listed separately

[154] Ibid.
[155] ANOM, C/4/22, f. 367r, Poivre to Praslin, 30 November 1767. On the establishment of hospitals, see also ANOM, E 362, the personal file of Charles de Saint-Mihiel, *médecin à l'Isle de France*, and ANOM, E 81, the file of Chevrillon, chirurgien-major des hôpitaux du Roi à l'Isle de France.
[156] AD Loire, FGB 15 J 8, 'Mémoire de Pierre Poivre sur l'état dans lequel il a remis la colonie d'île de France à son successeur', 23 August 1772, paragraph 10: 'Les hôpitaux'.
[157] Ibid. On the rivals Poivre and Fusée-Aublet, see Spary, 'Nutmegs and Botanists'.
[158] Besides the MA, OA 127, 'Liste des noirs de l'habitation de Monplaisir', n.d. possibly between 1772 and 1776, see also MA, OA 127, 'Etat et appréciation des noirs et régresses achtées par le Roy provenant l'habitation de Monplaisir', which lists Malay and '*noirs bambaras*'.
[159] MA, OA 127, no. 42, 'Liste des noirs de l'habitation de Monplaisir', n.d. possibly between 1772 and 1776.

and not shown as having any particular role. While enslaved people were generally given a French first name, some birth names also appear, such as the Mozambican Marimounou, Charimounou, and Maérouvria (or Mazouveira). The Malagasy enslaved people had to work as domestics, fishers, or guardians, and were listed with their own names: (phonetically transliterated into French spelling) Farlahé, Oudouda, Lahémanou, and Conlouvoulou.[160] The enslaved people's rank was determined by the tasks they carried out: the higher the rank, the more the privileges – individual accommodation perhaps, or being allowed to retain their own names. In Monplaisir the enslaved people had separate accommodation, 'the camp of the Black people': eleven huts located along the main path through the garden, along with some other small buildings and a kitchen.[161] The enslaved people cultivated the plants not only for their masters but also for themselves.[162]

Monplaisir helps to explain the transmission of plant exchange and information with other gardens, such as Cossigny's garden at Palma, a *habitation* he had bought in 1764 in Plaines Wilhems, in the west of the island. Cossigny told his enslaved people, both female and male, to take specimens to Commerson, who lived in Monplaisir. In January 1770, for example, Cossigny sent one of his enslaved people to Commerson with a bed of soil containing some (unidentified) live plants.[163] Commerson enjoined Cossigny not to rebuke his messenger for being late back, because the man had waited at Commerson's door for an hour until he was awake, and Commerson had then asked the man to stay while he wrote a message back to Cossigny. Commerson also asked Cossigny to tell his Black porters to bring things concerning Commerson to him straight away instead of, for instance, putting plants in a vase of water to leave them until they were picked up.

[160] Ibid.

[161] MA, OA 127, 'Etat de l'habitation de Monplaisir ainsi qu'elle se comporte suivant levalution faite par habitans experts et les offres faites par Cere même habitant en payant comptant', 17 September 1772. See also BnF Cartes et Plans, 220/Div1/P2, a surviving plan of Monplaisir, 1738. Though it must refer to the early establishments under La Bourdonnais, it also indicates the accommodation for enslaved people. On the enslaved people's accommodation after the Revolution, see MA, OA 127, f. 73, Céré, 'Mémoire sur le Jardin Nationales de l'Ile de France, adressé au Général Decarn le 1rt. Nivose an IV', 1796.

[162] MA, OA 127, 'Etat de l'habitation de Monplaisir ainsi qu'elle se comporte suivant levalution faite par habitans experts et les offres faites par Cere même habitant en payant comptant', 17 September 1772; On the enslaved people's accommodation after the Revolution, see MA, OA 127, f. 73, Céré, 'Mémoire sur le Jardin Nationales de l'Ile de France, adressé au Général Decarn le 1rt. Nivose an IV', 1796.

[163] Commerson to Cossigny, 3 January 1770, printed in Cap, *Philibert Commerson*, 122–3, see also letter 2, which Commerson ends with 'Unfortunately your Black is coming back, ready to depart', 125–6, and letter 12, in which Commerson speaks of Cossigny's '*negrèsse*', which underlines that enslaved women also carried out these tasks, 153–4.

As a reward for the enslaved people's good service, Commerson explained to Cossigny that 'there is always a cup of *eau de vie* or wine for them'.[164] Throwing a twist into the master–slave relationship, Commerson sought to purloin other settlers' enslaved people by offering them a reward for their services. While it is arguable that this more personal relationship with enslaved people served Commerson's own purposes, he did also made suggestions for improving the enslaved people's life on the island in general; for example, he asked his friend Cossigny for cinnamon wood, as it was suitable for making shoes, saying that the enslaved people always hurt their (bare, presumably) feet.[165]

Commerson and Cossigny had a vivid correspondence about knowledge of plants and specimen exchange. In this correspondence, the name of Commerson's slave Joseph, mentioned earlier, appears frequently. Most of the time, however, in Commerson's letters Joseph is only referred to as 'my Black'. Even so, the two had a friendly relationship, and Joseph could move about the island freely; he sometimes took messages for Commerson on his way to the market, and one night he slept at a wigmaker's.[166] These instances exemplify the trust the two had in each other.

Above all, Commerson's correspondence reveals that he relied on Joseph's plant knowledge, which made him of value to other naturalists.[167] For instance, when Cossigny sent samples of a '*Panicum orientale, Dactylon semine napi*', it is recorded that Joseph recognised it from his own country, and he ate the stem.[168] Although *Dactylon* is believed to be native to Turkey or Pakistan, varieties are native to numerous regions in Africa, India, Sri Lanka, Israel, Afghanistan, and Madagascar – so Joseph's ethnic origin remains unclear.

Cossigny's model treatment of enslaved people took place in his garden at Palma.[169] Here, Cossigny wrote, his enslaved people had two breaks per day, when they were fed breakfast and lunch, with time off from their work.[170] Furthermore, he stressed that his enslaved people were fed not only maize, manioc, and camanioc (a type of manioc native to the Antilles and Guiana), but also potatoes and cambers; the latter was, as mentioned in Chapter 3, a sort of yam or 'air potato'; it was only rarely grown on the

[164] Commerson to Cossigny, 22 September 1770, printed in Cap, *Philibert Commerson*, p. 136.
[165] Commerson to Cossigny, 2 September 1770, ibid., 147. [166] Ibid., 135–6.
[167] After Commerson's death, Sonnerat asked for Joseph. See ANOM, E 372, Maillart's letter to the Minister de la Marine, 31 August 1775.
[168] Commerson to Cossigny, n.d., printed in Cap, *Philibert Commerson*, 124.
[169] Wanquet, 'Joseph-François Charpentier de Cossigny et le projet d'une colonisation "eclairée" de Madagascar à la fin du XVIIIe siècle', 71.
[170] Charpentier de Cossigny, *Lettre à M. Sonnerat*, 32.

island.[171] Three more type of potato were planted at Palma: (1) a white one, the smallest and starchiest; (2) a red one from China, the biggest and the most precocious yet the least tasty; and (3) a type which Cossigny named 'les patates jaunes de Malaga' ('the yellow potatoes of Malaga'). The latter were normally served only to the Europeans because, Cossigny explained, the enslaved people found them too sweet.[172]

Two important facts can be derived from this potato business. First – as I have explained in Chapter 3 – there was a clear relationship between the slave population of Mauritius and the cultivation of certain crops. Some of these were introduced exclusively to feed the thousands of enslaved people who lived there. Second, there is a clear indication that the enslaved people working in the habitations and the gardens were fed differently from those working in construction, and so on; the enslaved people in the private gardens and habitations were given maize, potatoes, yam, and manioc, and were allowed about two hours' break each day.

Given that Cossigny was familiar with the Malagasy names of certain plants, he must have learned them from the enslaved people. There are further indications that Malagasy knowledge was put to practical use at Palma, and the usage of certain herbs was observed; for instance, Commerson described a herb called *anguîvi*, which he had also seen in the garden at Palma, and observed that his Malagasies used it to make a herbal infusion, while Cossigny explained that his Malagasies cultivated it and used it to make broth.[173] On a more general note, Cossigny also stated that the enslaved people in the Mascarenes were treated better than those in the Antilles; they were less strictly supervised, particularly at night – and, most importantly, they were given regular breaks during the working day. Sunday was a day off: if enslaved people had to work that day, they were paid and given some liquor.[174]

So, in some parts at least of Mauritius, the enslaved people were not only treated with a degree of latitude but were also appreciated for their skills and plant knowledge. The Malagasies in particular possessed knowledge of local flora, which could, as Lissa Roberts puts it, 'prove either a blessing or a curse. Knowledge of plant-based toxins provided a subversive weapon against oppression.'[175] The example of *songo* that I discussed in Chapter 3 illustrates these power relations. To cite a different example, Céré asked his female domestic worker for her advice

[171] Ibid., 34. [172] Ibid.
[173] BCMNHN, MS 888, f. 12, 'Notte de quelques plantes et fruits de Madagascar et de leurs propriétés', 1769.
[174] Charpentier de Cossigny, *Lettre à M. Sonnerat*, 34.
[175] Roberts, 'Le Centre de Toutes Choses', 331.

about *nourrouc* (*corallodendron*).[176] In a letter to Cossigny of 27 August 1778, he explained that his old enslaved woman relieved her asthma with it (indeed, it is still used for asthma today). She had suffered from asthma for more than forty years and used the leaves of the plant rather than the flower – as Céré had wrongly assumed – to make a herbal infusion.[177] Similarly, Cossigny explained in a letter to Céré that Hubert had learned in Réunion to treat venereal diseases using a tisane made of the decoction of benzoin bark and a substance from a tree called 'link', which healed wounds immediately.[178] Though this plant did not grow in Mauritius, Cossigny aimed at testing the benzoin bark for his enslaved people in Palma, who apparently were not familiar with the practices in Réunion. Making herbal infusions from plants seems to have been normal practice among the enslaved people in both gardens, where enslaved people prepared remedies for their own consumption.

Rama and Hilaire: Two Important Slave Gardeners

In his official writings, Poivre insisted that he never actually forced his enslaved people to work, but would instead allocate their tasks to them according to their abilities.[179] The instructions for the new manager of Monplaisir (*Instruction du roi pour le sieur régisseur de l'habitation du roi à Monplaisir*), which must have been written shortly before or after Poivre's departure from Mauritius in 1773, stressed that the director of the garden should ensure that enslaved people were, when ill, treated appropriately and given the necessary conditions to recover in.[180] Yet it also stated that enslaved people had to be subordinated to their masters in order to achieve their fidelity.[181]

Clearly the balance of power could – and did – shift, dependent as it was on economic factors. This section argues in favour of the role of interpersonal relations and the appreciation of the enslaved people's work in environments like small *habitations* and their gardens. This is particularly evident in a pair of striking cases. The first refers to the 'Black gardener', Charles Rama of Monplaisir; I learned about him in a pamphlet by

[176] Charpentier de Cossigny, *Moyens d'amélioration*,131. I thank one of my anonymous reviewers for pointing me to this reference.

[177] Céré to Cossigny, 27 August 1778, printed in *La Revue rétrospective de l'Ile Maurice* 1953 (4): pp. 296–7.

[178] Cossigny to Céré, 29 August 1778, ibid., 344.

[179] AD Loire, FGB 15 J 8, 'Mémoire de Pierre Poivre sur l'état dans lequel il a remis la colonie d'île de France à son successeur', 23 August 1772.

[180] AD Loire, FGB 15 J 3, 'Instruction du roi pour le sieur régisseur de l'habitation du roi à Monplaisir', n.d.

[181] Ibid.

Poivre's successor, Maillart-Dumesle. Its text consisted of a letter that purported to have been written by the head gardener, Rama, replying to the instructions that Poivre had, on his retirement in 1772, handed to his successor. 'Rama', referring to himself as a gardener, wrote that he had heard about Poivre's treatise, in which he had been referenced several times.[182] Although Maillart-Dumesle used his pamphlet as a tool to ridicule Poivre, the fact that Maillart-Dumesle helped himself to Rama's name as the fictional writer of the letter suggests that he, Maillart-Dumesle, was well aware of Rama's important role in the garden and his close relationship with his master, by attributing and even acknowledging the plant knowledge Rama possessed.

Rama's name would suggest that he had come from South Asia, and the Mauritian archives revealed that he was in fact Bengali.[183] According to Poivre's *mémoire* from 1772, Rama not only possessed specialised knowledge about the cultivation of nutmeg and cloves, but also was able to instruct others as to their cultivation. Before his departure, Poivre had handed Rama detailed instructions and a catalogue of every tree and plant that grew in Monplaisir.[184]

A list of the gardens' enslaved people, sixty-five in total, provides further insights: Rama, an 'excellent gardener', worked as head gardener in Monplaisir and is listed as Malabar.[185] Digging deeper, we learn that Rama and his wife Catherine (also Bengali), together with their daughter, were freed on 15 October 1776.[186] The document makes it clear that the reason for his emancipation was his excellent work. A month later, Poivre's successor, Maillart-Dumesle, commented on Rama's emancipation in a letter to the Minister of the Navy and the Colonies, saying that this favour would certainly have an effect on the king's other enslaved people. He assured that Poivre had been asking for Rama's emancipation for so long that he could not ignore it.[187]

The crucial point that I am trying to make here is that it was the good work of Charles Rama – and Poivre's satisfaction with it – that determined Rama's, hence his family's, status in the garden. His rank as head

[182] AD Loire, FGB 15 J 11, 'Objections de Rama, jardinier noir esclave de l'habitation de Monplaisir, au mémoire de Pierre Poivre' [libelle de Jacques Maillard-Dumesle], 12 August 1774.

[183] MA, OA 75, f. 135, 'Registre pour servir à l'enregistrement des actes de liberté accordée à des esclaves', 29 December 1768–5 February 1785.

[184] AD Loire, FGB 15 J 8, 'Mémoire de Pierre Poivre sur l'état dans lequel il a remis la colonie d'île de France à son successeur', 23 August 1772.

[185] MA, OA 127, no. 42, 'Liste des noirs de l'habitation de Monplaisir', n.d. [1772?].

[186] MA, OA 75, f. 135, 'Registre pour servir à l'enregistrement des actes de liberté accordée à des esclaves', 29 December 1768–5 February 1785. When emancipated, Charles Rama was 40, his wife Catherine 25, and their daughter Marie 7 years old.

[187] ANOM, F/3, art. 89, ff. 42v–43r, Maillart to the minister, 8 November 1776.

gardener afforded him certain privileges. He supervised twenty-two enslaved gardeners, mostly from Mozambique and Fohi, the 'small Malabar'. Like Conlouvoulou (the Malagasy livestock keeper), Rama and his family had a separate 'hut made of palisades'.[188] In the 'occupation' column of the list of Monplaisir's enslaved people, Rama is listed as a gardener, whereas those of his wife Catherine and their daughter Marie are blank.[189] Clearly, Rama and his family enjoyed higher status in the *habitations* than the gardeners he supervised; they are listed as a family unit, together with the enslaved people responsible for housework.[190] There is strong evidence that sixty-five enslaved people in the private, quite small, institution of Monplaisir enjoyed the latitude given them: they lived with their families and some were allowed to keep their original names (or to name themselves?).

Rama and his family were emancipated four years after the end of Poivre's tenure. Upon his departure, the *habitation* of Monplaisir was sold, along with all the land, livestock, and furniture – and eighty-seven enslaved people, for 1,000 livres each, plus a baby boy, François, for no payment. Poivre had freed none of them, although (or rather because?) the gardeners and cultivators had 'talent', as he explained, and were essential for his long-term plans for Mauritius after his departure.[191] Hence, he had also sold Rama for 2,400 livres (more than the double a regular garden slave), his wife Catherine for 800, and their daughter for 200.[192] Yet, as mentioned above, Rama and his family were later freed, along with Poivre's private domestic servants, the Guinean Julien Gueblé, his wife Marie, their son Louis (aged 6), and the Creole Jean (of Indian origin).[193] Furthermore, it is testified that in 1768 Françoise Poivre's personal Bengali servant Rita had arrived, as a free domestic servant, on the *Duc de Praslin* from India.[194] Poivre and his wife had prior to their departure ensured that their personal enslaved people, with whom they must have had a close relationship, would be freed.

[188] MA, OA 75, f. 135, 'Registre pour servir à l'enregistrement des actes de liberté accordée à des esclaves', 29 December 1768–5 February 1785.
[189] MA, OA 127, no. 42, 'Liste des noirs de l'habitation de Monplaisir', n.d. [1772?].
[190] Ibid.
[191] MA, OA 127, 'Etat de l'habitation de Monplaisir ainsi qu'elle se comporte suivant levalution faite par habitans experts et les offres faites par Cere même habitant en payant comptant', 17 September 1772.
[192] Indeed, MA OA 127 contains several pricing lists which were negotiated. See 'Etat et apréciation des noirs et negresses achtées par le Roy provenant l'habitation de Monplaisir'.
[193] MA, OA 75, ff. 135, 89–91, 'Registre pour servir à l'enregistrement des actes de liberté accordée à des esclaves, 29 December 1768–5 February 1785'.
[194] Ibid., entry from 1 August 1772.

Rama's name also appears in the correspondence of the royal Jardin des Pamplemousses, which Monplaisir became after it was sold, complete with its enslaved people, to the Crown in 1779.[195] Perhaps Rama stayed on in the garden as a free man, like Bernardin's former slave Duval when his master left Mauritius. Further, some enslaved people might have decided to stay in Monplaisir after their emancipation, and if so that indicates that it had become a home to them, or that they could not find a better situation.

The second case is that of Hilaire. In 1778, when Joseph Hubert, the head gardener of the acclimatisation garden in neighbouring Réunion, came to see Céré's work in his *habitation* of Belle Eau, he brought his talented Hilaire with him. Céré was so impressed by Hilaire's knowledge that he wanted to buy him, but Hubert refused to give up such a knowledgeable and accomplished person, explaining that Hilaire performed 'difficult tasks in the habitation and the garden'. He would, however, consider the proposal if the king offered him the equivalent of a thousand 'écus in piastres' – but on one condition: Hilaire himself had to agree to the deal.[196]

Two important aspects are stressed in this context. First, Hubert obviously appreciated Hilaire's work and skills to such an extent that he even refused to sell him for the king's advantage. Second, as far as Hubert was concerned, Hilaire could not be sold against his will.[197] So in the end, keeping Réunion as his home base, Hilaire travelled between Mauritius and Réunion in October 1778, transporting specimens and young plants to be transplanted, including clove, cacao, and Chinese apricot.[198] Most importantly, he had developed a highly unconventional method of grafting clove plants which promised rapid propagation. During his visit to Pamplemousses in September 1778, Hilaire proposed using *jamrosa* as stock when grafting cloves – a method 'against the principles in France' – and which Céré 'would never have thought of', as he exclaimed in a letter to Hubert.[199]

[195] Céré to Cossigny, 25 April 1779; see Ly-Tio-Fane, *Triumph of Jean Nicolas Céré*, 44. Céré complained that Hubert did not send tobacco and coffee to Mauritius as he had promised. According to Hubert's notes, he cultivated 'red coffee'. According to Hubert's gardening journal, dated between August 1783 and December 1784, he cultivated cloves, nutmeg, mangosteen, cacao, litchi, dak grapes, avocado, small peas, cabbage, and quite a few other species. Tobacco and coffee are not even listed. He might have abandoned their cultivation (ADR 4 J 62, Hubert 'Journal de mon jardin', n.d.).

[196] Ly-Tio-Fane, *Triumph of Jean Nicolas Céré*, 43–4, 145–6; see Roberts, 'Le Centre de Toutes Choses', 331.

[197] Hubert to Céré, 1 October 1778, transcribed in Ly-Tio-Fane, *Triumph of Jean Nicolas Céré*, 146.

[198] See letter from Céré to Hubert, 10 October 1778, transcribed ibid., 147.

[199] Ibid, 43.

Plant grafting is a horticultural technique that joins two plants, to reap greater benefit than from each of the two plants alone. A 'rootstock' variety is selected for its desirable root system, and it is intended to grow together with the stem ('scion'), inserted into a slit in the rootstock bark. If this has been done correctly, the tissues join ('inosculation') and the scion grows. Both stock and scion must be supported during this process, as the joint is weak. Hilaire stayed in Mauritius for at least a month in order to give Céré the benefit of his knowledge practice. Because Hilaire's method of propagating cloves was so promising, Céré suggested conducting experiments with different stocks for nutmeg, as suggested by another enslaved individual. But despite their hopes, the cloves dried up after a little less than a fortnight. So Céré continued grafting nutmeg and experimenting with oranges, which had also looked promising, for one and a half months, and, more successfully, grafting peaches onto pears.[200]

A similar case in the context of nineteenth-century Réunion is that of the 12-year-old gardener slave Edmond Albius (1829–80) in Saint-Denis.[201] He had discovered a quick and easy way of pollinating vanilla orchids by hand; they had been introduced to Réunion from Mexico in the 1820s but due to a lack of the correct pollinating insects did not propagate there. With his method, Edmond Albius revolutionised the cultivation of vanilla, and helped Réunion break into the lucrative international vanilla trade. Both Hilaire and Edmond Albius conducted experiments using methods that matched the plants' needs, and taught the European naturalists their strategies.

Conclusions

Following recent important works, this chapter has endeavoured to centre the agency of the enslaved people within Western botanical knowledge. By the second half of the eighteenth century, the population living on Mauritius had become very diverse, with its voluntary European settlers and the many forced immigrants from Africa, the Indian Ocean littoral, and elsewhere. This varied population created a huge potential for the ongoing meshing and adaptation of different knowledge traditions from across the world. This became particularly evident to me on analysing dynamics of different knowledge traditions, focusing on the paucity of

[200] Céré to Hubert, 10 October 1778, transcribed in Ly-Tio-Fane, *Triumph of Jean Nicolas Céré*, 44, 145–6.

[201] Jennings, 'Cartels et lobbies de la vraie vanille'. On popular history accounts of Edmond's discovery, see Ecott, *Vanilla*; *Chérer, La vraie couleur de la vanille*; Céré became interested in vanilla from 1784 onwards, but he did not succeed in cultivating it.

knowledge possessed by the French settlers. Their applied knowledge was experiential and heavily influenced by environmental conditions, leading to a creolising knowledge base which constantly evolved.

I examined Poivre's attitudes to slavery, and as a result I have concluded that he promoted his idea to cultivate spices by using enslaved people as a means to his own ends. Because spices were easy to grow, he argued, their cultivation needed less space and fewer enslaved people. He thereby hoped to kill two birds with one stone: enslaved people would not be forced to perform hard labour, and both lives and money would be saved because fewer enslaved people would be required. Hence, there was a deep relationship between economic motives and the treatment of enslaved people.

Poivre's vision of Mauritius was not anti-colonial, nor was it radical, nor abolitionist. His attitude towards slavery was highly ambiguous. He used his anti-slavery sentiments as a justification for his economic interests because 'free hands' worked more efficiently than those who were unfree. While a slave still remained a slave according to law, no matter how liberally they were treated, the treatment of enslaved people cannot be detached from political economy; what might seem at first glance like sentiment or even pity was in fact closely linked to economic considerations. It becomes clear that the colonial government was ultimately more adept at ensuring the successful acclimatisation of plants than that of the enslaved people on whose labour those plants depended; the enslaved people died in large numbers throughout the French colonial rule of the island.

Naturalists such as Commerson and Bernardin, and garden directors such as Fusée-Aublet, Poivre, Céré, Cossigny, and Hubert, appreciated the enslaved people's work and skills related to plant knowledge, so this sometimes led to high rank. Enslaved people such as Rama and Hilaire played a crucial role in the development of plant knowledge – but not only them: lower-ranked enslaved gardeners of African, Malagasy, Indian, and South East Asian origin – men such as Duval and Joseph – would enrich the islands' gardens through their everyday tasks as well as through their own ways of knowing and using the plants. But despite the valuable assistance of the enslaved people such as Hilaire and Rama and their revolutionary propagation methods, much of their embodied knowledge either disappeared during the time when the histories of Mauritius were written, or has never even entered the European discourse until now, despite the fact that during their lifetime the two men were so valued, and so highly praised, for their skills.

The Mauritian gardens were intimate worlds with their own social dynamics, in which the masters and the enslaved people often had strong and positive interpersonal relations. The French colonists embraced the enslaved people's knowledge of the preparation of certain plants for nourishment and remedies. And in turn the enslaved people's possession of their plant knowledge provided them with opportunities for social mobility; when they impressed their masters with their plant knowledge, this could lead them not only to high rank and privilege but even to freedom, as in the case of Rama.

More research needs to be done, however, on what 'freedom' meant in practice; and it would be good to know whether Rama was ever able to rebuild a life on the island or elsewhere, maybe continuing to make use of his experience with plants.

5 The Cross-Cultural Quest for Spices in South East Asia

In October 1777, the Parisian weekly newspaper *Affiches, annonces, et avis divers* published an article about Poivre's adventurous search for spices in South East Asia where he had 'to land in foreign lands, to deal with distrustful, wary and greedy people'.[1] Archival sources relating to encounters between the French and local populations in the South East Asian islands and their port cities, however, reveal quite the opposite. In fact, it was friendly relationships with people of South East Asia that were the key to making the search for spices possible, and indeed successful. Also, in the official reports, the names of South East Asian contributors were erased.[2] But while those important actors have vanished from printed sources, they are omnipresent in first-hand written testimony, as Neil Safier, for example, observed in the case of the *Encyclopédie* and South American expeditions.[3] Likewise, both contemporary printed accounts and later narratives about Poivre's apparent success in the spice project ignore the interplay and the importance of the many local relationships in that success; that neglect has made a serious contribution to a simplistic and above all Eurocentric portrayal of the acclimatisation of nutmeg and cloves in Mauritius.[4] But, in fact, as the Maluku Islands were the key to the Dutch spice monopoly, they were heavily defended against all comers, and Poivre in his explorations had to rely on the help of a range of actors from diverse backgrounds and ethnic groups.

The purpose of this chapter is to identify at least some of the individuals who have disappeared from the Eurocentric narratives about the French spice quests and to acknowledge their agency and contribution as opposed to their passivity and/or oppression. In this process, I focus on the role of the intermediaries and the suppliers omitted from the mainstream histories of the French spice quest, aware meanwhile of the asymmetries involved in

[1] *Affiches, annonces, et avis divers* (Paris), n. 42, 15 October 1777, 168.
[2] Poivre's report to the Academy, *HARS* 1772, 56–61.
[3] Safier, *Measuring the New World*, 9.
[4] Malleret, *Pierre Poivre*; Ly-Tio-Fane, *Mauritius and the Spice Trade*, 2 vols.; de Fels, *Pierre Poivre ou l'amour des épices*; Le Gouic, 'Pierre Poivre et les épices'; Piat, *L'île Maurice*.

the cross-cultural encounters. For some time now scholars have argued that Europeans were clearly not automatically superior – especially in sites of negotiation and in contact zones and when they were relying on intermediaries. In fact, it now seems to be accepted that 'imperial knowledge' could not have been created without cultural intermediaries.[5]

This chapter consists of two main sections, bringing to light first the informal relationships between the non-Europeans and the Europeans in the Indo-Pacific world, and second the local transmission of information in and around the islands of South East Asia. I look into the Spanish–French collaboration of the 1750s and trace the development of these relations, in that they formed the basis of a wide and deep exchange of knowledge, information, and materials on a secret Indo-Pacific island network in the late 1760s and 1770s. I endeavour to show that that the French researches required from the very outset an informal and cross-cultural network of Indo-Pacific islanders, Chinese traders, Dutch residents in Indonesia, and Spanish and Portuguese allies in the Philippines and Timor.[6] I question the constructs of 'nation' and 'imperial' and the general basic assumptions about successful European endeavours around the world. This approach facilitates the decentring of the history of science and of empire, especially in social and economic terms.

I argue two major points. First, that the spice quests were in fact decentralised ad hoc, and that much successful business negotiation operated outside imperial governance and formal conventions. Second, that relationships and patronage in the Malay–Philippine region were of crucial importance to the success of the French trade. I explore in particular the extent to which South East Asian islanders, as well as Spanish, Dutch, and Portuguese colonial settlers – all using a mixture of languages, promises, allegiances, and bribes – decided to cooperate (or not) with the French actors. I not only claim that these cross-cultural encounters were the key to the French attempts, but I also stress the profit-making activities of local cultures by exploring the interdependence of the French actors with the South East Asian islanders. This view encourages an understanding of the asymmetrical power relations, and reveals some at least of the injustices of European imperialism. More importantly, this approach helps us examine the meaning of the European presence from the viewpoint of South East Asian islanders and how they sought to turn the various conflicts between the European colonisers to their advantage.

[5] Pratt, *Imperial Eyes*; Schaffer et al., *Brokered World*; Raj, *Relocating Modern Science*.

[6] For the Spanish–French interaction and the spice quests in the Philippines, see de la Costa, 'Early French Contacts with the Philippines'; Nardin, 'La France et les Philippines sous l'Ancien Régime'. Poivre's biographer Malleret also offers superficial insights into Poivre's contacts in South East Asia (*Pierre Poivre*, ch. 7).

Poivre and the CIO (1748–1755)

To begin with, Pierre Poivre had in 1748 planned to smuggle cloves and nutmeg out of the Maluku Islands to Mauritius in order to turn the island into a cultivation centre for the French spice trade. But the Maluku Islands, a coveted spice monopoly belonging to the Dutch, were very well defended. However, during his earlier travels in the region in the early 1740s, Poivre had been given a tip that Chinese and Malay merchants were trading fresh seeds and seedlings in Manila, in the Philippines.[7]

He had spent several years as a novice in Cochinchina with the Société des Missions Etrangères (1742–3) and as he had travelled in many parts of Asia as well, during which time he had been able to closely observe local practices and gather secret information about the cultivation of spices in Indonesia.[8] So 1748, he made a dual proposition to the CIO: first, to establish commercial relations with the king of Đàng Trong (Cochinchina) and, second, to buy nutmeg and cloves in Manila and ship them to Mauritius, to propagate them there.[9] In order to break the Dutch spice monopoly, he was convinced that a twofold strategy was required: the establishment of a depot in Cochinchina, and the cultivation of spices in the French Indian Ocean outposts.[10] His proposal was welcomed by the CIO, partly at least due to the interest in Burma taken by Joseph François Dupleix (1695–1763), the CIO's governor-general in India from 1742 to 1754, who was willing to provide ships; although Dupleix did not put much energy into the idea of cultivating spices, he was very interested in commercial relations with Cochinchina,[11] in order to trade gold, sugar, iron, timber, indigo, silk, white pepper, and copper, among other things.[12]

Dupleix felt sure that the CIO could outdo its rivals. Indeed, he was nothing less than exuberant when, in 1746, he explained to his Tamil *dubash* ('man of two languages', i.e. translator or agent), Ananda Ranga Pillai, that 'the English Company is bound to die out. It has long been in an impecunious condition, and what it had to its credit has been lent to

[7] On the global dimension of the European trading companies and Asia in a comparative perspective, see the recent volume by Berg et al., *Goods from the East*.
[8] On Cochinchina at the time, see Li Tana, *Nguyen Cochinchina*.
[9] BnF, NAF 9341, 'Instructions de la Compagnie des Indes Orientales à M. Poivre. Approuvées par M. de Machaux, contrôleur général des finances', 30 September 1748.
[10] Kennedy, 'Anglo-French Rivalry in Southeast Asia', 200.
[11] See for instance Pillai, *Private Diary of Ananda Ranga Pillai*, vol. 6, 155.
[12] BnF, NAF 9377, f. 38r, 'Instructions du conseil supérieur de Pondichéry au Sr. Poivre, chargé de l'Expédition pour la Cochinchine', 5 July 1749.

the King, whose overthrow is certain. The loss of its capital is therefore inevitable, and this must lead to collapse.'[13]

Dupleix had a similar opinion of the Dutch VOC: 'In like manner; the Dutch Company is destined to share a similar fate. Its expenses continue to be enormous, whilst its trade has considerably decreased.'[14] But his arrogance blinded him.

In the CIO's instructions of 5 July 1749, its directors – Dupleix, de St Paul, Guillard, Le Maire, Boyelleau, and Friell – set out their plans for Poivre's travels.[15] The instructions provide very detailed information on how, where, and when Poivre was to go, but in the event this plan was useless. He met with insurmountable difficulties due to the weather; at the very start, he arrived two months later than planned, owing to the monsoon.[16] The reason why the instructions were so detailed could be that Friell in particular was deeply suspicious of Poivre's proposal; unlike the other CIO signatories, Friell could not see the overall advantages of relations with Cochinchina. The precision of the instructions helped persuade him to sign, and can be understood, in its control over the mission, as some form of security. In fact, the control remained theoretical. The instructions had been intended to regulate the processes, but their implementation remained a matter of local conditions – in this case, environmental ones in particular – as history shows, starting with natural conditions and the lack of European vessels.

Over a year later, in August 1750, Poivre landed in Canton, where he made arrangements with a Chinese merchant. He bought goods made for the Manilan market, and, due to absence of any CIO ships , took the Portuguese Santa Rita to Manila, where he went ashore in late May 1751.[17] At that point, he still had no credibility, in that the CIO officials were still dubious about his plan to buy spices in the Philippines. They had in fact only approved his venture when two Spaniards who lived in Mindanao (the second largest island of the Philippines), stopping off in Mauritius en route to Spain, had confirmed that the idea was in fact

[13] Bayly, *Empire & Information*, 6; Pillai, *Private Diary of Ananda Ranga Pillai*, vol. 2, 81. On Ananda Ranga Pillai and the role of dubashes in French India, see Marquet, 'La médiation des dubashes'.

[14] Pillai, *Private Diary of Ananda Ranga Pillai*, vol. 2, 81.

[15] BnF, NAF 9377, ff. 36r–41r, 'Instructions du conseil supérieur de Pondichéry au Sr. Poivre'.

[16] BnF, NAF 9377, f. 55r, 'Rapport de la mission du sieur Poivre'. On the instructions Governor David received regarding Poivre's spice quests, see BnF, NAF 9341, f. 291v, Friell to David, 25 October 1749.

[17] BnF, NAF 9377, ff. 54r–67v, 'Rapport de la mission du sieur Poivre a la Cochinchine'. At least two copies of the said report exist. The original by Poivre is held at the BCMNHN, MS 312 and MS 575. This document was transcribed, Cordier, 'Voyages de Pierre Poivre de 1748 jusqu'à 1757'.

feasible, in that the Dutch regularly traded their spice seedlings in the Philippines.[18] This is a good example of the confidence in local experience notwithstanding any differences in national identity.

In any case, once the mission had been approved, Dupleix, hoping for great profits, gave Poivre a secret letter, offering him 20,000 piastres for twenty-five nutmeg trees and twenty-five clove trees from Manila.[19] Dupleix gave a copy of the letter to Carvaillo, a Spanish merchant who had come from Pondicherry to Manila: he was instructed to receive the spice plants.[20] But by the time Poivre disembarked in Manila in late May 1751 (two months late), Dupleix's letter had become public knowledge.[21] This made Poivre's mission infinitely more challenging, because he was now in fear of the Dutch seeking to protect their spice monopoly.[22]

He was lucky enough, nevertheless, to come across some merchants from the Maluku Islands and Borneo on the Manila riverside. He hoped to receive information about spices from these traders and was able to buy 300 fresh seeds in their shells from a Chinese merchant, some even with their mace (the seed's reddish covering, also used as a spice) intact. According to Poivre's report, he was able to make a good dozen of the seeds germinate, later repeating his success with another thirty-two seeds from the same merchant.[23] Chinese merchants and their role as traders for goods from the Dutch-owned Moluccas can be traced throughout the VOC's annual reports, but the idea that Poivre could actually make the seeds germinate is probably false and should be viewed as self-promotion.[24] Nutmeg was – and still is – extremely difficult to germinate, which is what makes his report so implausible.

Spanish–French Relations and the Philippines

Although Poivre's search for nutmeg was going well, his search for cloves was proving more difficult. As he explained in his report to the CIO, clove seeds were not to be sourced from traders, as they sold only the mature

[18] De la Costa, *Early French Contacts with the Philippines*, 407. For the actual approval, see ANOM, C/1/2, f. 3r, 'Délibération de la Compagnie des Indes à l'assemblée du 25 juillet 1748'. On the instructions set up by the CIO, see BnF, NAF 9341, f. 294v, 30 September 1748.

[19] On the great hope of the spice project, see also ANOM, C/4/7, Bouvet to the Secret Committee, 30 April 1753.

[20] Cordier, 'Voyages de Pierre Poivre de 1748 Jusqu'à 1757'.

[21] BnF, NAF 9224, ff. 149–54.

[22] BnF, NAF 9377, ff. 57–8, 'Rapport de la mission du sieur Poivre'. See also ANOM, C/2/82, f. 284r, Dupleix to the CIO, 3 October 1750.

[23] BnF, NAF 9377, f. 58r, 'Rapport de la mission du sieur Poivre'.

[24] Schooneveld-Oosterling, *Generale Missiven*, vol. 7.

plants. So he decided to collect the seeds from the Moluccas themselves. During his earlier travels in the region, he had learned that nutmeg and cloves grew on some islands of the group which had not been colonised by the Dutch, so were left undefended.[25]

It was important, of course, to keep his plans as secret as possible. Since ships sailing to the Moluccas were few and Poivre had no official support, he turned to the Spanish governor-general of the Philippines, Don Francisco José de Ovando (1750–4).[26] Poivre, careful not to reveal to Ovando that his mission to challenge the Dutch spice trade had become public knowledge, persuaded Ovando to sign an order to Juan González Del Pulgar, the governor of Zamboanga (a district of Mindanao). This was to provide an armed ship with Mindanao islanders on it to take Poivre to the Spice Islands, and to ask the Mindanao chiefs to instruct the islanders to search for spices once they had landed.[27]

Del Pulgar expressed great interest in the enterprise – but Ovando then learned that Poivre's mission had become public. So, Ovando, acutely aware that a Spanish–French collaboration would challenge the Dutch, changed his plans: now, officially, the search for spice plants was to be carried out in his own interests.[28] He was playing a risky diplomatic game; while he could not risk open conflict with the Dutch, he was nevertheless interested in challenging their spice monopoly.

His underlying motivations would have been economic. We can see this from a letter from Poivre to the CIO dated 30 September 1751, in which he explained that the Spanish market and the colonies of the Philippines had suffered greatly after losing a galleon out of Acapulco, on the far side of the Pacific.[29] The Manilan government was in constant need of money from the Spanish Crown, and its relations with the Dutch and the English were deteriorating. Moreover, Ovando must have flirted with the idea of acquiring high-quality spices in other parts of the region, but although spices such as cinnamon were native to some of the Philippines, the plants grew high in the mountains – regions controlled by Indigenous people, who would not

[25] Cordier, 'Voyages de Pierre Poivre de 1748 Jusqu'à 1757'.

[26] On the Spanish presence and commercial relations in Asia in general, see Rodao García, *Españoles en Siam* (1540–1939); Yuste Lopez, *Emporios transpacíficos*; Fradera, *Colonias para después de un imperio*.

[27] In the handwritten French source, Pulgar is spelled Pulgarre. On Ovando and Pulgar, see AGI, FILIPINAS 457–70, 'Duplicados de cartas del gobernador de Filipinas (1750–54)' and AGI, FILIPINAS 385–7 'Correspondencia de los gobernadores con la via reservada (1746–1762)'. Del Pulgar was nominated in 1743 (AGI, FILIPINAS 342, L.11, ff. 155r–158r).

[28] BnF, NAF 9377, f. 58v, 'Rapport de la mission du sieur Poivre'.

[29] BnF, NAF 9224, ff. 169r–174r, Poivre to M. de St Priest, Manila, 30 September 1751.

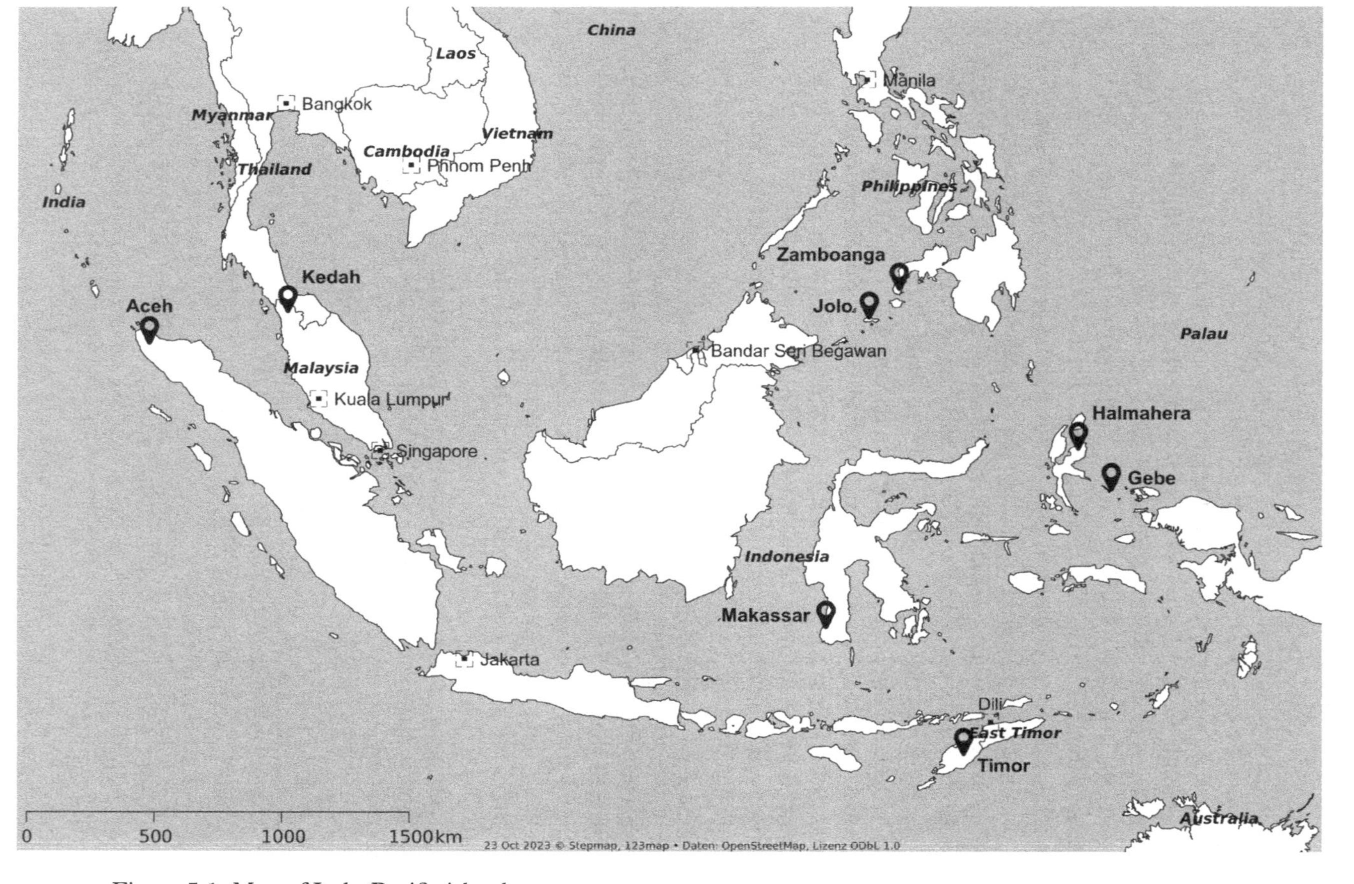

Figure 5.1 Map of Indo-Pacific islands

permit the Spanish colonisers to harvest them.[30] This is a cogent example of the Europeans' lack of power in the face of local resistance. Another example is the struggle between the Dutch and the kingdom of Kandy (in Ceylon) over the search for cinnamon: the Dutch had to ask the king for permission to harvest this spice, said to be best in Kandy.[31]

Then in the later 1750s, the Spaniards attempted to introduce spices to the Philippines for a financial boost. Nicolás Norton Nicols, a naturalised English physician who had lived in Spain for many years, and seeking his fortune in Manila, noted that one could do in Mindanao 'come en la Isla de Mauricio' ('as on the island of Mauritius'), that is to say, introduce and cultivate spices, such as cinnamon and pepper, and timber such as ebony, along with other commodities, because the soil on both islands was similar.[32] He proposed setting up a direct trade route between Spain and the Philippines in addition to China, the Coromandel and Malabar coasts, Bengal, and even Persia. But as we read that he then expressed his sincere concerns that the islanders of Mindanao, and particularly Muslims ('los moros') in the region, had blocked his enterprise, it becomes clear yet again that local populations had a degree of power over the colonisers.[33] Later, in the 1780s, the Spanish botanist Juan de Cuéllar hoped to capitalise on the plan to cultivate spices, especially cinnamon, from both Ceylon and China. He was, however, troubled by the fact that the sultans of Mindanao could easily put pressure on the project. He saw the need to enter into diplomatic relations with them, particularly Sultan Suibar Salaniad, as can be traced through his correspondence with the Spanish authorities.[34]

[30] On the struggle between Spain and Mindanao islanders in relation to the cultivation of spices, see, for instance, AGI, FILIPINAS 199, N. 7, Nicolás Norton Nicols 'El comercio de Manila y las Convenicias y Beneficio de las Islas Phelipinas', 2 September 1757; and AGI, FILIPINAS 39, containing a map entitled 'La Ysla de Mindanao, dedicada al Rey Nuestro Señor Don Fernando VI qui Dios guarde, por Don Nicolás Norton Nicols', 1757.

[31] On the resistance of islanders on Sri Lanka more generally, see Wickremesekera, *Kandy at War.*

[32] Nicolás Norton Nicols, 'El comercio de Manila y las Convenicias y Beneficio de las Islas Phelipinas (a Su Magestad Quien Dios guarde), Madrid, 2 September 1757; AGI, FILIPINAS 199, N.7. 9, 'Petición de licencia para pasar a Filipinas de Nicolás Norton Nicols', 29 October 1757. On Nicols, see Fish, *Manila–Acapulco Galleons*, 473. Nicols is also spelled Nichols; Nicolás, Nicholas, Nicholás. See also the map entitled 'La Ysla de Mindanao', 1757, AGI, FILIPINAS 39. I thank Ana Hernández Callejas (AGI) for providing further information on Nicols.

[33] AGI, FILIPINAS 199, N. 7, Nicolás Norton Nicols, 'El comercio de Manila y las Convenicias y Beneficio de las Islas Phelipinas', 2 September 1757. 'Moro' is the commonly accepted collective designation for all Muslim people in the southern Philippines (Pelras, *Bugis*, 2).

[34] AGI, FILIPINAS 723, Cuéllar's correspondence. On Cuéllar, see Díaz-Trechuelo Spínola et al., *La expedición de Juan de Cuéllar a Filipinas.*

Let us return to the Spanish–French collaboration in the Philippines relating to Poivre's expedition. Since the Spanish governor-general could no longer provide a ship for Poivre, he gave him permission to order one from Mauritius, but only on condition that it anchored in the port of Manila under an Asian flag rather than a French one.[35] Why so? According to one document, voyages to Manila, said to be the most lucrative in the region, were prohibited to all Europeans unless their dealings related to the Spanish Crown or the Portuguese from Macao.[36] All other Europeans wanting to trade in the Philippines were obliged to arrive under an Asian flag.[37] The reason for this law must have been rooted in Dutch–Spanish commercial competition: the VOC was not permitted to trade goods from its colonial possessions in the region.[38] In 1751, the Dutch observation of Spanish movements must also have played a role in the flag convention, in order to keep the Spanish–French collaboration secret. In trade, different identities were often quite fluid and national identities on the ground flexible, so even when these rules existed and even when they were officially required, this did not necessarily mean that they were actually implemented.

For instance, when Poivre anchored in Manila his captain, Du Bois, flew a Moorish flag.[39] Maybe, though, that mistake worked out in Poivre's favour. He explained in a letter to the Secret Committee in September 1752, while in Manila: 'Since Dupleix's attempt in 1750, the interests of the Company regarding the spices are so well known that it was impossible for me to embark without everyone being aware of the subject of my voyage, and after all, there are Dutch here.'[40]

There is evidence that VOC officials had indeed heard rumours of Frenchmen with Moorish accomplices in the Dutch colonies. A VOC report dated 30 December 1752 mentioned this, albeit in reference to Ceylonese cinnamon rather than the Maluku Islands or the Philippines.[41] In that respect, this seems surprising that – at least as far as the annual reports reveal – the Dutch did not have any further information about Poivre's activities in the region, nor that they believed it was necessary to act at this early stage. Yet, after Poivre collaborated with the Spaniards and Malays in 1755, a VOC report from 31 December 1755 makes an explicit reference to a certain 'Frans capitain De Poivre' in the region. He was said

[35] BnF, NAF 9377, f. 58r, 'Rapport de la mission du sieur Poivre'.
[36] Malleret, *Un Manuscrit Inédit de Pierre Poivre*, 32–3. The original manuscript transcribed by Malleret is held in the BM Lyons as MS 1094.
[37] Ibid. [38] This is also explained in Poivre's account, ibid., 56. [39] Ibid.
[40] ANOM, C/4/7, Poivre to the Secret Committee, 10 September 1752.
[41] Schooneveld-Oosterling, *Generale Missiven*, vol. 7, 245, report by Jacob Mossel, 30 December 1752.

to have been in Manila and Timor, where he was negotiating with the Portuguese. The VOC feared further French attempts on the Indonesian islands and the west coast of India.[42]

Eventually, in March 1752, probably with the help of the governor Del Pulgar, Poivre found two Spanish ships; these were under the command of a Malay captain who Poivre considered to be 'a man of capacity and trust'.[43] He took them to the island of Jolo (one of the Sulu Islands, between Mindanao and Borneo). While Poivre continued his mission to collect plants – which was now even more problematic, because he had arrived in the wrong season, just before the monsoon – his collaborator Del Pulgar died in 1752, to be replaced as governor of Zamboanga in February 1753 by Francisco Domingo Oscote.[44] He had a friendly relationship with the people of Mindanao, so was very important as a mediator for Poivre's mission. Poivre claimed that he had persuaded Governor-general Ovando to make Oscote governor of Zamboanga.[45]

Whereas Poivre received help from his Spanish collaborators, he got nothing but silence from the CIO. The ships he had ordered from Mauritius never appeared. Disappointed, and unwilling to wait for the CIO's reaction to his complaint, Poivre embarked on a vessel taking nineteen nutmeg plants to Pondicherry, whence he later sent plants to Mauritius before heading back there himself in December 1753.[46] He had been assured that the one nutmeg plant from Mindanao was the 'true spice plant' of the best quality. He now needed official approval of the plant's originality.[47] Mauritius's new governor and CIO agent, Bouvet, was however not in favour of an official report – perhaps he was jealous – so the plant was introduced into Mauritius as 'mangosteen'.[48] Poivre's biographer, Louis Malleret, assumes that French governors were the puppets of the CIO director, Duvelaer, who was sceptical about the acclimatisation of spices in Mauritius.[49]

[42] Ibid., 614–15, report by Jacob Mossel, 31 December 1755.

[43] ANOM, C/4/7, Poivre to the Secret Committee, 10 September 1752.

[44] AGI, FILIPINAS, 464, N.4, 'Duplicado de carta del marqués de Ovando sobre gobernador de Zamboanga', 5 February 1753. In the French source, Oscote is spelled Oscotte. Poivre paid Oscote 2,000 piasters in advance (BnF, NAF 9377, f. 60r, 'Rapport de la mission du sieur Poivre').

[45] ANOM, C/4/8, Poivre to the Secret Committee, 10 January 1754. See also on this diplomatic coup, Nardin, 'La France et Les Philippines sous l'Ancien Régime', 12–14.

[46] ANOM, C/4/8, Poivre to the Secret Committee, 10 January 1754.

[47] See in particular on this, Spary, 'Of Nutmegs and Botanists'.

[48] BnF, NAF 9377, f. 60v, 'Rapport de la mission du sieur Poivre'. See also on the five plants ANOM, C/4/8, Bouvet to the minister, 10 January 1754 and ibid, Poivre to the Secret Committee, 10 January 1754.

[49] Malleret, *Pierre Poivre*.

Undeterred by the lack of French support, Poivre made use of a local network. He returned to Manila in May 1754, but this turned out to be a bad moment; war had broken out between the people of Jolo and the Spanish colonisers. In addition, Pedro Manuel de Arandía Santisteban (1754–61) had replaced Ovando as Spanish governor-general.[50] The Spanish–Jolo conflict seriously impacted Poivre's mission; for example, when he arrived back in Manila after three and a half months at sea, it stopped him from corresponding with his collaborator Oscote in Zamboanga. Five months later, Poivre's patience came to an end. He had his ship repaired – it had been significantly damaged during the passage from Port Louis to Manila – and sailed to the port of Calderia in Mindanao. From there, he sent news to Oscote and heard back from him; it turned out that the war had made it impossible for him to come to Manila as he had promised. He reported that he had successfully collected some fresh cloves and nutmeg in the Moluccas – together with native islanders, he stated – but now, after a long voyage back to the Philippines, the plants were too old to be cultivated.[51]

Poivre considered going back to the Moluccas himself, but he had neither a ship nor a reliable captain who knew the region well enough. So, he approached a Dutch captain who had served his country for fifteen years before changing his allegiance to the Spaniards in the Philippines. But even when Poivre offered the captain 1,000 piastres, he did not accept because he was anxious about being caught by his compatriots.[52] Fortunately for Poivre, Oscote promised to continue the collaboration with him once the war was over and he could collect spices with people from Jolo and Mindanao.

After the disappointment with Spanish authorities, in April 1755 Poivre looked for allies amongst the Portuguese in Timor, the only Europeans permitted to trade in Manila.[53] The governor of Timor, Manuel Doutel de Figueiredo Sarmento (in office 1751–9?) was, unsurprisingly, interested in the spice project, and immediately sent two 'naturals' off around his island on a search for nutmeg plants matching the description given by Poivre.[54] The men found some small plants in the

[50] ANOM, C/4/9, letter 21, Poivre to the Secret Committee, Mauritius, 15 November 1755. On the Spanish–Jolo conflict, see AGI, FILIPINAS 706 and 707.

[51] ANOM, C/4/9, letter 21, Poivre to the Secret Committee, Mauritius, 15 November 1755; BnF, NAF 9377, f. 62v, 'Rapport de la mission du sieur Poivre'.

[52] ANOM, C/4/9, letter 21, Poivre to the Secret Committee, Mauritius, 15 November 1755.

[53] On Timor–Portuguese relations, see de Matos, *Timor Português*; Rony and Siqueira Wiarda, *Portuguese in Southeast Asia*; Jarnagin, *Portuguese and Luso-Asian Legacies*.

[54] BnF, NAF 9377, f. 63v, 'Rapport de la mission du sieur Poivre'.

western part of the island – but Poivre identified these as nutmeg greatly inferior to the 'true' variant.[55] He was acutely aware of the lower quality of the plants, inferring that this was caused by Timor's climate, which was different from that of the islands of Ambon and Banda where the 'true' spices grew (See Figure 5.3).[56] Poivre explained to Sarmento that the plants his men had found were the sort of nutmeg he had seen growing wild in the forests of those islands, of no better quality than that found in Timor.[57]

Despite the low quality of the plants, Poivre took them on board and signed a contract with Sarmento, promising 14,000 piastres for twenty 'true' nutmegs and twenty clove plants.[58] From 1755, Sarmento continually reasserted this collaboration, finally sending a letter to Poivre in Lyon in 1761 by the way of Macau that the spices had been brought by Makassar traders and were ready to be picked up in Timor.[59] Later, when Poivre returned to the Indian Ocean as intendant of the Mascarenes, he planned to go back to Timor. He originally wanted to send *Le Vigilant*, commanded by Captain Trémigon, to Timor, where the captain was to pick up nutmeg and cloves. Poivre wrote to Praslin in November 1767 that Sarmento had promised in a letter dated 1760 that the plants were ready to be picked up.[60] However, because of the tensions between Poivre and Governor Dumas, who was not in favour of the spice mission, Poivre was obliged to send a private boat, *L'Utile*, commanded by Yves Cornic, to Timor in early 1768 instead.[61] The voyage of *L'Utile* generates many unanswered questions. Poivre explained in his instructions to Cornic that he had signed an agreement with Sarmento in 1755, referring to the deal explained above. The agreement had included a payment for the spice plants which Poivre, however, was unable to raise, given his financial situation and the lack of French support; he could not pay Sarmento as he had promised, particularly in the face of Mauritius's debts.[62] So it could be that Sarmento broke his promise because Poivre had never paid up. Moreover, *L'Utile* never made it back: nearly all the crew died from tropical fever, and the ship sank in a terrible storm in early

[55] Ibid.
[56] BCMNHN, MS 1756, 'Extrait de quelques conversations avec M. Poivre en 1758'.
[57] As Poivre explained to Malesherbes, ibid.
[58] BnF, NAF 9377, f. 63v, 'Rapport de la mission du sieur Poivre'.
[59] ANOM, C/4/22, ff. 122r–124v, 'Copie des instructions secrètes données par M. Poivre aux Srs Cornic et D'Olbel, capitaine et lieutenant du vaisseau particulier L'Utile'.
[60] ANOM, C/4/18, Poivre to the minister, 30 November 1767.
[61] ANOM, C/4/22, ff. 122r–124v, 'Copie des instructions secrètes données par M. Poivre aux Srs Cornic et D'Olbel, capitaine et lieutenant du vaisseau particulier L'Utile'.
[62] Ibid.

summer 1770.[63] There were also rumours that Cornic was poisoned on the orders of the 'Black king'.[64]

Secrets and the Nutmeg's Identity

The French were searching for the particular type of nutmeg, the 'true' one, native to the Moluccas, whose identity had caused the botanical debate between Poivre and his French rivals in the 1750s.[65] When the Spanish governor announced that 'true' nutmeg had been found in the Philippines, the great excitement caused by the news was brought to a swift end when Provost identified the live plants as not 'true';[66] the variety in question was a different and inferior one, with features that distinguished it from the 'true' one: oval leaves, a felt-like fruit and, most importantly, an absence of the 'quality of the true spice'.[67] In response, Poivre continued his search for the 'true' spice and included descriptions of nutmeg and cloves in his secret instructions for Provost in 1772. He explained that the only visible – in fact it was tactile – feature identifying the 'true' species was the leaf, which could be easily ground between the fingers, producing an aromatic scent. This was a feature which the 'false' nutmeg did not have.[68] The easiest way to distinguish between the species was to examine the seeds when they were still in their mace, covered with husks. There is evidence relating to the inferior quality of the Philippine nutmeg which apparently naturally grew in Mindanao.[69] These debates about the 'true' spice were a constant feature of the French attempts to cultivate the spices. In his instructions to Provost during the second major spice quest in 1768, Poivre stated that 'the long or round nutmegs are of the same good quality'.[70] Yet, in observations he made in 1752 while in Manila, he stated that although the long and the round ones

[63] As reported by Trémigon. ANOM, C/4/25, ff. 125r–129r, Trémigon's journal.

[64] 'Requête de Maturin Cornic au Duc de Praslin sur la perte du vaisseau l'Utile commandé par son frère, Cornic Le Jeune', 20 July 1770, transcribed in Ly-Tio-Fane, *Odyssey of Pierre Poivre*, 70; see also Le Gouic, 'Pierre Poivre et les épices', 110–11.

[65] Spary, 'Of Nutmegs and Botanists'.

[66] As confirmed by Commander Trémigon in his journal, ANOM, C/4/25, ff. 125r–129r, 126r, 'Extrait du journal de M. de Trémigon relativement à l'acquisition qu'il a fait des plants de muscadier et de géroflier'.

[67] Funke, 'Muskatnüsse', 72.

[68] ANOM, C/4/22, f. 127r, 'Description abrégée du muscadier et du géroflier pour servir à mettre les Srs Trémigon et Provost dans le cas de n'être pas trompés dans le choix des plants de ces deux espèces d'arbres', 4 February 1772.

[69] AGI, FILIPINAS 723, 'Nuez moscada fruto algo largo de la Provincia de la Lafuna de Bay en las Islas Filipinas cogida en el año de 1789'.

[70] ANOM, C/4/22, ff. 119r–121v, 'Instructions secrètes pour Mrs. Trémigon, lieutenant des vaisseaux du Roi, commandant la corvette Le Vigilant et Provost subrécargue sur ledit bâtiment pour le voyage de Quéda et d'Achem', 4 February 1768.

were similar and of equal quality, the long ones were more oily and aromatic.[71]

Even during Poivre's tenure as intendant of Mauritius between 1767 and 1772, he continued the secret spice quests through this patronage network without informing the governor of the Mascarenes.[72] Poivre mounted four expeditions, ordering the French naval officer Provost to negotiate with the local people on his behalf. These expeditions were not easy to set up and Poivre had to move money around in an attempt to balance the books. Sometimes there was not enough money to send a second vessel, as, for instance, Provost had requested in August 1769.[73] So Poivre sent a letter to Provost authorising him to ask the Spanish governor for cash; in the letter Poivre had promised, on behalf of the French Crown, to reimburse the Spanish government.[74] Given the delicate financial circumstances and secrecy of the mission, this letter was in fact duplicitous; the French Crown was not in fact willing to invest money in the spice project and Poivre was well aware of it. Before Provost revealed the real purpose of his mission in the region – he explained that he was anxious about Spanish jealousy – he gave the Spanish governor in Manila a letter stating that he was on an astronomical expedition. He fabricated the story that they were exploring the islands of Bali and Lombok in order to create better maps of the region for ships coming from China, particularly in wartime.[75] As they had a French astronomer on board, this lie was plausible … the mission continued to be a matter of secrets, games, and lies.

Transit Spots and the Role of Local People

Shortly after Poivre's appointment as intendant, in 1768, *Le Vigilant* went to what is now peninsular Malaysia (see Figure 5.1), where Provost and Trémigon were to collect spices.[76] They considered using Kedah – like Manila in the 1750s – as a trading station, in the hope that this would improve relations with local people. In fact, in the secret instructions to Trémigon and Provost, Poivre had explicitly stated that the real reason for *Le Vigilant* to go to Kedah was to make contacts with expert navigators

[71] ANOM, C/2/285, f. 159r, Poivre, 'Observations sur le muscadier et principalement sur la culture de cet arbre', Manila 12 February 1752.
[72] See ANOM, C/4/22, Poivre to Praslin, 9 January 1768.
[73] ANOM, E 343, ff. 219r–228v, Provost to Praslin, Manila, 13 January 1770.
[74] Ibid., ff. 229r–230r, Authorising letter, 25 September 1769.
[75] Ibid., ff. 219r–228v, Provost to Praslin, Manila, 13 January 1770.
[76] ANOM, C/4/22, Poivre to Praslin, 9 January 1768.

from Celebes (now Sulawesi, see Figure 5.3), who had an 'interloping trade of spices' and probably had fresh seeds and live plants of both nutmeg and cloves as well.[77]

These experienced mariners were the Bugis, and they had a powerful community in Makassar (see Figure 5.1), whence their maritime commerce had developed after the fall of Makassar in 1666–7.[78] They are a South East Asian people belonging to the great family of the Austronesian people and whose homeland is south Celebes.[79] Its best-known harbour city, Makassar, was home to many Bugis from the eighteenth century onwards. There is, however, a difference between the Bugis and the Makassar people, and they speak distinctly different – albeit closely related – languages.[80] The Bugis and the Makassar had been smuggling goods from the Dutch Spice Islands from before the seventeenth century. Combined with the kingdom of Goa's active maritime trade based on Maluku spices, these two peoples imposed 'a very serious threat to the monopolistic ambitions of the Dutch'.[81] The Bugis themselves traded nutmeg with the people of the Papuan Islands (New Guinea), where according to Provost the 'true' nutmeg could also be found. The Papuans sold nutmeg to the Bugis, and the Bugis sent nutmeg – in its mace – to China, the Coromandel Coast, and Bengal.[82] This type of nutmeg, as the botanists Michel Adanson and Bernard de Jussieu explain in their report from 1773, was similar to the one that Provost had found in Manila, but it had no aroma.[83]

Besides the Bugis, French deserters who sought to return to France played a significant role in helping Provost avoid the Dutch in the region. When he was gathering information about the Dutch trading post at Saway (in the north of Ceram, now Seram, one of the main islands of the Moluccas), he stopped in the village of Wayen, whose inhabitants told him that Saway was not well protected: there were only twelve soldiers, most of them French. Foreign ships were prohibited from anchoring in Dutch territory. So, heading to Saway, Provost pretended that he urgently needed fresh water and refreshments for his crew, and gave some presents to the VOC officer in charge, who turned a blind eye to the regulations.[84]

[77] Ibid., ff. 119r–121v, 'Instructions secrètes pour Mrs. Trémigon, lieutenant des vaisseaux du Roi, commandant la corvette le Vigilant et Provost subrécargue sur ledit bâtiment pour le voyage de Quéda et d'Achem', 4 February 1768.

[78] Pelras, *Bugis*, 254, On the Bugis's sea trade and navigation, see ibid., 254–68; on local trading networks in the Maluku Sea, see Andaya, 'Local Trade Networks in Maluku', 71–96.

[79] Pelras, *Bugis*, 1. [80] Ibid., 15. [81] Ibid., 265.

[82] Report by Adanson and de Jussieu, 17 February 1773, in Ly-Tio-Fane, *Triumph of Jean Nicolas Céré*, 120.

[83] Ibid.

[84] ANOM, C/4/27, Provost's journal (May 1769–June 1770), attached to Poivre's letter to the minister, 20 July 1770.

In Saway, an old French soldier who had helped to mediate between the French and another local informant of the island (who was afraid of the French soldiers and did not appear in front of the French) said that in exchange for the information he could give, he wanted to send his 20-year-old son back to the French colonies. Provost agreed to the deal since he expected that the young man and his knowledge would be useful for the spice garden in Mauritius.[85] A report by the Mauritian governor Desroches confirms that the young man and another Frenchman who grew up in the Moluccas were admitted to Mauritius, where they were employed in the spice gardens.[86] But although Provost had assumed that the young men could provide useful knowledge about spice cultivation, it turned out that they were not as helpful as he had hoped, as will be revealed in Chapter 6.[87]

Poivre and Provost established diplomatic relations with several high officials in both the Moluccas and the Philippines during their voyages. On the island of Jailolo (now Halmahera, see Figure 5.1), the largest island of the Moluccas, for instance, when it seemed that the Dutch forces were threatening the French, Provost was offered 3,000 men as reinforcements.[88] And, earlier, during his travels in the 1750s, Poivre, on visiting the island of Jolo, had succeeded in gaining the favour of its Sultan, Mohamad Alino Dien.[89] Poivre's biographer Dupont stated that the Sultan saw Poivre as a 'father'.[90] This French–Jolo alliance was reinforced in 1770, when the Sultan nominated the French king as his immediate protector in a letter which he (or rather one of his diplomats) wrote in Spanish (see Figure 5.2).[91] Provost pointed out that it was because of Poivre's friendship with the Sultan of Jolo – 'a prince with lots of *esprit* and knowledge' – that he was well received on the island.[92] Poivre added in a letter to Praslin in 1771 that if the French Crown aimed to establish a colony in the Philippines, it was of great importance to form an alliance at the same time with the Sultan

[85] Ibid.
[86] ANOM, C/4/26, ff. 168v–179r, Desroches's report to the minister, 22 July 1770.
[87] ANOM, C/4/29, ff. 22v–23r, Poivre to the minister, 22 August 1771.
[88] 'Journal de Sonnerat', transcribed in Ly-Tio-Fane, *Odyssey of Pierre Poivre*, 96–100. The Dutch gathered their forces on the island of Jailolo (also spelled Jilolo or Gilolo, present-day Halmahera), the biggest island of the northern Moluccas.
[89] See ANOM, C/4/22, ff. 247r–259r, 'Extrait du journal de M. de Trémigon'. This extract, together with Provost's journal (ff. 260r–263r), are attached to Poivre's letter to the minister, dated 18 December 1768.
[90] Ibid.
[91] ANOM, C/4/27, f. 43v, Sultan Mohamad Alino Dien to Louis XV, Jolo, February 1770. See also Dupont de Nemours, *Notice sur la vie de M. Poivre*, 46. In Spanish, the Sultan is known as Mahamat Alimudin.
[92] Provost's journal (May 1769–June 1770), attached to a letter from Poivre to the minister, 20 July 1770, ANOM, C/4/27. See also AGI, FILIPINAS 706, N. 4, f. 3, 'Año de 1752. Copia a la Relacion del expediente formado con el motivo a la expedicion de Jolo'.

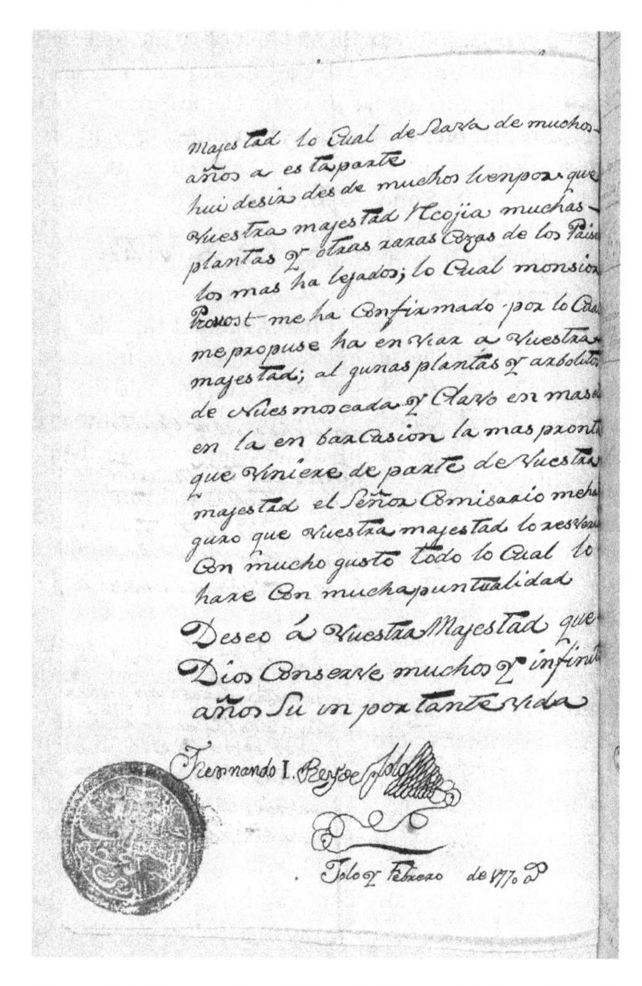

majestad lo Cual de Rarla de muchos
años a esta parte
hui desiar des de muchos tienpo por que
vuestra majestad Ycojia muchas
plantas y otras raras Cozas de los Paise
los mas ha lejados; lo Cual monsion
Provost me ha Confirmado por lo Cual
me propuse ha enviar a Vuestra
majestad; al gunas plantas y arbolitos
de Nues moscada y Clavo en mas
en la en barCasion la mas pronta
que viniere de parte de Vuestra
majestad el Señor Comisario meha
guro que Vuestra majestad lo resVeru
Con mucho gusto todo lo Cual lo
haze Con mucha puntualidad
Deseo á Vuestra Majestad que
Dios Conserve muchos y infinit
años Su im portante vida

Fernando I. Rey de Jolo

Jolo y Febrero de 1770

Figure 5.2 Extract from Sultan Mohamad Alino Dien's letter to Louis XV, Jolo, February 1770, Archives nationales d'outre-mer (France), C4/27, f. 44

of Jolo. Indeed, Mohamad Alino Dien was a reference point for other local island kingdoms.[93]

It seems highly likely that the Sultan would indeed have entered diplomatic relations with the French in order to challenge the Spaniards in the region; the Spanish–Moro conflict was playing itself out in a long struggle

[93] ANOM, C/4/27 f. 40r, Poivre to Praslin, 21 July 1771.

between Jolo and Manila. It eventually led to the arrest of Mohamad Alino Dien, and ended with his conversion to Christianity in 1749, after which he became King Fernando I.[94] The conflict was a tricky affair for the French, especially in that it hindered their collaboration with the Spanish in Mindanao in the search for nutmeg. Disappointed by their European counterparts, Poivre and Provost decided to prioritise relations with the local populations who, willing to collaborate with them, also wanted to intensify the conflict between the various Europeans in South East Asia.

The French had formed alliances in the region from their very first mission in the Moluccas. In June 1768, Provost laid the foundations for a strong Franco-Malay collaboration in Kedah. Poivre had given explicit instructions to Provost and Trémigon to negotiate a price for the fresh spices with the Bugi pirates who traded nutmegs in the Malaysian peninsula, and to offer them some money to seal the deal.[95] Poivre also authorised Trémigon and Provost to offer them rifles, gunpowder, lead bullets, sheet iron, and cannons, to be stored for the time being in Mauritius, in return for plants and seeds. Poivre stressed that the two men should not make promises they could not keep, because the Bugis, who were easily offended, would then not keep their own promises.[96]

Material exchange was at the heart of these Eurasian agreements. Relying on the help of the old and respected dervish, Hadé Hachem, in today's Indonesia Provost managed to negotiate informally with Dinck Poudony, a Bugi from Celebes. He was to gather spices for Provost, who planned to collect them a year later, in July 1769. Provost and Trémigon held a secret meeting with Poudony in Hadé Hachem's house, where they came to an agreement using basic Portuguese and Bahasa Malay.[97] The Frenchmen signed a contract promising the Bugis money and gunpowder in exchange for spices.[98] They agreed in addition to cover the expenses of Poudony's voyage to Selangor, while Hadé Hachem was to engage more captains in the Malay Peninsula.[99] With this deal, Provost felt sure that he could achieve a breakthrough by the following year. However, according to the sources, Dinck Poudony never returned to the meeting point in Kedah

[94] On the Spanish–Moro conflict, see AGI, FILIPINAS 706 and 707 as well as FILIPINAS 463, N.6. See also AGI, FILIPINAS 334, L.15, ff. 300r–303r, Philip V of Spain to Amiril Mahomenin Campsa of Tamontaca, 1744.

[95] ANOM, C/4/22, ff. 119r–121v, 'Instructions secrètes pour Mrs. Trémigon, lieutenant des vaisseaux du Roi, commandant la corvette Le Vigilant et Provost subrécargue sur ledit bâtiment pour le voyage de Quéda et D'Achem', 4 February 1768.

[96] Ibid.

[97] ANOM, C/4/22, f. 247v, Trémigon's journal. Hadé Hachem is also called Hadé Hassen in the sources.

[98] Ibid., f. 249r. [99] Ibid., f. 249v.

in 1769. The reasons for this remain unclear, but it forced the French to enter further relations deep in the Maluku Sea.

Poivre clarified in the secret instructions to Provost that if Hadé Hachem – this 'religious man' considered as 'friend of the nation' – succeeded in engaging sailors in Mindanao or Makassar, Provost was authorised to give him gifts.[100] Similarly, on the island of Gebe, Provost entered relations with another 'Mohammedan priest, an intelligent man', named Bagousse.[101] Desroches described him as a young priest who was particularly attached to Provost: he asked for certain gifts, and Provost hoped to motivate him by promising even bigger presents when time was running short.[102] As is clear from this account, the French spice quest was not only a cross-cultural but also a cross-religious affair, and the Muslims were an important reference point.

Spies, Collectors, and Their Island Networks

By offering gifts to local people or paying them to bring young plants and seedlings to trading centres outside Dutch territory, the French actors were relying heavily on local collaborators, and many of the South East Asian islanders worked as spies and transmitters. Information about Dutch vessels travelled rapidly on the larger islands and from one island to another.[103] Local allies were not only important in providing the French ships with fresh water and food, but also as spies, as in Gebe in late February 1772, where the French were informed that Dutch vessels were patrolling the region.[104] Provost described the Gebes as having 'a pleasant character' and that 'one does not have to be afraid of anything if one is careful that one does not look at their women because they are very jealous'.[105] Provost understood the need to show respect for people's customs, and ordered his men to follow his example. He skilfully used his knowledge about the islanders in order to achieve his goals, as during the negotiations with the Sultan of Patani, described below.[106]

[100] ANOM, C/4/28, Poivre's secret instructions to Coëtivi and Provost, 24 June 1771.
[101] ANOM, C/4/29, f. 245r, Provost, 'Instruction secrète pour M. Cordé commandant la corvette du Roi le Nécessaire, au cas de séparation', 25 December 1771.
[102] ANOM, C/4/26, ff. 168r–179r, Desroches to Praslin, 22 July 1770.
[103] On the travel of information between South East Asian islands, see especially Yoo, 'Wars and Wonders'.
[104] See 'Journal de Sonnerat', in Ly-Tio-Fane, *Odyssey of Pierre Poivre*, 96–100.
[105] ANOM, C/4/29, ff. 242r–245r, Provost, 'Instruction secrète pour M. Cordé commandant la corvette du Roi le Nécessaire, au cas de séparation', 25 December 1771.
[106] ANOM, C/4/26, f. 168r, Desroches's report to the minister, 22 July 1770.

The Dutch made enquiries into the French movements and questioned the islanders of Patani and the islanders of Gebe in September 1771.[107] But the islanders remained loyal to the French, because they felt that with them they had the best deal, and they could use the conflicts between Europeans in the region to their advantage. Conflicts were also used as diplomatic coups. For example, the Muslim head of Makassar had much earlier established a colony in Timor under Portuguese protection.[108] Those political partners were interested in weakening their Dutch enemy in the region. But, then, in came the French, so the South East Asian islanders hoped to encourage French interests against the Dutch. On the other, the French hoped to interest the islanders in getting rid of the Dutch. In the 1750s, when Poivre landed on the island of Buton (where he hoped to collect spices and set up local relations), the Buton islanders would not even let the crew off the ship. This was because the islanders had been at war with the Dutch, and although Poivre's interpreter spoke Malay, he could not convince the Butons that these European-looking men were not Dutch.[109] While Poivre's mediator was trying to negotiate with the islanders, 200 to 300 armed men arrived, shouting at the 'Dutch'. Poivre and his crew were left with no option other than to depart.[110]

Provost was aware that when it came to receiving information about the Dutch he had the Gebe islanders on his side: nevertheless, in the port of Pulofaux (in the south of Gebe, see Figure 5.3), in order to secure the flow of information, he still had to make promises to them and present them with gifts.[111] The French hoped to accelerate the conflict between the Dutch and the islanders just as islanders sought to manipulate the French into getting rid of the Dutch. Poivre clearly said so when, in a letter to Praslin on 16 June 1768, he wrote that he felt the Makassars would surely collaborate with the French in order to see the Dutch chased off their lands.[112] As the chiefs of Moa (the biggest of the Leti Islands, between Leti and Lakor) explained, they were at war with the inhabitants of Jailolo, and they were keen to collaborate with any European explorers

[107] Ly-Tio-Fane, *Triumph of Jean Nicolas Céré*, 25.

[108] BCMNHN, MS 1756, 'Extrait de quelques conversations avec M. Poivre en 1758'. However, when Provost returned to Timor in 1769, no trace of the Portuguese possession could be found on the island because the inhabitants had destroyed all the Portuguese buildings in Lifao.

[109] As Poivre explained in a letter to the Secret Committee, ANOM, C/4/9, letter 21, Poivre to the Secret Committee, 15 November 1755.

[110] Ibid.

[111] ANOM, C/4/30, f. 144r, Provost's report on the mission to the Moluccas, June 1771–June 1772. On the encounters with the Gebe islanders and detailed promises the French made to them, see Ly-Tio-Fane, *Triumph of Jean Nicolas Céré*, 26–7.

[112] ANOM, C/4/22, no. 66, Poivre to Praslin, 16 June 1768, secret letter.

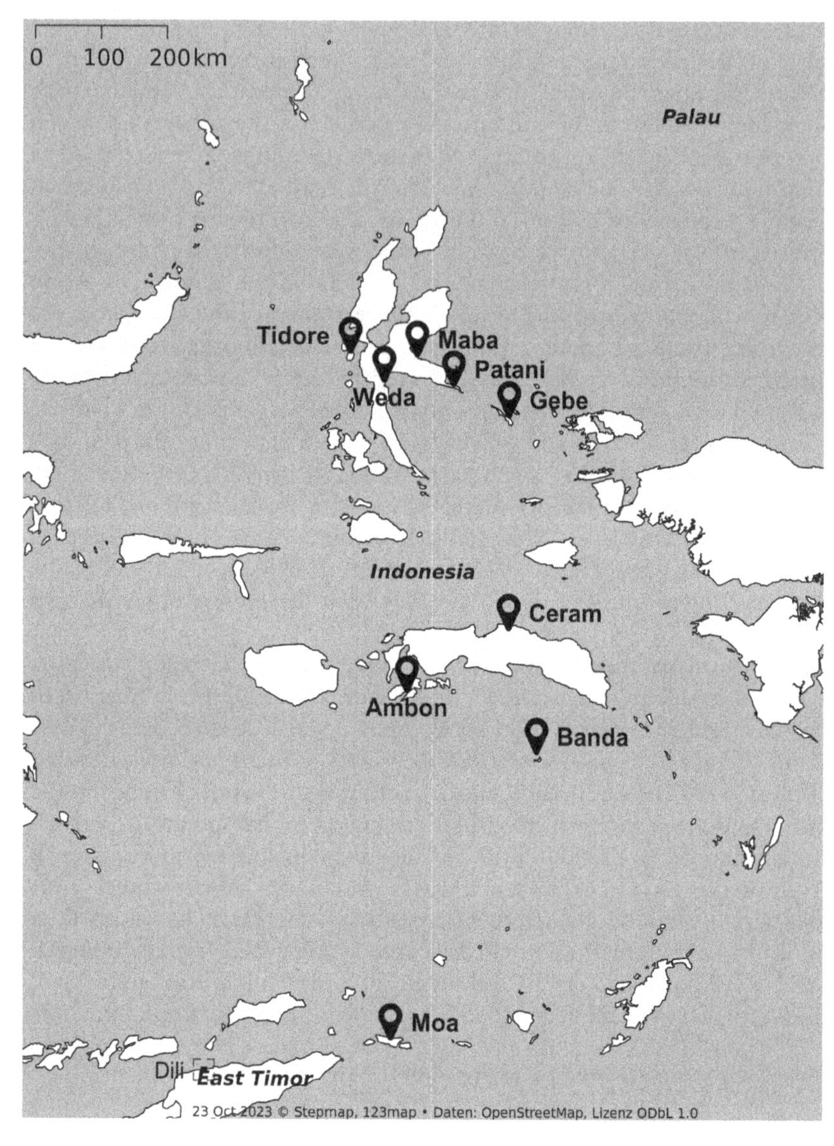

Figure 5.3 Map of the Maluku Sea

who had not tried to take their lands. Collaborating with the French was thus in their interest for two reasons. First, they could offer their services in exchange for money and goods. Second, by entering commercial relations, they could pit the competing Europeans against each other in the hope of getting rid of the ones who threatened them.

In the remainder of this chapter, I will provide several examples of this duplicity. To start with, in 1770 – that is, during Poivre's third attempt to collect nutmeg and cloves in the Moluccas – a local informant told him that the Dutch had destroyed the nutmeg and clove trees on all the islands except Ambon, which was very well defended. His informant was another old French soldier, who had been living in the Moluccas for more than thirty years.[113]

In Ceram the same year, Commander D'Etcheverry (another French agent searching for spice plants) encountered a Dutch turncoat who did not hide his hatred of the Dutch. He revealed that nutmeg of as fine a quality as that in Ambon could be found in Gebe.[114] The fact that the Dutch had not managed to completely destroy the nutmeg trees in the region was very interesting news for the French. The Dutchman knew the region very well and even sold some maps to D'Etcheverry, who was negotiating in his rusty Dutch. The Dutchman was very careful that D'Etcheverry was not glimpsed by the '*noirs malais*' ('Malay Black people') who worked in his house, quite possibly because they would have sold their knowledge to the Dutch.

With this new information about Gebe, D'Etcheverry, with Provost, set sail for the island. The Gebe islanders, who hated the Dutch for destroying their nutmeg, on seeing that the vessel was not a Dutch one, offered refreshments to the French while waiting for the island's ruler. Like the Sultan of Jolo, the ruler of Gebe was very interested in becoming an ally of the French in order to 'deliver himself entirely from the tyranny of their current masters [the Dutch]', which he graphically demonstrated, ripping a Dutch flag apart and replacing it with a French one in a grand ceremony.[115] D'Etcheverry explained to Provost that their collaboration must nevertheless remain secret in order not to challenge the Dutch openly.

The crucial point here is that the islanders always received something in return for working as spies and collecting spice plants. The price rose according to the difficulty of finding the plants: it was higher when the

[113] 'Extrait du voyage fait en 1769 et 1770 aux Isles Philippines et Moluques', printed in Poivre, *Mémoires d'un Botaniste et Explorateur*, 158–61.
[114] Ibid.
[115] 'Extrait du voyage fait en 1769 et 1770 aux Isles Philippines et Moluques', printed in Poivre, *Mémoires d'un Botaniste et Explorateur*, 161.

islanders had to look for the plants in the mountains or in foreign territory. The inhabitants of Ceram's mountains were called the 'Alphours': they were said to know 'no manners, no religion, no authority', and even the Dutch did not succeed in taming them.[116] This was the reason why nutmeg and cloves still grew in their lands: the Dutch did not dare enter, so could not destroy the spices there.[117] The islanders used the French to help get rid of the Dutch and to gain money and goods. In return, the Dutch ferocity in safeguarding their clove and nutmeg monopoly led to more violent clashes through the mid-eighteenth century.

These diplomatic games were frequently played out: for instance, at the end of the sixteenth century, the Sultan of Ternate had welcomed some newcomers – the Dutch, as it turned out – as allies in his fight against the Portuguese. The latter had established themselves in Ternate in 1512 and had been bullying the islanders ever since.[118] In his study *The World of the Maluku*, Leonard Andaya observed in relation to the Spaniards and the Dutch in the region that:

> both the Europeans and the Malukans sought allies with one another to achieve their respective goals, though their alliances were never more than fragile accommodations because of the radically differing perceptions of the 'enemy'.[119]

These diplomatic games continued through the centuries, but studies on the Spice Islands neglect the French–Maluku alliances, where a similar mechanism arose, as this chapter shows. It is important to understand that the islanders were in a powerful position precisely because the French actors were so very dependent on them. As the French–Patani alliance highlights, Provost was desperate to bring nutmeg and cloves to Mauritius and finally set up commercial relations with the king of Patani. The king offered to bring a large quantity of nutmeg to the island of Gebe, where Provost was supposed to meet him.[120] The king had the advantage of Provost who, if he wanted to take the spices to Mauritius, had no option other than to accept the king's offer. The king, keenly aware of his favourable position, asked the French ships to fly the flag of the French king and to send French forces to the region in order to defend themselves and Patani against the Dutch. Provost was entitled to sign this

[116] ANOM, C/4/26, ff. 168r–179r, Desroches to Praslin, 22 July 1770. Notably, Desroches refers to the island of Jailolo but was possibly confused in this, since the 'Alpours' were on Ceram.
[117] Ibid. [118] Widjojo, *Revolt of Prince Nuku*, 1.
[119] Andaya, *World of the Maluku*, 151.
[120] ANOM, C/4/29, ff. 242r–245r, Provost, 'Instruction secrète pour M. Cordé commandant la corvette du Roi le Nécessaire, au cas de séparation', 25 December 1771, and ANOM, C/4/26, ff. 168r–179r, Provost's report about the expedition attached to a letter written by Desroches to Praslin, 22 (24?) July 1770.

'treaty of defensive alliance' (*'traité d'alliance défensif'*).[121] The French dependence on local networks for the spice quest empowered the islanders, and the French were ever aware of the risks of losing them as allies.

Local Hierarchies of South East Asian Island Rulers

The sources also allow us to examine the relationship and connections between the inhabitants of several islands in the region, especially between important individuals. French sources confirm that the Sultan of Jolo was regarded not only as a sovereign but also as a legislator and a prophet by various people.[122] The islanders of Mindanao, the Moluccas, and Celebes would never take any action without consulting him.

The local hierarchies had to be respected, as shown in some more examples. Because the Dutch had largely destroyed the nutmeg of Gebe, its ruler had made secret arrangements with the ruler of Patani, a kingdom in the south-east of Jailolo. Patani's ruler ordered his people to go deep into the forests to look for the spices. Because the rulers of Gebe and Patani were allies, the French were welcomed by the chief of Patani and were able to explore the island without any difficulties. They found nutmeg, which D'Etcheverry loaded on to his vessel – but it was much harder to find cloves, and Bagousk Hundes, the *'grand prêtre de la loi de Patanie'* (the 'grand priest of the law of Patani'), was dispatched on this challenging enterprise.[123] It was not until D'Etcheverry had given up all hope of his return that Hundes returned with his cloves. D'Etcheverry, offering gifts to both the rulers of Patani and Gebe, promised to continue the collaboration, and indeed the Patani dynasty extended its alliance with the French for some time. In their initial encounters in 1770, Provost had flattered the king by asking him to report on the curiosities of his kingdom, particularly spices, to which the king of Patani had responded, according to a French source: 'we speak with open hearts, you see how much I am fond of being a friend of the French . . . I do not ask for anything better than subordinating myself to your king'.[124]

However, in 1771/2, during the last attempt to gather more spices, when Provost and Coëtivy, commander of *L'Isle de France* (accompanied by *Le Nécessaire*, commanded by Cordé), visited Patani in early March 1772, the sultan was busy putting down a village uprising in Gebe. So, he sent his wife, the queen, to apologise and to promise to continue the alliance.[125] While the

[121] ANOM, C/4/26, f. 173r, Desroches to Praslin, 24 July 1770.
[122] See 'Journal de Sonnerat', in Ly-Tio-Fane, *Odyssey of Pierre Poivre*, 96–100.
[123] Bagousse is also spelled Bagouce.
[124] ANOM, C/4/26, ff. 168r–179r, Desroches to Praslin, 22 July 1770.
[125] See the secret instruction for Provost and Coëtivy, dated 24 June 1771; and *Instructions particulières pour M. de Coëtivi*, dated 24 June 1771, both ANOM, C/4/28 and the secret

queen was sent on this mission, Provost, upon his return to the Patani kingdom, employed Bagousk, whom he had befriended in 1769, during his first voyage.[126] As a man given the confidence of the Patani dynasty, Bagousk aimed to collect spices in the territory of the kingdom of Maba, in the east of Jailolo. Here, rumours said that the king of Maba captured him.[127]

Some islanders – and it remains unclear who they were – tried to turn this tricky situation to their advantage: one man came to the French announcing himself as the king of Maba. He proposed setting Bagousk free if the French purchased the (few) spice plants he had with him, a deal that the French did not accept, as they felt his behaviour was suspect.[128]

Maybe the man was trying to take the French for a ride and make some money: whatever his reasons, this anecdote brings to light two important aspects of the relationship between the local people and the Europeans. First, the islanders knew that they could offer their spices in exchange for money. Second, when the 'king of Maba' had arrived, he had been very hesitant – as described by Sonnerat's journal – because French soldiers had not threatened him when he came on board, but had let him pass.[129] This, along with other examples in this chapter, provides evidence that it was reasonably normal for local people to visit French vessels in port. The sources are silent about what really happened. Eventually, however, Bagousk and the true Maba king, accompanied by a larger number of Maba men, arrived with a large supply of spice plants and seventy-three sea turtles, which were given to Provost in exchange for fabric ('brasses de toile').[130]

As mentioned earlier, the French not only gave presents to the islanders so that they would work as spies, but also generously rewarded them in the search for spices. On the island of Moa in March 1772, the islanders offered their service to the French by finding a good anchorage for *Le Nécessaire* and providing refreshments.[131] Once the ship had anchored, the chiefs arrived in order to negotiate the price for their help in the search for spices. As the chiefs of Moa explained, they were at war with the inhabitants of Jailolo. But the Jailolos – subordinated to Patani – had become an ally of the French in 1770. Gebe was subordinated to the Patani king, and the Patani kingdom was subordinated to the sultanate of Tidore (an island west of Jailolo). The sultan of Tidore was, however, allied to the Dutch – and he was one of the most important authorities in the region. Although Jailolo had a good relationship with Tidore, he

instructions to Cordé, 8 June 1772, ANOM, MAR B/4, ff. 313r–315r; 'Journal de Sonnerat', in Ly-Tio-Fane, *Odyssey of Pierre Poivre*, 98.

126 'Journal de Sonnerat', in Ly-Tio-Fane, *Odyssey of Pierre Poivre*, 98.

127 Ibid., 99. 128 Ibid. 129 Ibid. 130 Ibid. 131 Ibid., 98–9.

would no longer subordinate his kingdom to their authority but rather to the French Crown.[132] Many years later Prince Nuku (1725/35?–1805) of Tidore launched a rebellion against the Dutch, defeating the VOC in close collaboration with the English in 1801.[133]

Conclusions

This chapter has, with Chapter 2, set out the sociocultural interactions between Europeans and non-Europeans in relation to the French attempts to bring useful plants to Mauritius. Because of the French attempts to acclimatise spices on the island being well documented and available in the archives, I was able to trace the developing relationships and patronage in the Malay–Philippines region.

The newspaper article mentioned at the beginning of this chapter portrays the French quests for spices in a totally different way from that shown throughout this chapter. The spice quest was in fact a matter of cross-island networks that are invisible in the printed accounts of the time. These island networks extended all over the Indo-Pacific, involving French, Filipino, Malukan, Malay, Dutch, Spanish, and Portuguese agents, among others. While this chapter reveals important points relating to the Franco-Spanish entanglements in Asia, more importantly it broadens our perspective on the multidirectional and multicultural components of an enterprise that has traditionally been portrayed as French alone. The actors actually used a mixture of languages, promises, and allegiances. Their encounters would be determined by spontaneous possibilities and random limits, and were often carried out by intermediaries, who would secretly trade spices such as nutmeg in different parts of South East Asia. In acquiring the plant materials they desired, French agents turned to a network of local informants and local people who offered their services in exchange for money and other goods.

It is especially important to realise that this was far from a one-way process. This chapter, focusing on the asymmetries of Eurasian encounters, has moved away from a European perspective by seeking to understand what the trade as a whole meant for the local populations. Because the French had to rely on the islanders who were aware of that fact, the initially perceived power relations shifted: the islanders in fact had power over the French. All the parties involved sought

[132] ANOM, C/4/29, ff. 242r–245r, Provost, 'Instruction secrète pour M. Cordé commandant la corvette du Roi le Nécessaire, au cas de séparation', 25 December 1771; and ANOM, C/4/26, ff. 168r–179r, Desroches to Praslin, 22 July 1770.

[133] Widjojo, *Revolt of Prince Nuku.*

alliances with one another to achieve their goals. Communication, language, negotiations, and secret diplomacy were the key factors in these interactions. The French attempts to collect spice plants met with the degree of success they did only because of their interdependence and interactions with other European informants and above all, local South East Asian islanders.

6 Materials, Environment, and the Application of Knowledge

In 1805 an article on the eighteenth-century French attempts to establish a nutmeg trade appeared in a German economics periodical. The author of the article, a Herr Funke, elaborated on the difficulties of nutmeg propagation:

> In the past, one believed that the nutmeg could not germinate through human hands at all but that it had to be ingested by a native bird instead, which would, without digesting it, excrete it. This falsity was naturally rooted in the fact that purchasable nuts do not germinate, and a bird indeed swallowed the same in their fresh condition.[1]

The Maluku people did indeed believe that humans could not propagate the nutmeg. But in fact if the seeds are fresh enough, they can germinate without being passed through a bird, and this fact was known to the eighteenth-century French actors in the Indo-Pacific. Poivre explained in his secret instructions to his collaborators in the Moluccas that nutmegs:

> are not proper to sow and germinate unless they are freshly picked from the trees which carry them. Already after eight or ten days after their harvest, they are no longer able to germinate unless they are conserved in earth.[2]

Thus, the successful cultivation of nutmeg required efficiency, fast action, and precise know-how relating to the plant material. But these challenges in nutmeg cultivation remained largely absent from the published accounts of the eighteenth-century French spice project.[3]

While Chapter 5 took a sociocultural approach to the quest for nutmeg, this chapter analyses the material culture associated with the application of knowledge, ranging from the cross-oceanic transportation to the planting of spices in foreign soil. I examine in detail the interplay of human and natural factors in the application of knowledge. In doing so, I focus on the ongoing creolisation of knowledge in an island space where there was no pre-existing or predominant form of knowledge relating to the cultivation

[1] Funke, 'Muskatnüsse', 72.
[2] ANOM, C/4/22, Poivre's secret instructions to Trémigon and Provost, 4 February 1768.
[3] See for instance 'Observations Botaniques', *HARS* 1772, 56–61.

of a foreign plant. Early modern historians, using praxeological approaches, have argued that an exploration of the various practices will help us to understand history from different angles and to reinterpret existing great narratives which were, in fact, much more complex.[4] In contrast I revisit the dynamics of knowledge production and colonial practices more generally, by analysing the application of plant knowledge as a practice. By engaging with the actual transmission of plants, seeds, and knowledge, I am thus contributing to the reinterpretation of the complexity of material culture, cultivation, and environmental conditions.[5]

Nature itself created restrictions on human hopes of turning it to their purposes. Consider the plant material's passage from the Asian islands to Mauritius, a voyage that was extremely risky for the plants. Scholars have examined such cross-oceanic specimen voyages in a range of colonial contexts, from the Americas to the Pacific, focusing on the differences between the transport instructions and their implementation.[6] Harold Cook observed in the context of the Dutch East Indies, 'the successful transportation of live plants over long distances by sailing ship took elaborate preparation and continuous attention'.[7] In their work on specimen transport in the British and French Atlantic, Christopher Parsons and Kathleen Murphy pointed to the importance of the ocean; they argued that it was a space where existing methods for preserving curiosities and specimens proved increasingly insufficient.[8] Moreover, Dániel Margócsy made the point that:

> shipping costs mattered to *curiosi*, and influenced the concern of natural history collections, because of the commercial mindset of many collectors. Immediate profit might not always have been their primary concern, but they were loath to suffer financial losses when trading specimens.[9]

[4] Freist, *Diskurse – Körper – Artefakte*; Brendecke, *Praktiken der Frühen Neuzeit*; Chateauraynaud and Cohen, *Histoires pragmatiques*; Freist, 'Historische Praxeologie Als Mikro-Historie'; Cerutti, 'Histoire pragmatique, ou de la rencontre entre histoire sociale et histoire culturelle'; Lepetit, *Les formes de l'expérience*.

[5] The most recent studies on cross-oceanic transportation being Easterby-Smith, 'Recalcitrant Seeds'; Parsons and Murphy, 'Ecosystems under Sail'; Klemun, 'Live Plants on the Way'.

[6] Easterby-Smith, 'Recalcitrant Seeds'; Parsons and Murphy, 'Ecosystems under Sail'; Klemun, 'Live Plants on the Way'; Rigby, 'Politics and Pragmatics'; Mariss, '*World of New Things*'; Kury, 'Les instructions de voyages'; Klemun, 'Live Plants on the Way'; Allain, *Voyages et survie des plantes*; Bourguet, 'Measurable Difference'; Lacour, *La République naturaliste*, 105–33. On ships as scientific instruments, see Sorrenson, 'Ship as a Scientific Instrument in the Eighteenth Century'; Adler, 'Ship as Laboratory'.

[7] Cook, *Matters of Exchange*, 325.

[8] Parsons and Murphy, 'Ecosystems under Sail'. Few scholarly works are primarily concerned with the instructions handed to travelling naturalists. Some exceptions are Kury, 'Les instructions de voyages' and Collini and Vannoni, *Les Instructions Scientifiques*, 15–54.

[9] Margócsy, *Commercial Visions*, 38.

The costs were huge, and if anything went wrong all the work could be lost. So, the appropriate packing and transport of the plants was as important as their collection itself. However, specimens too often arrived damaged, if alive at all, because of the long time when the sailors and other members of the crew either could not be bothered to take proper care of them, or had neither the time nor the knowledge to do so.

During the eighteenth century, practical questions on how to make the plant material survive long voyages were of abiding concern to naturalists. Many instructions on the plants' transportation and cultivation circulated in European correspondence and in the acclimatisation gardens. In Paris, for example, the botanist Antoine de Jussieu observed that it was necessary to ensure that the illustrations and descriptions of seeds and live plants showed how they had to be stored so that they would arrive alive and ready for transplantation.[10] Yet, as Nigel Rigby pointed out: 'The failure rate in all these methods was high.'[11] And as this chapter will show, so too was the failure rate of cultivation methods, whether based on European or non-European theoretical knowledge. In Mauritius, in the 1770s, the nutmeg did not propagate in the way the administrators had expected. Acclimatisation was a matter of clear explanation of the plants' needs and the successful implementation of the techniques required to carry out the explanations. Although Indigenous knowledge from South East Asia about fruit-bearing and non-fruit-bearing trees arrived with the plants in Mauritius, the horticulturalists there faced many and various practical challenges when seeking to apply this natural knowledge.

One of the main problems turned out to be that not all nutmeg trees bore fruit. This was because, as we now know, nutmeg is a dioecious plant: it has distinct male and female specimens. While male trees are important for pollination, it is only the female that bears fruit, so in order for a commercial venture to be profitable there needs to be a large preponderance of female trees.[12] But the cultivators were not necessarily aware of this factor. In addition – more pertinent to my argument – they did not actually have the means to implement it. While the Indo-Pacific classification systems had enabled the practical application of cultivation knowledge and reproduction in Mauritius, they did not promise success. Both European and non-European theoretical knowledge faced a reality

[10] BCMNHN, Jus 3, ff. 95r–105r, Jussieu, 'Des Avantages que nous pouvons tirer d'un commerce littéraire avec les botanistes étrangers', n.d.

[11] Rigby, 'Politics and Pragmatics', 84.

[12] As instructions on cultivating nutmeg from 1789 reveal, trees only blossomed after seven or eight years (BCMNHN, MS 42, file 2, 'Instructions pour servir aux 30 caisses et 105 barigues d'arbres plantes et graines remises au Sr. Martin pour le Jardin royale de Plantes à Paris', 31 January 1789). On Martin's mission, see Easterby-Smith, 'Recalcitrant Seeds'.

check when those cultivators tried to put their knowledge in action. In addition, it is misleading to assume that all techniques would have travelled from Europe to the colonies and vice versa. As I will illustrate through the example of grafting, certain methods were developed independently in different parts of the world.

The inability to understand the plants' needs had a serious effect on their cultivation. Misinterpretations of instructions were frequent and could easily lead to a plant's demise. Poivre was deeply occupied in providing detailed instructions on the appropriate transportation and cultivation of nutmeg (and cloves), and his concern is reflected in the primary sources of the colonial and botanical archives. His detailed instructions, together with their practical implementation in the Indo-Pacific islands, form the heart of this chapter. First, I engage with the cross-oceanic transport on ships, considering the vessels in accordance with recent claims as to their being important spaces for an in-depth analysis of material practices. The successful transportation, cultivation, and eventual acclimatisation of the plant specimens was not only a matter of appropriate climate conditions and soil, but also one of mastering the relevant practices, including when, where, and how to plant the material, as I analyse in the second part. Third, I focus on the dioecious nature of the nutmeg: it was only after many years that the French naturalists began to understand why it was so difficult to propagate nutmeg using the seed's sex as the criterion. This chapter, then, rather than concluding that the whole project was a success or failure, sets out the reasons why it progressed so slowly and how contemporaries endeavoured to explain the lack of reproduction.

Tools, Containers, Packing Methods

Let us begin at the very start of the transplantation process: digging up the specimens. For live plants in the Moluccas, French actors brought tools (unfortunately, the source at hand is imprecise as to their type) to dig up the fertile plants so that they were fresh and could be boxed for transportation to Mauritius.[13] Provost showed the islanders how to use the tools for the best result. It is not clear whether these islanders had experience with digging up and transplanting plants or indeed had any more general horticultural knowledge. He instructed the workers to pay particular attention to the roots and to leave them surrounded with their native earth; this is reflected in the Malay–French vocabulary attached to the

[13] ANOM, C/4/29, ff. 242r–245r, Provost, 'Instruction secrète pour M. Cordé commandant la corvette du Roi le Nécessaire, au cas de séparation', 25 December 1771.

secret instructions for his collaborator Cordé: 'I require them with their earth' (*'Minta dangan tanair'*/*'J'en demande avec leur terre'*) as well 'with their roots' (*'Diar dangan acana, yangan bri poutousse'*/*'avec leurs racines'*).[14] These examples are a reminder of the importance of language, and show that the French actors had to learn local expressions in order to communicate with islanders.

The specimens then had to be packed and prepared for transportation. The type of container was crucially important, and over the eighteenth century it varied considerably, depending on the type of specimen (i.e. seed or young plant). For instance, in 1772, some nutmeg and clove trees sent to members of the Académie des Sciences for examination left Mauritius in '4 jars filled with eau de vie'. Dried branches and berries were sent in a separate package.[15] Live plants were normally packed in wooden boxes, or very occasionally, in porcelain vases; this was the case for nutmeg from China, when in 1769 Father Galloys planted some germinating seeds in a porcelain vase filled with earth.[16] But since porcelain often proved too fragile, wooden boxes were more typical.

The type of timber used for boxes, the construction of those boxes, and the type of earth used was also of concern to the naturalists. In 1763, the naturalist François-Etienne le Juge recommended that the boxes should be one inch longer than the plants packed in them.[17] According to Poivre's recommendation from 1768, the boxes were to be made of 'strong boards'.[18] However, in instructions for Provost, dated 1771, Poivre then stipulated the complete opposite: he instructed Provost to make some boxes of light wood while in Manila.[19] They were to be narrow, like the boxes used for weapons, and at least 6 feet long. They had to be long enough so that it would not be necessary to cut the plants: 'The best thing would be to have boxes which are big enough so that we are not obliged to cut the stalk of the plant, the nutmeg and the clove in particular.'[20] So the instructions for the containers could be paradoxical and even contradictory.

[14] ANOM, C/4/29, f. 245r, Provost, 'Instruction secrète pour M. Cordé commandant la corvette du Roi le Nécessaire, au cas de séparation', 25 December 1771.

[15] Report by Adanson and Jussieu, ADS, PV 1773, f. 34v.

[16] ANOM, C/4/25, Poivre to Praslin, 18 February 1769.

[17] Le Juge, 'Manière de transporter les jeunes plantes de toutes sortes d'arbres', 185–6.

[18] ANOM, C/4/22, 'Mémoire sur la manière de conserver les arbres et les graines pour les transporter par mer', 4 February 1768.

[19] ANOM, C/4/28, f. 138r, 'Instructions secrètes communes à M. de Coëtivi, enseigne de vaisseau, commandant la flûte du Roi, L'Isle de France, et M. Provost, commissaire de la Marine, chargé d'une mission particulière', 24 June 1771.

[20] ANOM, C/4/22, 'Mémoire sur la manière de conserver les arbres et les graines pour les transporter par mer', 4 February 1768.

Once the boxes had been constructed, the method of placing each specimen was to be carefully considered, depending on its type; boxes were filled with earth whose type differed from one set of instruction to another. Placing live plants in loose good earth, not too dry and not too moist, reduced the risk of them drying out. The soil was to be 'light and not too moist or too dry', and the best soil for the operation was the sandy black earth from the island of Gebe.[21] The importance of good earth echoes Poivre's own experience: 'The earth must be pure, without any mixing of compost and especially not rotten wood or non-rotten foreign wood or roots.'[22] While some preparation methods refer to loose earth, other instructions stipulate that the plants should come in a 'thick earth', which was preferable because it could keep fresh for longer.[23]

The exact placement of the materials was just as important as the packing material. Nutmegs were to be placed in beds of earth alternately like a chessboard, with each bed of earth being about 3 to 4 inches thick.[24] The earth, not dry or too wet, was to be worked before a little by hand before the seeds were placed in it, without touching each other, then covered with a second layer of earth (another 3 inches), on which another layer of seeds was placed. The same procedure was to be conducted until the box was full.[25] The last layer of earth had to be a little thicker than the previous ones. On top should be placed a layer of lime ('chaux') mixed with water. To this layer, sugar ('gros sucre') should be added, probably for conservation reasons.[26] It was also necessary to bear in mind the different seasons of the plants and to ensure that only seeds of plants that fruited at the same time of year should be mixed.[27]

There is no indication, however, that despite the precision of these instructions these materials were actually used or methods actually applied when transporting seeds from South East Asia to Mauritius.

Similarly, clove berries were put in boxes filled with earth and arranged alternately in layers of 2 inches of earth (like a chessboard, as for nutmegs).[28] According to another method, clove seeds were to be put in open boxes, filled with 'good earth', loose and not firm, and then

[21] ANOM, C/4/29, f. 245r, Provost, 'Instruction secrète pour M. Cordé commandant la corvette du Roi le Nécessaire, au cas de séparation', 25 December 1771.
[22] Ibid.
[23] BCMNHN, MS 280, 'Instruction sur la manière de cultiver avec succès des plants et graines de gérofliers et muscadiers, apportés des Isles Moluques par M Provost commissaire de la marine'.
[24] Ibid.
[25] ANOM, C/4/22, Poivre, 'Mémoire sur la manière de conserver les arbres et les graines pour les transporter par mer', 4 February 1768.
[26] Ibid. [27] Ibid.
[28] ANOM, C/4/29, ff. 242r–245r, Provost, 'Instruction secrète pour M. Cordé commandant la corvette du Roi le Nécessaire, au cas de séparation', 25 December 1771.

planted at a distance of 4–5 inches from each other, covered with just one layer of earth. To ensure that water could flow and the seeds would not 'swim', the bottoms of the boxes had to be pierced, and the holes then had to be covered with oyster shells (or something similar) before the earth was added, so that the holes would not become blocked.[29] The following method to keep them alive could also be tried: to mix them, especially those which lost their ability to bud easily, with coarse-grained sugar and 'put them in a vial or flask and form the base with a mastic [putty]'.[30] This method differed from the one described above, emphasising that caulk was required so that air (through the joints of the boxes but also the boards) could not enter and spoil the seeds. It was not desirable for the seeds to germinate either too much or not at all: if they were to arrive in good condition there was very little room for manoeuvre.

As for seeds, careful attention was to be paid to live plants, which were to be placed horizontally in boxes containing 4–5 inches of earth mixed by hand. If there was space, a second layer of plants should be added, separated from the first by earth: the last layer had to be earth. During the voyage, the plants would not require anything, but it was advised to water them – with fresh water of course – but on the open sea that was a rare and precious commodity.

Seedlings were especially fragile. During their transport from the Moluccas to Mauritius in 1771, the clove trees were too young and too small (less than 2 feet high) and thus suffered badly; for successful transportation, they should have been at least 4–5 feet high.[31] The instructions had said that the boxes were to be stored below and never exposed to sunlight. Plants which germinated too much could be cut slightly.[32] The latter is in direct contradiction to the instructions above, in which it was advised not to cut plants and that one should pay attention to the length of a box in order to avoid cutting them.

On more than one occasion, the preparation instructions were directly contradictory, and thus sometimes even dangerous for the plant material. For instance, in the boxes, the plants were to be embedded in alternating layers and the boxes filled with earth so that the cover could only be fixed on with effort. This meant in practice that the plants were squeezed together. When new specimens arrived in 1772, two different types of boxes were used: 500 clove trees (and twenty-eight nutmeg plants) arrived in thirty-six open boxes covered only with some wire netting, while another 500 clove and nutmeg trees were placed between two

[29] Ibid., f. 244r. [30] Ibid.
[31] ANOM, C/4/29, ff. 22v–23r, Poivre to Praslin, 22 August 1771.
[32] ANOM, C/4/22, Poivre, 'Mémoire sur la manière de conserver les arbres et les graines pour les transporter par mer', 4 February 1768.

beds of earth in closed boxes wrapped in tarred cloth.[33] Nearly all the plants died from lack of air. For the same reason, half of the nutmeg died when placed in *closed* barrels filled with earth. Plants would easily die when covered with heavy earth in sealed boxes, and Poivre must have been aware of this. In instructions from 1771, he explicitly referred to the fact that young/small plants must be transported in *uncovered* boxes filled with native earth and protected against rats, the sun, and the wind by trellises of iron and bamboo.[34] Then in 1763 Le Juge stated that the boxes must never be closed entirely: it was vital to leave a gap of two to three inches for ventilation, extremely important in the tropical heat.[35] But in practice these important details were not adhered to.

On Board: Sun, Ventilation, Wind, and Water

Because vessels were themselves complete ecosystems, with changing weather conditions, air, light, salt water, heat, and plagues (of insects, rats, and so on), the precise placement of any specimen would help determine whether or not it survived the trans-oceanic voyage. Plants stored on deck could easily be damaged by seawater or overexposure to sunlight, while those below would be short of air and light.[36] Over time, naturalists developed what Parsons and Murphy call an 'environmental science of ships' – extensive knowledge practice of handling the environmental factors on ships correctly – and the successful transport of a plant was subject to its mastery.[37]

Ideally, naturalists or other experts would travel on the ships with the plants to take care of them. In practice, however, this was rarely possible. Instead, transportation was often determined by spontaneous decisions and supervised or carried out by actors who could not be bothered, or were unable for a multiplicity of reasons, to take care of the plants. Some sources even highlight that the packing itself could be chaotic. A list of specimens sent to Paris from Mauritius in 1772 shows that the boxes were numbered '18°1, 18°2, 18°3, 18°4' as well as one 'open box' and that the list corresponded to the contents of the boxes.[38] However, it remains an

[33] Tessier et al., *Encyclopédie méthodique*, vol. 4, 199.

[34] ANOM, C/4/28, f. 138r, 'Instructions secrètes communes à M. de Coëtivi, enseigne de vaisseau, commandant la flûte du Roi, L'Isle de France, et M. Provost, commissaire de la Marine, chargé d'une mission particulière', 24 June 1771.

[35] Le Juge, 'Manière de transporter les jeunes plants de toutes sortes d'arbres', 185–6. On Turgot's methods on live plant transport and placing and a discussion in France, see Turgot, *Mémoire instructif*, 165–205.

[36] The point is also made by Parsons and Murphy, 'Ecosystems under Sail', 523.

[37] Ibid.

[38] BCMNHN, MS 277, IV, 'Nottes sur la nature et la prospirité de quelques vegetaux de L'Isle de France faittes a l'occasion d'un envoi d'iceux en France en fevrier & mars 1772. Notte des Graines Envoyées a Mr Le Monnier', 1772.

open question how the individual seeds could be identified, particularly in light of a curious note, perhaps written by Commerson: 'We could not write the botanical names on the items, which were insufficiently drawn because the shipment did not pass our eyes.'[39] There are further indications that this shipment was chaotic, in that the list also contains a section on '*articles oubliés*' ('forgotten items').[40] These examples contribute to the argument that transportation was a matter of local and ad hoc decision-making.

In 1768 Poivre stated that the seeds were quite easy to transport and that 'pure earth', neither dry nor humid, should be supplied, in a properly closed box: the seeds were not to touch each other but were to be placed between 'beds' of earth. This was 'the only attention one must pay', Poivre claimed, particularly when a box was to be stored in the great cabin where they were normally safest and where they were 'exposed to the light from the windows'.[41] But the great cabin had restricted space, and other boxes and barrels had to be stored in places where environmental conditions might even be harmful to the plant material. Once the plants were on board, it was important to them to be stored without being exposed to too much wind or direct sunshine; they would then require only a little water, a scarce commodity on ships. Fresh water was certainly an issue on vessels crossing salt waters, which also attracted the attention of the naval minister; he stated that barrels of fresh water must be stored on vessels for this purpose.[42] In fact, too much or too frequent watering would also kill the plants, Poivre explained. So, it was best to put them in the ship's storerooms, where they would not require water during the first two months of the voyage.

The lack of air, however, concerned the botanists more than the lack of water. It seemed that nutmeg trees would die from lack of air because they were too packed too tightly in the barrels, not because of a lack or an overdose of water. The question of air was always a tricky affair, as shown below. In 1767, Poivre set out two distinct methods to transfer live plants by sea from China. He put pressure on Father Galloys to secure appropriate transport, and to pack the plants so that they could be replanted in the colonies immediately upon their arrival.[43] The first method, for plants and trees, required stable boxes with iron clamps so that the earth would not be loose. The samples were to be packed in horizontal layers in beds, separated by layers of earth.[44] The second method was to keep the roots

[39] Ibid. [40] Ibid.

[41] ANOM, C/4/22, 'Mémoire sur la manière de conserver les arbres et les graines pour les transporter par mer', 4 February 1768; Rigby, 'Politics and Pragmatics', 93.

[42] Ly-Tio-Fane, *Odyssey of Pierre Poivre*, 137.

[43] ANOM, C/4/18, f. 19v, Poivre 'Instructions pour Mr L'abbé Galloys'. [44] Ibid.

in their original rootball surrounded with '*mousses*', which should be covered with straw or rags, frequently dampened: they should be arranged in boxes by plugging one part of the root in earth and folding their branches so that the plants could breathe.[45]

But in direct contradiction to these careful considerations regarding ventilation, live clove trees, with their roots, were to be embedded in boxes wrapped in cloth to keep them humidified (but lacking air) then stored in the depot or cargo hold (where there was no light, either).[46] When live plants were stored in the great cabin, they would not be damaged by the vessel's movement on the open sea. But plants stored elsewhere needed to be protected not only from the weather but also from the pitch and roll of the ship, which could move them and break them. Some of the boxes were to be kept open in order to guarantee ventilation – but at the same time they had to be protected from rats; a tall order. Other boxes were to be completely closed and stored in the cabin just below the poop deck (the highest level) on the starboard side.[47]

The failure rate of all these methods was nonetheless very high. When in 1772 seeds and plants arrived in Mauritius, out of the 12,000–13,000 seeds half of them, having fermented in the ship's hold, had died. The other half were germinating.[48] A great number of clove buds also arrived, most of them germinating and some even with seminal leaves. The live clove and nutmeg trees were green, fresh, and budding, but the clove trees had suffered seriously during the voyage and in the end failed to acclimatise.[49]

In light of the complex techniques, most of which did not work in practice, transportation was a very hard nut to crack. The methods were very experiential, hence sometimes even paradoxical. Then the ever-greater pressure to take larger quantities of plant material from the Moluccas to Mauritius was yet another reason for the instructions not being followed.

[45] Ibid.

[46] ANOM, C/4/29, ff. 242r–245r, Provost, 'Instruction secrète pour M. Cordé commandant la corvette du Roi le Nécessaire, au cas de séparation', 25 December 1771. Later, Poivre's successor Maillart, for instance, conducted experiments in preserving clove berries by steaming them. ANOM, F/3, art. 89, ff. 42v–43r, Maillart to the Minister, 8 November 1776.

[47] ANOM, C/4/28, ff. 138v–139r, 'Instructions secrètes communes à M. de Coëtivi, enseigne de vaisseau, commandant la flûte du Roi, L'Isle de France, et M. Provost, commissaire de la Marine, chargé d'une mission particulière', 24 June 1771.

[48] ANOM, C/4/30, f. 303r, Commerson's report, 8 June 1772; and ADS, PV 1773, ff. 32v–37r, Report by Michel Adanson and Bernard de Jussieu, 17 February 1773. See also Tessier et al., *Encyclopédie méthodique*, vol. 4, 198–9.

[49] ANOM, C/4/30, f. 303r, Commerson's report, 8 June 1772; and ADS, PV 1773, ff. 32v–37r, Report by Michel Adanson and Bernard de Jussieu, 17 February 1773.

The optimisation of transportation of plants around the globe developed only slowly over the course of the eighteenth century: the question of appropriate boxes in particular was a most difficult problem. It was only in the first half of the nineteenth century that their transport was facilitated by the Wardian case. In 1829, the amateur naturalist Nathaniel Bagshaw Ward (1791–1868) invented a sort of terrarium, made of glass, in which live plants could thrive.[50] Until then, it was not only the environmental conditions on ships that endangered the live plants, but the boxes themselves. Boxes and packing methods thus went hand in hand with the environmental conditions on ships which, like the packing methods, were difficult to optimise.

Cultivation, Environment, and the Adaptation of Knowledge Practice

In this section, I focus on the complex processes of acclimatising nutmeg in Mauritius alongside the introduction of knowledge and people that sometimes accompanied it. The French actors were aware of the potential value of embodied knowledge, and Provost, for example, encouraged the transport of knowledgeable people from South East Asia to Mauritius. For the cultivation of nutmeg, he took two young 'Creole'[51] men from the Maluku Islands, one of them of French ancestry, to Mauritius. He had hoped that they could assist with the cultivation of spices even though they had apparently not worked as cultivators in the Maluku Islands. It turned out, however that the two did not have much agricultural knowledge, and they overwatered some young nutmeg plants. They had done this because they believed that the plants required humidity, but, as Poivre explained in a letter to Minister Praslin, nutmeg trees with collateral roots required only a little water, particularly when they were young.[52] Clove seedlings, on the other hand, required a lot of humidity: but even though the boys had watered those correctly, the plants, which had suffered during their transportation, did not survive either. Since the young men had not known that they were sensitive to too much heat and sunlight, they had not planted them in the shade.[53] Although Poivre blamed the boys for their lack of knowledge it was in fact he who had failed to instruct them properly. It was wrong for him to assume that because the two had lived on an island where native nutmeg was cultivated, they would be familiar with the appropriate technologies. It is easy

[50] Klemun, 'Live Plants on the Way', 33.
[51] Here, 'Creole' clearly means locally born, for one at least of French ancestry. There is no further indication of a person of partial African descent, which might be the case.
[52] ANOM, C/4/29, ff. 22v–23r, Poivre to Praslin only, 22 August 1771.
[53] Ibid.

to imagine, as well, that forced migration and homesickness would have left these islanders completely uninterested in tending similarly dislocated plants.

On 22 September 1770, Commerson wrote to Cossigny de Palma: 'Everyone has the same complaints about the nutmeg as you, but M. Poivre underlines nonetheless that one must not lose hope, and that in two months' time one will see it grow.'[54] According to Poivre, the French colonists were too incompetent and too impatient. He considered that too many of them were incapable of treating plants properly.[55] That is why at the start of the enterprise, until early April 1771, nutmeg and cloves were acclimatised in Poivre's garden alone. Then four months later they were distributed around the island.[56] Prior to that, Poivre had complained that small sprouting seeds were thrust into the ground with too much strength so that their sprouts and taproots were destroyed. In addition, because many of the seeds were planted too deeply, they only showed signs of germination after ten or eleven months.[57] Yet even so, confident of his plan to acclimatise nutmeg, Poivre asked for Praslin's patience; in ten years' time, he asserted, the nutmeg trade would flourish and challenge the Dutch, just as he had envisaged.[58]

The colonists' impatience and apathy had deeper reasons – reasons related to the terrible food situation in Mauritius, as Chapter 3 shows – and their objections sprang, too, from the obvious fragility of the project as a whole. When in 1772 Provost called at Mauritius with a vessel loaded with nutmeg and clove seedlings, the inhabitants were lukewarm about the chances of success. Father Galloys, who described the arrival of spice plants in Port Louis, reported sarcastically 'no one applauded ... there was not a single man shouting "Long live Pierre Poivre and the great Provost should be crowned"'.[59]

Galloys provides several explanations for this lack of enthusiasm. He stated that it was the lack of previous success that had discouraged the inhabitant; they had come to believe that these trees could not grow successfully in a climate so different from that of their original habitat.[60]

[54] Commerson to Cossigny, 22 September 1770, printed in Cap, *Philibert Commerson*, letter 6.
[55] ANOM, E 197, Galloys to Praslin, 14 August 1769, and Poivre to Galloys, 16 February 1769, printed in Laissus, 'Note sur les manuscrits de Pierre Poivre'.
[56] Galloys to Le Monnier, 6 April 1771, printed in Laissus, 'Note sur les manuscrits de Pierre Poivre', 47, and ANOM, C/4/29, ff. 22v–23r, Poivre to Praslin only, 22 August 1771.
[57] ANOM, C/4/22, letter no. 66, Poivre to Praslin, 16 June 1768, secret letter.
[58] Ibid.
[59] Galloys to Le Monnier, 20 July 1772, printed in Laissus, 'Note sur les manuscrits de Pierre Poivre', 50–1.
[60] Ibid., 51–2.

People in need of basic necessities were not likely to be enthusiastic about spices which could not feed them and whose success was, in their opinion, the wishful thinking of the intendant.

In fact, a summary of Poivre's observation on the cultivation of the spice plants in Mauritius, probably dated 1772, makes it clear that only a small area had been given over to the cultivation of nutmeg and cloves. Because the island itself was so small and its cultivable land had to be prioritised for staple crops, there was little land left for commercial crops such as spices.[61] Thus, the spice project and the agricultural enterprise hindered each other.

Finally, in 1772, Poivre printed his 'Instructions on the Manner of Successfully Planting and Cultivating Plants and Seeds of Clove and Nutmeg, and Distributed Them to the Colonists'.[62] The instructions were based on the observations Poivre had probably made himself in 1752, and perhaps also those by Provost between 1771 and 1772. Poivre had spent time observing the agricultural skills and methods of the Maluku islanders and other parts of the Indo-Pacific during his travels in the 1740s. He explained in a letter to the Minister of the Navy and the Colonies that he intended to practise this knowledge in his acclimatisation garden according to the cultivation methods he had observed for cloves on Ambon and nutmeg on Banda.[63]

He had also derived knowledge from written sources on the cultivation of Maluku spices. In particular, the Dutch naturalist G. E. Rumphius's *Het Amboinsch Kruid-Boek*, the first account on nutmeg in Dutch territory, on Ambon in particular.[64] *Het Amboinsch Kruid-Boek* was translated into Latin as *Herbarum Amboinense* (1750) but the circulation and the editions of the book are a matter of further research. *Het Kruid-Boek* (or rather its Dutch–Latin edition) became an important point of reference for naturalists both in France and overseas. It is unknown which edition Poivre accessed, but the actors in Paris would surely have consulted the 1750 Dutch–Latin edition, which I elaborate on below.

Poivre's instructions show that a plant could not be successfully cultivated if it were kept in isolation; instead, it required an interactive environment – one in which plants could help each other. It was in

[61] Poivre's summary is transcribed in Ly-Tio-Fane, *Odyssey of Pierre Poivre*, 109. It is probably dated 1772, Bibliothèque Angers, MS 612, f. 28r.

[62] AN, MAR G101, file 4, ff. 171v–175r, 'Instruction sur la manière de planter et de cultiver avec succès les plants et les graines de Géroflier et de Muscadier', 1772.

[63] ANOM, C/4/22, letter no. 66, Poivre to Praslin, 16 June 1768, secret letter.

[64] Rumphius, *Ambonese Herbal*. On how Poivre uses Rumphius, see ANOM, C/2/285, ff. 158r–162v, 'Observations sur le muscadier et principalement sur la culture de cet arbre', 12 February 1752.

Poivre's studies of plants in other settings that he began to imagine an environment for the nutmeg trees which could draw on the available resources, both human and environmental, in Mauritius. Precise precautions were necessary to ensure the necessary fresh air, shade, and protection from the cyclones of the south-west Indian Ocean.[65] He elaborated in particular on how other types of plants could be used to provide just the right amount of shade and shelter for successful propagation. When constructing a nutmeg stand, for example, the plot should be protected by an outer hedge of bamboo. Within that, a ring of trees such as the mango, the jackfruit, or the betel or coconut palm, and, in a third row, trees such as the orange, lemon, cinnamon, or coffee.[66] Like the nutmeg, these plants were also not native to Mauritius, and contemporaries tested transplanted crops for ways in which they could be integrated in a new ecosystem.

Cultivation methods were learned from experience with creolised ecosystems. In order to ensure that spice seedlings would be protected against the wind and the sun, Poivre recommended planting them with banana trees, to provide shade and shelter.[67] In the later 1770s, experience with cloves showed, however, that banana trees were wrong, because they provided too much shade.[68] (Prior to this, some instructions to colonists (possibly dated 1771/2) had stated that the banana tree was very suitable for cinnamon (which may have come from Ceylon) because cinnamon required lots of shade.) Clove seedlings were likewise to be protected from the burning sun with the careful placement of small leaves such as those of the tamarind.[69] As these instructions emphasised, horticulturalists were encouraged to experiment with what worked best for a range of spice plants only newly brought together.[70] Ultimately, the cultivators had to be sure that the plants intended to protect did not actually harm the nutmeg; for example, the bamboo should be made to grow outwards from the hedge instead of spreading inwards, encroaching on the nutmeg.[71]

[65] On cyclones in the south-west Indian Ocean and Mauritius under British rule, see Mahony, 'Genie of the Storm' and Grove, *Green Imperialism.*

[66] ANOM, C/2/285, f. 159r, Poivre, 'Observations sur le muscadier et principalement sur la culture de cet arbre', 12 February 1752.

[67] Ibid.

[68] Céré's instructions for the cultivation of the clove tree, 7 April 1779, transcribed in Ly-Tio-Fane, *Odyssey of Pierre Poivre*, 128.

[69] BCMNHN, MS 280, 'Instruction sur la manière de cultiver avec succès des plants et graines de gérofliers et muscadiers, apportés des Isles Moluques par M. Provost commissaire de la marine'.

[70] Ibid. [71] Ibid.

The soil was another crucially important factor in the colonial cultivation techniques. Nutmeg required 'fat' and humid soil.[72] After mixing it, you had to enrich it by using 'some other materials suitable as amendment', such as leaves. Although the soil in Mauritius appeared to be fertile, Poivre provided several justifications for improvement. For instance, when the soil became too dry, it was too sticky and required loosening so that water could flow more easily.[73] The nutmeg was to be sown in the rainy season, which in Mauritius was at a different time of year than in the Maluku Islands. According to the 'gifted gardeners of Banda', who Poivre may have observed himself when travelling – and surely noted by him from *Het Kruid-Boek*, since he used it as a reference point in the production of his own document – it was better to plant the seeds as much as two months before the rains, because the seeds could then germinate and take root prior to the downpours.[74] That way, the seeds could also grow their stalk more easily once the rains came. Otherwise, the young plants could easily drown.

Once the soil was prepared, the appropriate placement of the nutmeg was equally important. Seeds on germinating first develop a rootlet, then a stem. This, however, was not known to most of the settlers, who had little experience with agriculture. Even though the exact method of planting the seed in the soil would have been meticulously detailed, this was often misinterpreted. According to the instructions, the plant required a 2–3 foot bed of deep 'pure' earth, that is, without any compost. An accompanying drawing demonstrated that a partly germinated nutmeg should be planted horizontally with its long and thick root buried in the soil (see Figure 6.1).[75] A fully germinated seed, however, was to be planted as follows: it was not to be buried in the soil. Instead, only the rootlet was to be buried, and the seed then covered lightly with foliage.[76] The document explained that some settlers confused root and stem, so would plant the seed with the root on top and the developing stem in the soil.[77] This of course led to the seed's demise. Despite the most precise instructions possible, mistakes were frequent due either to ignorance, or a lack of

[72] ANOM, C/2/285, f. 159r, Poivre, 'Observations sur le muscadier et principalement sur la culture de cet arbre', 12 February 1752.
[73] BCMNHN, MS 280, 'Instruction sur la manière de cultiver avec succès des plants et graines de gérofliers et muscadiers, apportés des Isles Moluques par M Provost commissaire de la marine'.
[74] ANOM, C/2/285, f. 159r, Poivre, 'Observations sur le muscadier et principalement sur la culture de cet arbre', 12 February 1752.
[75] BCMNHN, MS 280, 'Instruction sur la manière de cultiver avec succès des plants et graines de gérofliers et muscadiers, apportés des Isles Moluques par M Provost commissaire de la marine'.
[76] Ibid. [77] Ibid.

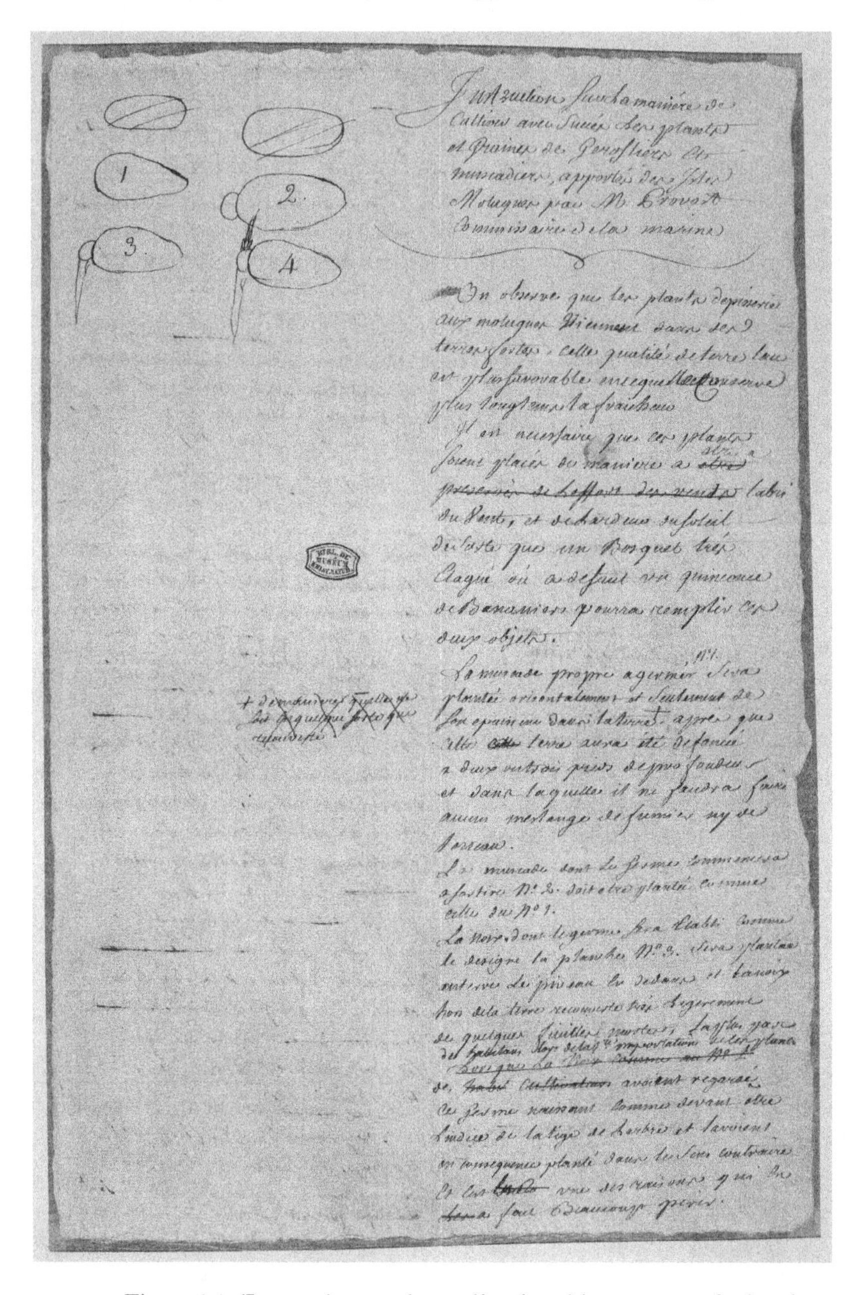

Instruction sur la manière de Cultiver avec Succès les plants et Graines de Gérofliers et muscadiers, apportés des Isles Moluques par M. Provost Commissaire de la marine

Figure 6.1 'Instructions sur la manière de cultiver avec succès des plants et graines de gérofliers et muscadiers, apportés des Isles Moluques par M. Provost commissaire de la marine' ['Instructions on how to successfully grow clove and nutmeg plants and seeds, brought from the Maluku Islands by M. Provost, Commissioner of the Navy'], *c.*1770, Muséum national d'Histoire naturelle, MS 280/1

experience, or carelessness, or misinterpretation, or a combination of any of these factors.

Two things were definitely required when growing nutmeg: cultivation know-how and above all patience. While the trees were maturing, the climate remained an omnipresent factor. Jean Nicolas Céré, in his role as head gardener of the Jardin des Pamplemousses after Poivre's departure in 1773, wrote a report describing how cyclones damaged or even uprooted the young plants – as did plagues of rats, one of which led to the loss of the biggest 'Creole' (as he called it) tree in 1779.[78] Then, on 4 November 1783, his report on the cultivation of spice plants explained that two nutmeg trees had died because of a '*vase mortière*' around their roots. Bad soil or mud might have harmed them, indeed killed them.[79] The colonial ecosystems were fragile, and the new trees were easily lost.

The difficulties the local actors had to face in transplanting spices from the Moluccas were overwhelming. They had to find the right method and the perfect conditions for the trees to grown in soil foreign to them. This is precisely where the tragedy of Mauritius's spice enterprise began: the island's climate and microclimates were very different from those of the Moluccas. In 1758, Poivre explained to the Académie's librarian Malesherbes that the perfect conditions for nutmeg were to be found not in the Mascarenes but in Madagascar.[80] Yet Poivre nevertheless felt sure that he could cultivate the spices regardless of the climate in Mauritius, so different from that of the Moluccas. He gave three reasons for this. First, Mauritius had the same climate as, and soil conditions similar to those of, Madagascar, where he had also found wild nutmeg of good quality. Second, the cinnamon he had introduced to Mauritius from Ceylon had not lost its aroma. Third, he stressed that other vegetables which naturally grew in the Moluccas also grew in the Mascarenes.[81] In later observations (probably dated 1772), he asserted that in Mauritius cloves and nutmeg were the only exchange commodities that could be grown profitably.[82]

These observations must be seen in the light of Mauritius's and Poivre's decline during his tenure. The competition between the colonies, and particularly the colonisation of Madagascar, reflected badly on the Mascarenes, and Poivre was doing his best to ensure that his spice project would be acknowledged and continued after his departure in 1773. The

[78] Céré's instructions for the cultivation of the clove tree, 7 April 1779, transcribed in Ly-Tio-Fane, *Odyssey of Pierre Poivre*, 128.

[79] Céré's report on the cultivation of spice plants, 4 November 1783 (ibid., 109).

[80] BCMNHN, MS 1765, f. 240r, 'Extrait de quelques conversations avec M. Poivre en 1758'.

[81] Ibid. [82] Poivre's summary is printed in Ly-Tio-Fane, *Odyssey of Pierre Poivre*, 91.

debate about the difference in climate was crucial yet paradoxical, and reflected French colonial politics; it went hand in hand with the political climate.

Particularly because of the climate and soil factors, the transplantation of young plants from one place to another was a difficult task. One aspect which was forever a problem was that that the soil had to 'rest', as a gazette article suggests on a more general note regarding cultivation in Mauritius. 'Resting' in this context meant that the soil had to be given time to recover from cultivation, which the author considered a self-regulating mechanism of nature determined by the season and the rains.[83]

From the outset, the French attempts to acclimatise nutmeg were tentative and experimental. An appreciation of the complications of environmental factors and the right manner of cultivation was central to the project. The cultivation methods, based as they were on Maluku experience, European systems and techniques, and the African and Malagasy ethnobotanical knowledge that travelled with the huge slave population, all had to be fundamentally adapted to the island's ecosystem. This required an understanding of each type of plant, along with its cultivation, soil, and climatic requirements. In acknowledging the creolised knowledge that produced Mauritius's nutmeg, I reveal the creolisation of the cultivation practices that adapted to environmental factors (including rain, sun, and storms), the island's soils, and horticultural techniques.

Maluku Knowledge in Transit: Making Sense of Plurality around 1770

An understanding of how the French actors tried to make sense of Maluku classification systems is important in order to make sense of the contemporary confusion about the identity of the nutmeg that, based as it was on the European interpretation, led to generalisation. In relation to botanical plurality, European knowledge production in taxonomy, generalised and universal though it purported to be, dismissed the complexity of the Maluku world. It becomes particularly evident through language that this complexity was erased by the summary, translation, and descriptions that came from a different world view of the nutmeg.

Historians have shown that among the French historical actors there was constant confusion between the different kinds of nutmeg; Emma

[83] 'Réflexions sur la nécessité prétendue de varier les cultures des mêmes terres', in *Annonces, Affiches et Avis divers pour le Colonies des Isles de France et de Bourbon*, 19 January 1774, 2.

Spary, for example, attributed the misperceptions in the identity building of the 'true' nutmeg in Mauritius in the 1750s as the source of a serious dispute between Poivre and Fusée-Aublet.[84] The essential problem in the later 1770s and 1780s, however, was not so much the different species of nutmegs and the discussion about 'true' or 'false' as the lack of recognition of the nutmeg's two sexes, which led to both misinterpretations and increasing practical challenges in its propagation. Even while the French drew on the Maluku classification as a source of knowledge – partly via *Het Kruid-Boek* – this knowledge created a tension between theory and the practical needs of the cultivators.

A 1773 report produced by two Parisian naturalists on nutmeg specimens sent from the Indian Ocean showed how the actors in Mauritius tried to make sense of the plurality of Maluku classification.[85] According to the report, in 1772 Poivre had sent a letter to Paris, together with some specimens for examination. Although I have been unable to locate that letter in the French or Mauritian archives, the resultant report opens a window onto the transfer of Maluku knowledge to Mauritius, where it was accumulated and appropriated.

The French botanists Michel Adanson and Bernard de Jussieu examined three different kinds of nutmeg which had been sent from Mauritius to the Académie royale des Sciences.[86] The naturalists officially approved the authenticity, value, and commercial quality of the nutmeg plants and clove trees that Poivre's collaborators had gathered on the Maluku Islands.[87] Poivre used the acknowledged expert judgement of the Academy's members in order to turn their findings into official knowledge and policy.[88] He needed to publicise his claim of introducing the 'true' nutmeg after several decades of uncertainty, and he pursued the Academy's approval and scientific justification in the hope of official recognition.[89] So this report was just as political as it was botanical.

According to Adanson and Jussieu's report, the specimens had been sent in glass containers, filled with alcohol, numbered, and labelled with Malay names. Most of them contained fruits without branches and were labelled '*Pala parampuan*', '*Pala lakki-lakki*', and '*Pala Lakki Parampuan*'.[90] These names, however, are not included in the Dutch–

[84] Spary, 'Nutmegs and Botanists'.

[85] ADS, PV, f. 32v, Report by Adanson and de Jussieu, 1773. Neither in the archives of the French Academy and the BCMNHN nor in AN was I able to find any surviving documents in relation to Ambonese taxonomy or the distinction of seeds more generally dated to 1772 when the specimens together with Poivre's letter must have been sent.

[86] ADS, PV, f. 32v, Report by Adanson and de Jussieu, 1773. [87] Ibid.

[88] On this point, see also Albritton Jonsson, *Enlightenment's Frontier*, 125.

[89] On voyages and scientific discourse, see also Linon-Chipon and Vaj, *Relations savantes*.

[90] ADS, PV, f. 32v, Report by Adanson and de Jussieu, 1773.

Latin edition of Rumphius's *Kruid-Boek*, which the Parisian naturalists and Poivre (when, back in the 1750s, it had been his source of knowledge) relied on.[91] Instead, somebody – probably Poivre himself – had written these names, possibly drawn from Maluku sources during the several spice quests conducted by Poivre and Provost.

As Bahasa Malay was so widely spoken on the islands of South East Asia, the French actors used its names for the nutmeg; this meant that the complexity of the linguistics and the botanical knowledge of the region was reduced, simplified, and generalised.[92] The Malay names went from the Maluku Islands to Mauritius and on to Paris.

'*Pala*' in Bahasa Malay means nutmeg, and that was the word used as the general term for that spice; the female nutmeg was *Pala parampuan*, the female long nutmeg *Pala Lakki Parampuan* and the 'wild' male or long nutmeg *Pala lakki-lakki*.[93] In Bahasa Malay, these names indicated the distinction between female and male, *perempuan* meaning 'female' or 'woman' and both *laki* and *laki-laki* meaning 'man' or 'male'. *Laki* also means 'long', perhaps in relation to the phallic symbol.

Whereas in Paris the misinterpretations led merely to theoretical debates, in Mauritius they created practical problems for the cultivators. The most damaging misinterpretation by the Parisian actors relates to the fact that Rumphius referred to the distinction of male and female trees but not – as the Parisian botanists seem to have believed – to the seeds. In other words, whereas Rumphius and the cultivators in Mauritius classified the sexuality of nutmeg by referring to the obvious differences in the trees, the Parisian botanists – who had never actually seen a nutmeg tree and working from *Het Kruid-Boek* alone – looked only at the seeds for their classificatory data when examining the specimens. They were so fixated on the seed specimens that they ignored the plant as a whole.

Finally, the question of priorities and purpose should be taken into consideration: the Parisians' classification was based on the commodity while the colonists' classification would have been based on the parts of the plant they worked with, in order to make it grow. This case, then, provides another illustration of the disruption of the knowledge transmission from the Maluku Islands and Mauritius to Paris.

Similarly, in *Het Kruid-Boek* (and its Latin translation), Rumphius provided Malay names that might have had a different meaning in local Maluku languages and were by no means used everywhere. Local Maluku

[91] Compare their report to Rumphius, *Herbarum Amboinense*, vol. 2, 14–18, 24–9.

[92] Bahasa Malay was widely spoken in the Maluku world and served, besides Portuguese and Dutch, as lingua franca in that part of the world. On Bahasa Malay, see Leow, *Taming Babel*.

[93] ADS, PV, ff. 35r–36r, Report by Adanson and de Jussieu, 1773.

classificatory nomenclature differed significantly from one island to the other, even though they were often closely connected in other ways.[94] Rumphius gave the Maluku names for wild nutmeg and distinct local classifications that were used on different islands: the Malay *pala laki-laki* (meaning male or wild nutmeg) was *pala fuker* in Banda, meaning 'mountain nutmeg'.[95] On Ambon, however, it was *pala utan* and *palala*, the latter word a combination of *pala* with *ala*, the nutmeg-eating bird I referred to at the opening of this chapter.[96] It appears to be what the Bandanese called *falour* and the Malays *burong pala*, and what we now we call Blyth's hornbill.[97]

However, Rumphius then explained that the *pala lacki-lacki* bore fruit. It might have been possible, then, that the wild nutmeg and inferior types were denoted as male, and the true and superior nutmegs were denoted as female (in recognition nonetheless that there were male trees) because other Maluku names do not actually make any reference as to whether a tree is male or female. The Malay names referred to sex, the Bandanese to the habitat, and the Ambonese to a bird.

Rumphius – well aware of local taxonomic diversity – also documented the local knowledge systems of the second half of the seventeenth century. These were, however, completely ignored by the eighteenth-century French actors in both the Indo-Pacific and Paris. They might have misinterpreted the Ambonese classifications in the first place, as they grouped the trees according to fruit-bearing (female) and non-fruit-bearing (male) trees and the names do not refer to the seeds (as the French actors apparently believed). These examples reveal the incredibly complex linguistic and cultural world of the Maluku Islands – a world the French actors (in Paris) were unaware of. Even while Poivre had clearly drawn on Rumphius in his descriptions of nutmeg cultivation, he omitted many of the classificatory nuances. It was in the documentation of such taxonomic difference that clues about the species and their relationships were hidden, seemingly because Poivre had misinterpreted them or had different priorities. This fact is worth repeating: even though Rumphius had provided a complex description of Maluku classificatory schema in *Het Kruid-Boek*, it was used neither by Poivre nor by the Parisian naturalists.

[94] See Yoo, 'Wars and Wonders'.
[95] Rumphius, *Herbarum Amboinense*, vol. 2, 24. See also the original held at Leiden University Library, BPL 314/2–3, Rumphius, *Amboinsch Kruidboek*, boeken I–XII, met het Auctuarium of Toegift BPL 314 – parts 2–3, ff. 19r–38v, on the nutmeg.
[96] Rumphius, *Ambonese Herbal*, vol. 2, 37, margins.
[97] Rumphius, *Herbarum Amboinense*, vol. 2, 20.

Yet Rumphius's description of the Maluku classificatory schema gives us important evidence that the concept of the sex of plants was commonly accepted in the seventeenth-century Maluku world – much earlier, then, than in France, as the next section will show. Different people in different contexts divide their natural environment into different units; the original cultivators had long recognised that the plants had their two sexes, and had successfully coped with the inherent practical challenges. Classification systems, whether framed as agricultural knowledge or folk taxonomy, should not be understood as a lesser form of knowledge than that which botany and science seek to become. As Stephen Jay Gould put it: 'In short, the same packages are recognized by independent cultures.'[98]

European knowledge was only one type of knowledge, and even though it borrowed from other systems it nevertheless had significant limitations; this is particularly visible as far as nutmeg was concerned, in that the taxonomic knowledge based on European traditions produced no benefits whatsoever in the colonies.

The Sex of Plants and the Limits of Reproduction

The crucial element in the plans to acclimatise foreign species for commercial gain was abundant plant reproduction. In Mauritius, however, this was never realised during the eighteenth century. Through the process of acclimatisation, the cultivators in Mauritius became increasingly aware of the challenges of propagating the plants. This section analyses the process of recognising these challenges, the ways the cultivators tried to make sense of them by drawing on Maluku knowledge, and how they hoped to overcome them eventually by introducing slavery.

In France, the discovery of plant sexuality was only slowly accepted and, as Antoine Jacobsohn argued, at the turn of the nineteenth century it created a tension between the botanical identification of new species, cross-pollination, and the practical needs of the farmers.[99] Scholars have suggested that until that point the sexual system of classification was largely ignored, which is not necessarily true.[100] The sexual character of plants was discussed in professional correspondence in the early eighteenth century, but at that time little was known about the sexual reproduction of plants, let alone widely accepted.[101] In the wider public sphere, commercial plant nursery catalogues used the Linnaean System from the

[98] Gould, *Panda's Thumb*, 207. [99] Jacobsohn, 'Seed Origins'.

[100] Ibid., 65–6; Hoquet, *Buffon-Linné*, 55; Taiz and Taiz, *Discovery*; Ruppel, *Botanophilie*, 131–43. See also more widely, Schiebinger, *Nature's Body*.

[101] See Williams, *Botanophilia*.

1770s onwards.[102] Further, adherence to the sexual system of classification was not a prerequisite to an understanding of the implications of the plants' two sexes: the former relates to the ways in which plant species were organised in relation to each other, as against the latter, involving as it did the practical questions of how the plants might be successfully cultivated. Even if the institutional botanists did not attach much importance to the fact that plants had sexes, the gardeners who worked for them, and who accrued more practical insights into plant cultivation, might well have done.

So, the theoretical sexual nature of the nutmeg was not necessarily well understood in Paris. The French actors in Mauritius with their practical insights became, however, increasingly aware of the fact that nutmeg bears monosexual flowers.[103] What we know today is that while female plants bear the seeds, the male plants, which do not bear the seeds, are nevertheless indispensable for pollination and consequently reproduction. So, like other dioecious plants, nutmeg is more difficult to propagate than monoecious ones. Further, as it is only the female trees that bear the seeds it was (and still, is of course) desirable to grow many more of the females. For all these reasons, growing enough trees for a commercial harvest was far from straightforward. French cultivators struggled to put their knowledge about sex in plants into practice to produce more plants and more seeds. The sex of nutmeg caused huge problems relating to its propagation in Mauritius, as this section shows.

In the following examination of the practical problems of dioecious plant breeding in Mauritius, I argue that Maluku knowledge did not necessarily contribute to its application. Here, only when looking through the lens of knowledge application does the practical problem relating to dioecious plants become evident. The ways in which cultivators in Mauritius interpreted and applied the Maluku knowledge are unclear, but it is clear is that the question of sex was crucial to the cultivators: Céré's 1783 report on nutmeg trees illustrated this dilemma. He had a male overpopulation – or, rather, a too-small female one, insufficient for extensive reproduction and consequently for sustainable harvesting.[104] Although he recognised that the sex of his trees was an insurmountable obstacle, he had no practical way of overcoming it, in that he could not plant female seeds.

[102] See Easterby-Smith, *Cultivating Commerce.*

[103] On a discussion of a plant's sex from Leibniz to Linnaeus and the classification in the *Systema Naturae*, see Drouin, *L'Herbier des Philosophes*, 110–23.

[104] Céré's report on the cultivation of spice plants, 4 November 1783, transcribed in Ly-Tio-Fane, *Odyssey of Pierre Poivre*, 100–12.

He was baffled by some fundamental questions. How did the tree the seed came from determine the sex of the tree that would grow from it? Did each seed have the capacity to create a male or female tree, or was there a way of ascertaining which seeds would grow into male plants or female? He proposed carrying out what turned out to be an impossible task: predicting the sex of an offspring from the seed available to the cultivator.[105] He worked on this problem from 1776 to 1783, before concluding that he had reached the limits of his practical knowledge (and his own capacity). At the same time, he decided to conduct further experiments with different types of nutmeg, in different shapes and forms. He argued that only by practical experimentation could observations 'enlighten us on the promptest way of reproducing this precious gift of nature'.[106]

He accepted that even though it was desirable to know which seeds would grow into female trees, it was impossible to establish that. Indeed, it is still true that the sex of a seed cannot be determined by visual examination, and it is impossible to tell if a plant is female or male until that plant blooms. Unfortunately, Céré's focus on cultivating nutmeg from seeds blinded him to experimentation with alternative methods.

An example shows that similar kinds of knowledge practice were developed independently in different parts of the world: the gardener Hilaire – as explained in detail in Chapter 4 – started experimenting on grafting clove trees, using jamrosa as stock. This argument goes hand in hand with the recognition of the sex of plants, as shown in the previous section.

Imitation Spices, Climate Debates, and Colonial Distribution

Poivre, meanwhile, because of his lack of success considered the idea of conducting experiments with 'imitation spices' such as the ravensara before his return to the Indian Ocean in 1767; he surely would have had these possibilities in mind when commissioning Rochon to collect these 'imitations'.[107] By the time Rochon was in Madagascar, in 1768, Poivre had long been pursuing his plans to domesticate nutmeg and cloves on the Mascarenes: he not only told Rochon to collect useful plants in general but also had a particular interest in the Malagasy plants that resembled nutmeg. These included the *malao-manghit*, which the Malagasies

[105] Ibid., 111. [106] Ibid., 112.

[107] The search for imitation spices was common in the early modern world. For the French case and the debate about the 'true' spice, see Spary, 'Nutmegs and Botanists'. For the Portuguese context, see Cardoso, 'Especiarias na Amazônia Portuguesa', 118. See also Romaniello, 'True Rhubarb?'.

considered to have the same virtues as the 'true' nutmeg, and the '*rarabé*' (the Malagasy rarabe tree), described as a sort of wild nutmeg whose seeds were used by the Malagasies to make oil for skin and hair, and a remedy to settle the stomach.[108] Other such plants included the '*bachi-bachi*', with an aromatic seed, stem, and blossom, the '*rharha-horac*', which was a wild but 'true' species, and the '*ravend-sara*' ('ravensara'), which united 'the scent of cloves, cinnamon, and nutmeg' and whose leaves and seeds produced better essential oil than the 'true' cloves. It was used to season stews.[109] In 1756, Poivre, visiting Madagascar, observed that the ravensara fruit was harvested when small because this resulted in the dried fruits being more aromatic: this was similar to the way cloves were harvested in the Moluccas.

Poivre reflected that a little experimentation on the ravensara might be of interest, to see whether this species could serve as a new spice; it had features similar to those of cloves.[110] So, in case nutmeg and cloves did not acclimatise in Mauritius, so in 1752 he suggested introducing ravensara from Madagascar. He argued that the climate of Mauritius was so very similar to that of Madagascar that the acclimatisation process for Ravensara was more likely to be successful than that for nutmeg and cloves.[111] This exemplifies the importance of the climate debates, which extended from the French Indian Ocean to the Atlantic colonies, as explained next.

From the early 1770s the Parisian political actors had demanded the distribution of spice specimens. Poivre's idea of turning the Mascarene Islands into a 'second Java' had been shared by his patrons, Choiseul and Praslin. But after Praslin had lost his position as Minister of the Navy and the Colonies in 1770, the secretary of the Ministry, abbé Terray,[112] began to put pressure on Poivre to tighten the links with French Guiana. Terray wrote a letter to Poivre ordering him to send spices to French Guiana, explaining the need to cultivate and propagate spices in the colonies; Terray felt that French Guiana had the same climate as that of the Moluccas.

Although Choiseul had planned to reinforce France's position in the West Indies with the help of Guiana, the project was only set in motion in 1773, under the incumbent, Pierre Étienne Bourgeois de Boynes

[108] Rarabe butter, produced from seeds, abundant in fatty acids and myristic acid, has entered today's beauty market as a rare ingredient for skin care.
[109] Rochon, *Voyages à Madagascar*, vol. 1, 272–4.
[110] BCMNHN, MS 1265, 'Raven-tsara'.
[111] ANOM, C/2/285, f. 158r, 'Observations sur le muscadier et principalement sur la culture de cet arbre', Manilla, 12 February 1752.
[112] Until 1771, when de Boynes was nominated Minister of the Navy and the Colonies.

(1718–83) – and then only after Terray had insisted on it, contacting Poivre directly in 1771. Terray instructed Poivre to send fresh nutmeg, cinnamon, and clove seeds, to him in Guiana, from where, in the expectation of successful acclimatisation, they would send plants to the other French colonies in the Atlantic.[113]

In 1770, in order to protect the secret distribution of the plants to the other parts of the world, Desroches and Poivre had written an ordinance declaring that stealing, damaging, and – the most heinous crime – exporting the plants and seeds of the newly introduced spices was treason.[114] There is evidence, though, that in 1770 the British captured a French shipment of nutmeg specimens: if this is true, these plants would probably have been transferred to the British colonial garden of Saint Vincent, which had been re-established in 1765.[115] Recently, it has been discovered that cinnamon and nutmeg had been introduced to either the Lesser Antilles or Venezuela as early as 1767. These facts have been extracted from a list drawn up by the French botanist Jean Baptiste D'Arnault, although it seems likely that he was incorrect about the nutmeg.[116]

While it remains questionable as to whether D'Arnault really had seen any nutmeg, he may well have seen cinnamon, which had been smuggled from gardens from Sri Lanka via Karikal (India) into the French colonial spice gardens of Mauritius in 1752 and was to be acclimatised on the French West Indian island of Guadeloupe in 1762. Indeed, with the new Minister of the Navy and the Colonies exerting pressure for an increase in the trade, nutmeg, cloves, and cinnamon were, with the relevant care instructions, imported to French Guiana and cultivated there.[117]

During Poivre's tenure, however, the nutmeg project was exclusively based in the Indian Ocean. In 1772, for instance, Provost sent nutmeg seedlings to Réunion: Governor Bellecombe thanked him for the plants that he had brought him from Jolo.[118] The nutmegs, however, did not germinate and Poivre sent new ones to *ordonnateur* Crémont, asking him to report on their development.[119]

In addition, the local administration of Mauritius, planning to expand the French territories in the Indo-Pacific, came up with the idea of taking the Seychelles; this was to be achieved by shipping to one of the archipelago's many islands eighty to one hundred men and women, together with their children and a food supply (corn, rice, salted fish, oil, and tobacco

[113] Terray to Poivre, dated 31 March 1771, ANOM, B/202.
[114] The ordinance is reproduced in Ly-Tio-Fane, *Odyssey of Pierre Poivre*, 94–5.
[115] Zumbroich, 'Introduction of Nutmeg', 156. [116] Ibid., 155.
[117] ANOM, C/14/40, ff. 223r–224v, Ternay to anon., 31 March 1771.
[118] BCMNHN, MS 357, De Bellecombe to Provost, 6 July 1772.
[119] ADR, Sér. C, f. 264r, Poivré to Crémont, 20 July 1772.

from Mauritius.[120] The Seychelles project was also to include the acclimatisation of spices; Poivre wrote to the minister in August 1771 that he planned to 'transplant a considerable quantity [of spices] in the Seychelles ... which lie in the same latitude as Banda'.[121]

When the idea of sending specimens to French Guiana arose, Poivre disapproved, because he hoped to keep the cultivation monopoly within the French Indian Ocean colonies. He may even have feared that the Mascarenes would be abandoned. In his observations on the introduction of spices in Mauritius, Réunion, and the Seychelles (1773), he argued that the project would not have the same effect on one of the Atlantic islands as keeping it exclusively on the Mascarenes.[122] First, he argued, the largest consumption of spices took place not in Europe but in Asia (China, Japan, the Persian Gulf, and the Red Sea). So, transporting the harvests of spices from the Atlantic all the way back round the Cape to the Indo-Pacific would not make sense. Second, even if the introduction of spice cultivation in the Atlantic could defeat the Dutch spice monopoly, the aim of turning Mauritius into a profitable colony could still not be attained; the island would be ruined forever and would remain a burden to the French Crown.[123]

This too may explain why the overwhelming problems of nutmeg cultivation on Mauritius never reached metropolitan ears at the time, let alone entered any published accounts. The acclimatisation of spices was a political project to ensure that France would continue to maintain Mauritius as a colony, and thus a French presence in the Indian Ocean world, and this influenced the way Poivre attempted to promote the economic potential of nutmeg cultivation on Mauritius. It also explains why he did not explain to his correspondents in Paris how difficult it was to acclimatise and cultivate nutmeg.

Conclusions

This chapter has considered the cross-oceanic transportation and acclimatisation of spices, as both live plants and seeds, showing the spice project as disruption rather than success, and analysing knowledge as a practice by which we can illuminate its adaptation according to new social, cultural, material, and environmental contexts. While the French

[120] BnF, NAF 9341, ff. 249r–254v, 'Mémoire sur l'établissement des Iles Seychelles'.
[121] ANOM, C/4/29, ff. 18r–21r, Poivre to the minister, 22 August 1771.
[122] ADL, FGB, 15 J 17, Poivre, 'Observations de monsieur Poivre sur l'introduction de la culture des épiceries aux îles de France, de Bourbon et de Seychelles', 12 September 1773.
[123] Ibid.

met with a degree of success in changing the environmental landscape of Mauritius, both intentionally, through the introduction of foreign crops, and unintentionally, through their deforestation of the island, it was particularly difficult for them to control the environment in spaces such as ships, where human influence was very restricted (as indeed it was, in fact, on the island itself, as well).

Even when live plants and seeds finally arrived in Mauritius, they did not promise immediate success. On the contrary: acclimatising nutmeg and cloves was a very long drawn-out process, always with only slow progress. Since the spice project had been intended to give the island a financial boost, the local administrators did not abandon it – Poivre, in particular, possibly also because of his personal ambitions and hoped-for personal cultural capital. These too were reasons why he so desperately tried to keep the project exclusively within the Indian Ocean.

I have visualised the challenges and fragility of the whole spice project by highlighting the barriers to progress – climate, weather, and soil conditions, as well as packing methods – as factors that contributed to the project's becoming ever-more uncertain and experiential. I have given consideration to maritime space, in particular the long voyages from the Moluccas to Mauritius, as part of the process. The project depended heavily on the environmental factors and the plant material itself, both during its transportation on ships and once on the island. Aboard ship, the proper treatment of the live plants and the seeds in boxes and barrels determined whether they arrived alive or dead; the methods were quite tentative and chaotic, even paradoxical, and often led to the death of the plants.

The spice project was a matter of intersections of plant knowledge based on Asian, African, and European systems. Cultivation knowledge which arrived in the colony was sometimes lost or wrongly explained, and the projects were doomed to fail if the correct conditions for acclimatisation were not achieved. These problems were both human-generated, when plants were handled in the wrong way, but would also occur naturally due to environmental conditions, many of which could not be controlled. The history of nutmeg acclimatisation is a case of an ongoing knowledge creolisation par excellence closely linked to the practical problems inherent within the creolisation of nature (that is non-native plants in an environment new to them).

Further, this chapter has presented the scant knowledge regarding the reproduction of nutmeg. Even while cultivation methods and the plants' needs were studied, there was a clear explanation why the nutmeg did not reproduce well: nutmeg is a dioecious plant but the French lacked female trees. The French actors could not tell whether a seed was female or male, so they could not plant more female seeds to increase the reproduction rate. It was the plant itself, into which they had put so much effort and hope, which limited their attempts. The spice project and Indo-Pacific knowledge transmission to and within Mauritius, including ruptures and pushbacks and leaps forward, can be seen as evidence that tropical plant knowledge was very experiential, and indeed experimental. Experience was lacking, and methods were developed by the initiation of the enslaved gardener Hilaire.

This chapter concludes by making three major points: first, techniques and methods were developed independently in different parts of the world, and were not 'transferred' from Europe to the colonies, or indeed vice versa. To illustrate this, I drew on the example of grafting. Second, the nutmeg project was neither a success nor a failure; instead, it was a grindingly slow process. Third, the Maluku classification systems served as sources for European naturalists, but those systems did not necessarily convey the complexities of the plants.

Looking at practices and the plant material as such, the spice project was a random, chaotic enterprise as opposed to a smooth and coherent project supported by the Parisian institutions. It was determined by the intersection of local possibilities and limits of both cross-oceanic transportation and cultivation based on the different kinds of knowledge originating in the Moluccas. It was tested further in Mauritius, where it remained fragile and tentative as part of both an intellectually and environmentally creolising process.

Conclusion

For the traveller to Mauritius today, one of the most interesting places to visit is the Pamplemousses Botanical Garden (also known as the Sir Seewoosagur Ramgoolam Botanical Garden) which as Monplaisir had originally belonged to Pierre Poivre. The beauty of the garden and its rich variety of plants are far from a reminder, though, of the harsh reality of life in the eighteenth century for the island's enslaved people, labourers, and settlers. In contrast, after ambling around the magnificent gardens you can enjoy a splendid meal at a 'creole' restaurant around the corner, offering a selection of the vegetables, fruits, and spices taken to the island during the course of its European colonisation and the slave trade, as I have examined in the previous six chapters.

In *Creolised Science*, I have endeavoured to write a comprehensive and inclusive history of Mauritius's botanical past during the course of the eighteenth century, drawing extensively on European colonial archives, and reading them closely. By focusing primarily on the plants themselves, I examined the asymmetrical hierarchy of knowledge claims and their practical implementation in the context of the cultivation of plants in Mauritius. In so doing, I was able to perceive the development of scientific knowledge in a colonial context, decentring European science in favour of integrating Asian/African knowledge – and that of the francophone Indian Ocean context in particular – into the history of science.

My aim was threefold: first, to stress the importance of 'local knowledge' within 'global' science; instead of seeing local knowledge as *opposed* to global science, I suggested considering the former as *an integral part* of the latter. So, local knowledge was not analytical but instead conceptual, describing the ongoing creolisation of knowledge traditions and related materials in eighteenth-century Mauritius. My second aim was to expose the discrete knowledge traditions of the colonists, the settlers, the labourers, and the enslaved people of Mauritius in the context of their cultivation of plants there, and to do this by tracing whence, how, and to what extent the plants and the relevant knowledge came from. With this geographical scope, my narrative, far from being limited to that small colonial island, reaches from the African mainland to China. I have nevertheless paid special attention to certain sites in the Indo-Pacific, and in Mauritius in particular.

While the historiography has mostly been centred on the French Atlantic colonies, my geographical scope has enabled a broadening of the horizons, to encompass the French expansion around the Indian Ocean (and into the Pacific). In addition, my reach has extended beyond that of the traditional French narrative in that I have used a cross-imperial approach, teasing out the many global connections, along with those of the Indian Ocean littoral and the South East Asian islands in particular.

Third – and at the heart of this monograph – I have striven to understand the creolisation of knowledge as a process within the island space, focusing on the practical challenges of applying knowledge. The island, without any Indigenous population, is arguably a case study par excellence to test the hierarchy of knowledge claims and their practical implementation in the context of the cultivation of plants. The processes of combining knowledges and techniques were in constant flux. When knowledge and plants were accumulated on the island, they underwent recontextualisation related to social, cultural, material, and environmental conditions. The knowledge applied to cultivation had to be adapted to the island's climate and environment.

The environmental problems the settlers faced generated barriers of a similar type in the acclimatisation of both their food crops and their spices. Nature was the force – both driving and restricting – that the island's human agents sought to manage. Too often, it was nature itself that limited natural productivity, and the settlers struggled, often in vain, with the complexities of managing it.

In addition, I have focused on the incomplete and uncertain knowledge in relation to practical agriculture and the appropriate handling of the plant material – in short, the limitations of European knowledge. Instead, I have emphasised the role of non-European knowledge and the ongoing creolisation within the island. This book has thus paid due attention to the ruptures and failures that characterised the processes of knowledge creation, and has explored the creolisation that resulted from cross-cultural and material interactions between the different people who, one way or another, found themselves living on the island.

Thinking along the lines of recent claims, and by embracing a microhistorical, praxeological approach, I have in this book provided detailed connected case studies that show how interactions came about, and how the processes of knowledge production operated. Clearly, the French colony of Mauritius was built on the interplay of sociocultural diversity, materials, and climate conditions which, in addition to human-generated causes and financial problems, exacerbated the tense situation in the colony.

Many of the often nameless subaltern actors who contributed significantly to Mauritius's botanical past have disappeared from the historiography, in which my research has relocated at least some of them. This book has brought to light the contributions of many hitherto generally neglected actors, such as enslaved people, Indigenous informants, cultural brokers, pirates, local kings, traders, religious intermediaries, and others from Africa and Asia. While shining a light on their agency, I have examined the island's cultivation activities by following how plants and knowledge were collected, applied, processed, and transformed in the eighteenth-century Indo-Pacific context.

In setting up the chapters, I focused on two types of 'useful' plants (of everyday and commercial value) and related knowledge while exploring the knowledge – sometimes embodied knowledge – that derived from distinct Indo-Pacific regions, ranging from the African mainland to China. Hence, for two chapters, I explored the gathering of knowledge in these regions. For the other chapters, and more importantly for my overall argument, I analysed what happened when this knowledge came to Mauritius and was put into action.

The new point made by this book has to do with the understanding of the application of plant knowledge and the related practical challenges in the handling of these temperamental plants in their new social and natural environment. Here, as I have argued throughout, when analysing knowledge as a practice, we can truly see the creolisation of intellectual and material inputs with all its facades: an ongoing, constitutive, and transformational process consisting of testing, contrasting, meshing, rejecting, interrupting, coexisting, blending, and rooting, and any combination of those.

In each chapter, I have teased out the nuances and subtleties of the creolising processes of plant knowledge production on, and in relation to, Mauritius in the mid-eighteenth century. Chapter 1 played a key role in setting out the historical context of French colonial rule within Mauritius, and inserted the history of science into that narrative. Chapters 2 and 3, analysing the collection of plants outside the island and their cultivation on it, focused on the settlers' attempts to increase agricultural productivity. In Chapter 2 I also examined the ways in which French actors obtained information about plant cultivation from their informants in Madagascar and East Asia. Returning to Mauritius in Chapter 3, I explored the settlers' efforts to make it a self-sufficient island by importing and naturalising plants (primarily foodstuffs and fodder), then I analysed the recontextualisation of plants and related knowledge within the island. Chapter 4 was about the contributions made by the enslaved people to the cultivation of plants and the application of knowledge relating to their uses. A key section on the enslaved gardeners Charles Rama and Hilaire was a reminder of the importance of enslaved people to the construction and application of natural knowledge – and a reminder, too, of the limits of European knowledge. Chapters 5 and 6 addressed the French attempts to cultivate spices in Mauritius, and I elaborated on the attempts in South East Asia to identify the agency of those non-European individuals searching for spices. Lastly, Chapter 6 provided a detailed explanation of the practical and political challenges the project organisers faced in introducing nutmeg and cloves into Mauritius. Its detailed praxeological approach to knowledge in action made it the key chapter on the creolisation of knowledge.

In sum, this book is an attempt to extract the narrative of Western botanical knowledge from its traditionally Eurocentric framework, and it does this by highlighting the crucially important roles of the forgotten actors, especially the non-European ones, in the creation and dissemination of that knowledge. The field is still dominated by studies of global movements of knowledge rather than its (potential) consolidation within European colonial territories, let alone knowledge-practice in the context of local cultures. In addition, much of the existing historiography has been concerned with questions about how, and in what forms, knowledge was transferred to Europe. Instead of analysing the global circulation of plant knowledge with a European endpoint, this book has focused on understanding the accumulation and practical implementation of horticultural knowledge within (or in relation to) this small colonial island in the middle of the Indian Ocean.

In a nutshell, *Creolised Science* has focused on knowledge-making within Mauritius rather than on knowledge transfer from the island to the French mainland, while seriously considering Indigenous knowledge both in its very own culture and/or in the imagined service to the French colony. In this way, it stands out from the field. By locating this study firmly within a colonial island space, I was able both to examine the (often tense) relations between local practices and colonial policies, and to explore the various connections and channels through which new information was transferred and new knowledge created, where it reached intellectual and practical limits. The specific intention of *Creolised Science* was – is – to definitively shift the history of colonial science away from its long-standing assumptions about the intellectual primacy of metropolitan centres over knowledge developed within the colonial peripheries. Hence, I conclude that the processes of creolisation must be considered a central aspect of the history of science.

Bibliography

Archival Sources

France

AD Loire
Fonds Bailly: 15J 3; 15J 15; 15J 16; 15J 17

AD Montauban
20J-130

ANOM Aix-en-Provence
Dépôt des Fortifications des Colonies: DFC IV, Mémoires 10, piece 38; DFC IV, Mémoires 10, piece 49; 22 DFC 4 A; 22 DFC 7
Colonial Series B: B 202
Colonial Series C: C/1/2; C/2/82; C/2/285; C/3/13; C/4/3; C/4/5; C4/7; C/4/9; C/4/17; C/4/18; C/4/19; C/4/22; C/4/23; C/4/25; C/4/26; C/4/27; C/4/28; C/4/29; C/4/30; C/14/40; C/14/54; C/14/68; F/5A/2; F/5B/1
Personal Series E: E/4; E/99; E/184; E/197; E/295; E/343; E/372

AN Paris
Series MARINE: MAR/1JJ/111; MAR/3JJ/353; MAR 3JJ/377; MAR/B/4; MAR/G/101

BCMNHN Paris
Jus 2; Jus 3; Jus 6; Jus 7; MS 42; MS 47; MS 277; MS 279; MS 280; MS 281; MS 299; MS 303; MS 308; MS 319; MS 357; MS 452; MS 453; MS 579; MS 580; MS 887; MS 888; MS 1265; MS 1319; MS 1327, MS 1765; MS 1904; MS 3378

BM Angers
MS 612; MS 613

BnF Cartes et Plans, Paris
220/Div1/P2; 220/Div5/P2

BnF Manuscrits, Paris
NAF 9341; NAF 9343; NAF 9344; NAF 9345; NAF 9347; NAF 9377; NAF 9408; NAF 22253

Brest, Service Historique de la Défense, Département Marine
MS 89

BSG, Paris
MS 2551; MS 1085

Electronic Enlightenment (www.e-enlightenment.com)
Correspondence of Bernadin de Saint-Pierre: BSP_013; BSP_0107; BSP_0121; BSP_0140; BSP_0143; BSP_0215

Institut de France, Académies des Sciences, Service Archives, Paris
69J 67/26
PV 1754; PV 1757; PV 1769; PV 1770; PV 1772; PV 1773
File Commerson; File Poivre

Mauritius

National Archives, Coromandel
Fonds Gouvernement Royal O: OA 97; OA 75; OA 116; OA 127
Fonds Gouvernement Impérial G: GA 91/4

The Netherlands

Leiden University Libraries Digital Collections
BPL 314/2–3, G. E. Rumphius, Amboinsch Kruidboek, boeken I–XII, met het Auctuarium of Toegift BPL 314 – parts 2–3, http://hdl.handle.net/1887.1/item:3442390

Réunion

Archives départementales de la Réunion, Saint-Denis (ADR)
4J series, Archives Joseph Hubert: 4J 14; 4J 15; 4J 61; 4J 62; 4J 65; 4J 68; 4J 70; 4J 72
Fonds Compagnie des Indes, série C (1665–1767): C 1476; 1477; 1478; 1479; 1480; 1481; 1482; 1573; 1575; 1577; 1579
Fonds Période royale (PR): 10C; 12C; 22C; 37C; 38C; 56 A

Spain

Archivo da Indias, Seville (AGI)

FILIPINAS 39; FILIPINAS 199; FILIPINAS 334; FILIPINAS, 342; FILIPINAS, 464; FILIPINAS 706; FILIPINAS 707; FILIPINAS 723

United KingdomBritish Library Manuscripts

India Office Records: IOR/E/4/862; IOR/H/111

European Manuscripts: Add MS 33765; Add MS 34450; MSS Eur D 707/1; MSS Eur D 707/4; MSS Eur 707/7c; MSS Eur D 707/10; MSS Eur D 707/12; MSS Eur/ORME OV 4, MSS Eur/ORME OV 325

National Archives Kew

HCA 30/290

Printed Sources

Primary

Académie des Sciences. *Histoire et mémoire de l'Académie des Sciences: guide de recherches*. Paris: Tec et doc-Lavoisier, 1996.

Adanson, Michel. *Histoire naturelle du Sénégal*. Paris: Claude-Jean-Baptiste Buache, 1757.

Annonces, Affiches et Avis divers pour le Colonies des Isles de France et de Bourbon. Port-Louis: Imprimérie Royale, 1773, 1774.

Anon. 'Auszug Aus der Lebensbeschreibung des Herrn Poivre, Ritters vom Heiligen Geistorden, und ehemaligen Interdanten der Inseln Isle de France und Bourbon'. *Hannoverisches Magazin* 27 (1789): 1089–98.

Anon. *True Report of the Gainefull, Prosperous and Speedy Voiage to Iaua in the East Indies, Performed by a Fleete of Eight Ships of Amsterdam: Which Set Forth from Texell in Holland, the First of Maie 1598, Stilo Nouo, Whereof Foure Returned Againe the 19. of Iuly Anno 1599. in Lesse Than 15. Moneths, the Other Foure Went Forward from Iaua for the Moluccas*. P.S., 1599.

Baissac, Charles, ed. *Folk-Lore de l'Île Maurice (texte créole et traduction française)*. Paris: Maisonneuve, 1888.

Bernardin de Saint-Pierre, Jacques-Henri. *Œuvres de Jacques-Henri-Bernardin de Saint-Pierre*, ed. Louis Aimé Martin, 2 vols. Paris: Lefèvre, 1838.

'Etudes de la nature'. In *Œuvres de Jacques-Henri-Bernardin de Saint-Pierre*, vol. 1, 21–128.

'Voyage à l'Ile de France'. In *Œuvres de Jacques-Henri-Bernardin de Saint-Pierre*, vol. 1, 129–518.

Buffon, Georges Louis Leclerc de. *Œuvres complètes de Buffon*, vol. 19. Paris: Pourrat, 1835.

Cap, Paul-Antoine. *Philibert Commerson naturaliste voyageur: Étude biographique*. Paris: Victor Masson et Fils, 1861.

Charpentier de Cossigny, Joseph-François. *Lettre à M. Sonnerat, Commissaire de la Marine, naturaliste, pensionnaire du roi, correspondant de son cabinet & de l'Académie Royale des Sciences de Paris, membre de celle de Lyon.* Palma: Imprimérie Royale, 1784.

Collini, Silvia, and Antonella Vannoni, eds. *Les instructions scientifiques pour les voyageurs: XVIIe–XIXe siècle.* Paris: L'Harmattan, 2005.

Commerson, Philibert. 'Sommaire d'observations d'histoire naturelle présenté au ministre qui, à l'occasion du voyage proposé de faire autour du monde par M. de Bougainville, demandait une notice des observations qu'y pourrait faire un naturaliste', 129–34. In *Les instructions scientifiques pour les voyageurs: XVIIe–XIXe siècle,* eds. Silvia Collini and Antonella Vannoni. Paris: L'Harmattan, 2005.

Cordier, Henri. 'Les correspondants de Bertin, Secrétaire d'État au XVIIIe siècle, IV. Pierre Poivre'. *T'oung Pao* 15, no. 3 (1914): 307–38.

'Voyages de Pierre Poivre de 1748 jusqu'à 1757'. *Revue de l'histoire des colonies françaises,* 1918, 5–88.

Cossigny Charpentier de, Jean-François. 'Treize lettres de Cossigny à Réaumur'. *Recueil trimestriel de documents et travaux inédits pour servir à l'histoire des Mascareignes françaises* 4 (1939): 168–96, 205–302, 305–16.

Moyens d'amélioration et de restauration, proposés au gouvernement et aux habitants des colonies. 3 vols. Paris: Marchant, 1803.

Cossigny de Palma, Joseph-François. *Essai sur la fabrique de l'indigo.* Paris: Imprimérie Royale, 1779.

Mémoire sur la fabrication des eaux de vie de sucres. Isle de France: Imprimérie Royale, 1781.

Delaleu, J.-B.-E. *Code des Isles de France et de Bourbon.* 2 vols. Isle de France: Imprimérie Royale, 1777.

Dupont de Nemours, Pierre-Samuel. *Notice sur la vie de M. Poivre, chevalier de l'ordre du roi, ancien intendant des isles de France et de Bourbon.* Philadelphia: Moutard, 1786.

Œuvres complettes de P. Poivre, précédées de sa vie. Paris: Fuchs, 1797.

Dürr, Michel. 'Quatre inédits de Pierre Poivre'. Mémoires de l'Académie des Sciences, *Belles-Lettres et Arts de Lyon* 7, no. 4 (2007): 210–32.

Funke, K. P. 'Muskatnüsse'. *Magazin Der Handels- und Gewerbskunde* 3, no. 1 (1805): 68–72.

Fusée-Aublet, Jean Baptiste Christophore. *Histoire des plantes de la Guiane françoise, rangées suivant la méthode sexuelle, avec plusieurs mémoires sur différens objets intéressans, relatifs à la culture & au commerce de la Guiane françoise.* 2 vols. Paris: Pierre-François Didot jeune, 1775.

Galaisière, Guillaume Joseph Hyacinthe Jean Baptiste Legentil de la. *Voyage dans les mers de l'Inde: fait par ordre du roi, à l'occasion du passage de Vénus, sur le disque de soleil, le 6 juin 1761, & le 3 du même mois 1769.* Paris: Imprimérie Royale, 1779.

Gennes de la Chancelière, Mattieu. 'Observations sur les Isles de Rodrigue et de France en Mars 1735. Réflexions sur ce qui peut tendre à l'accroissement du commerce'. *Recueil trimestriel de documents et travaux inédits pour servir à l'histoire des Mascareignes françaises* 3 (1933): 223–36.

Godeheu, N. 'Extrait du Journal de Godeheu, fait en 1754', *La revue rétrospective de l'Ile Maurice* 4, no. 31 (1953): 152.

Grant de Vaux, Charles. *The history of Mauritius, or the Isle of France, and the neighbouring islands: From their first discovery to the present time; Composed principally from the papers and memoirs of Baron Grant, who resided 20 years in the island.* London: Nicol et al., 1801.

Honoré Torombert, Charles Louis. 'Éloge historique de M. Poivre. 23 Juin 1819'. *Mémoires de l'Académie des Sciences, Belles-Lettres et Arts de Lyon* 8 (2009).

Juge, François-Etienne le. 'Manière de transporter les jeunes plantes de toutes sortes d'arbres, sans embarras, et sans dépense d'eau pour les arroser dans les vaisseau', 20 November 1763. *Recueil trimestriel de documents et travaux inédits pour servir à l'histoire des Mascareignes françaises* 4 (1939): 185–6.

La Bourdonnais, Bertrand-François Mahé de. *Mémoire historiques de B.F. Mahé de La Bourdonnais, gouverneur des îles de France et de Bourbon.* Paris: Pélicier et Chatet, 1827.

La Caille, Nicolas-Louis de. *Journal historique du voyage fait au Cap de Bonne-Espérance.* Paris: Guillyn, 1763.

Launay, Adrien, ed. 'Rélation de la persécution de Cochinchine en 1750. Par Mgr Lefebvre'. In *Histoire de La Mission de Cochinchine. Documents historiques* 2 (1728–71). Paris: Libraire orientale et américaine, 1924.

Le Gentil de La Galaisière, Guillaume-Hyacinthe-Joseph-Jean-Baptiste. *Voyage dans les mers de l'Inde.* Paris: Imprimérie Royale, 1779.

Lislet, Geoffroy. 'Notice sur le Voyage de M. de Crémont au Volcan de Bourbon, en 1772 (D'après le manuscrit autographe de Lislet Geoffroy)'. *Revue historique et littérature de l'Ile Maurice, archives coloniales* 3, no. 33 (1890): 361–5.

Lougnon, Albert, ed. *Correspondance du conseil supérieur de Bourbon et de la Compagnie des Indes (1724–1750)*. Saint-Denis (Réunion): G. Daudé, 1933.

Ly-Tio-Fane, Madeleine. *Mauritius and the Spice Trade: The Odyssey of Pierre Poivre.* Vol. 1. Port Louis: Esclapon, 1958.

Mauritius and the Spice Trade: The Triumph of Jean Nicolas Céré and His Isle Bourbon Collaborators. Vol. 2. Paris: Mouton, 1970.

Malleret, Louis, ed. *Un manuscrit inédit de Pierre Poivre: Les mémoires d'un voyageur.* Paris: Ecole Française d'Extrême-Orient, 1968.

Niort, Jean-François, ed. *Code noir.* Paris: Dalloz, 2012.

Pillai, Ananda Ranga. *Private Diary of Ananda Ranga Pillai: Dubash to Joseph François Dupleix, a Record of Matters Political, Historical, Social, and Personal, from 1736 to 1761*, ed. J. F. Price. 12 vols. Madras: Government Press, 1904.

Pingré, Alexandre-Gui. *Voyage à Rodrigue: Le transit de Vénus de 1761, la mission astronomique de l'abbé Pingré dans l'océan indien*, eds. Sophie Hoarau, Marie-Paule Janiçon, and Jean-Michel Racault. Saint-Denis (Réunion): SEDES Université de la Réunion, 2004.

Poivre, Pierre. *Mémoires d'un botaniste et explorateur*, eds. Denis Piat and Jean-Claude Rey. La Rochelle: la Découvrance, 2006.

Travels of a Philosopher: Or Observations on the Manners and Arts of Various Nations in Africa and Asia. From the French of M. Le Poivre, Late Envoy of the King of Cochin-China, and Now Intendant of the Isles of Bourbon and Mauritius. Glasgow: Robert Urie, 1770.

Voyages d'un Philosophe: Nouvelle édition à laquelle on a joint une notice sur la vie l'auteur, deux de ses discours aux habitants et au Conseil Supérieur de l'Isle de France et l'extrait d'un voyage aux Isles Moluques, fait par ses ordres, pour la recherche des arbres à épiceries. Yverdon, [1768] 1796.

Raynal, Guillaume-Thomas. *Histoire philosophique et politique des établissements et du commerce des Européens dans les deux Indes.* 6 vols. Amsterdam, 1770.

Histoire philosophique et politique des établissements et du commerce des Européens dans les deux Indes. 3rd edn, 5 vols. Geneva: Pellet, 1780.

Rochon, Alexis-Marie. *Nouveau voyage à la Mer Du Sud, commencé sous les ordres de M. Marion.* Paris: Barrois l'aîné, 1783.

Voyage à Madagascar et aux Indes orientales. Paris: Prault, 1791.

Voyages à Madagascar, à Maroc et aux Indes orientales. 3 vols. Paris: Prault and Levrault, 1801.

Rumphius, Georgius Everhardus. *Herbarum Amboinense/Het Amboinsch Kruid-Boek.* Vol. 2. Amsterdam: Meinard Uytwerf, 1750.

The Ambonese Herbal: Being a Description of the Most Noteworthy Trees, Shrubs, Herbs, Land- and Water-Plants Which Are Found in Amboina and the Surrounding Islands according to Their Shape, Various Names, Cultivation, and Use; Together with Several Insects and Animals; [for the Most Part with the Figures Pertaining to Them; All Gathered with Much Trouble and Diligence over Many Years and Described in Twelve Books], ed. Eric Beekman. New Haven, CT: Yale University Press, 2011.

Schooneveld-Oosterling, J. E. ed. *Generale Missiven van Gouverneurs-Generaal en Raden Aan Heren XVII der Verenigde Oostindische Compagnie. Dl. XII: 1750–1755,* vol. 7. Den Haag: Instituut voor Nederlandse Geschiedenis, 2007.

Sonnerat, Pierre. 'Description du Cocos de l'Île de Praslin, vulgairement appelé Cocos de Mer', *SE* 7 (1776): 263–66.

Voyage à la Nouvelle-Guinée: dans lequel on trouve la description des lieux, des observations physiques et morales, et des détails relatifs à l'histoire naturelle dans le règne animal et le règne végétal. Paris: Ruault 1776.

Voyage aux Indes orientales et a la Chine, fait par ordre du Roi, depuis 1774 jusqu'en 1781. Tome 1 /: dans lequel on traite des mœurs, de la religion, des sciences & des arts des Indiens, des Chinois, des Pégouins & des Madégasses; suivi d'Observations sur le Cap de Bonne-Espérance, les isles de France et de Bourbon, les Maldives, Ceylan, Malacca, les Philippines & les Moluques, & de recherches sur l'histoire naturelle de ces pays. Paris: chez l'Auteur, 1982.

Tessier, Alexandre-Henri et al., *Encyclopédie méthodique. Agriculture.* Vol. 4 [*Dactyle-Hyssope*]. Paris: Panckoucke et al., 1787–1821.

Turgot, Etienne-François. *Mémoire instructif sur la manière de rassembler, de préparer, de conserver, et d'envoyer les diverses curiosités d'histoire naturelle: auquel on a joint un mémoire intitulé Avis pour le transport par mer, des arbres, des plantes vivaces, des semences, & de diverses autres curiosités d'histoire naturelle.* Lyon: Jean-Marie Bruyset, 1758.

'Lettre aux auteurs du Journal de l'Agriculture, du Commerce et des Finances, au sujet des Colonies'. *Journal de l'agriculture* 5 (1766): 32–40.

Varro, Marcus Terentinus. *Varro on farming = M. Terenti Varronis Rerum rusticarum libri tres, translated, with introduction, commentary and excursus by Lloyd Storr-Best.* London: Bell, 1912.

Secondary

Adler, Antony. 'The Ship as Laboratory: Making Space for Field Science at Sea'. *Journal of the History of Biology* 47, no. 3 (2013): 333–62.

Adolphe, Harold. *Les Archives démographiques de l'Ile Maurice, registres paroissiaux et d'état civil, 1721–1810*. Port-Louis: Impr. commerciale, 1966.

Agmon, Danna. *A Colonial Affair: Commerce, Conversion, and Scandal in French India*. Ithaca, NY: Cornell University Press, 2017.

Aguirre, Robert D. *Informal Empire: Mexico and Central America in Victorian Culture*. Minneapolis: University of Minnesota Press, 2005.

Alberts, Tara, and David Irving, eds. *Intercultural Exchange in Southeast Asia: History and Society in the Early Modern World*. London: I.B. Tauris, 2013.

Albritton Jonsson, Fredrik. *Enlightenment's Frontier: The Scottish Highlands and the Origins of Environmentalism*. New Haven, CT: Yale University Press, 2013.

'Natural History and Improvement: The Case of Tobacco'. In *Mercantilism Reimagined: Political Economy in Early Modern Britain and Its Empire*, eds. Philip J. Stern, and Carl Wennerlind, 117–33. New York: Oxford University Press, 2014.

Aldrich, Robert, and Kirsten McKenzie. 'Why Colonialism?'. In *The Routledge History of Western Empires*, eds. Robert Aldrich and Kirsten McKenzie, 3–13. Hoboken, NJ: Taylor & Francis, 2013.

Allain, Yves-Marie. *Voyages et survie des plantes: au temps de la voile*. Marly-le-Roi: Editions Champflour, 2000.

Une histoire des jardins botaniques: entre science et art paysager. Versailles: Éditions Quae, 2012.

Allen, Richard B. 'Economic Marginality and the Rise of the Free Population of Colour in Mauritius, 1767–1830'. *Slavery & Abolition* 10, no. 2 (1989): 126–50.

Slaves, Freedmen, and Indentured Laborers in Colonial Mauritius. Cambridge: Cambridge University Press, 1999.

'The Mascarene Slave-Trade and Labour Migration in the Indian Ocean during the Eighteenth and Nineteenth Centuries'. *Slavery & Abolition* 24, no. 2 (2003): 33–50.

The Mascarene Slave-Trade and Labour Migration in the Indian Ocean during the Eighteenth and Nineteenth centuries. London: Frank Cass, 2003.

'Free Women of Colour and Socio-Economic Marginality in Mauritius, 1767–1830'. *Slavery & Abolition* 26, no. 2 (2005): 181–97.

'A Traffic Repugnant to Humanity: Children, the Mascarene Slave Trade and British Abolitionism'. *Slavery & Abolition* 27, no. 2 (2006): 219–36.

'The Constant Demand of the French: The Mascarene Slave Trade and the Worlds of the Indian Ocean and the Atlantic during the Eighteenth and Nineteenth Centuries'. *Journal of African History* 49, no. 1 (2008): 43–72.

European Slave Trading in the Indian Ocean, 1500–1850. Athens, GA: Ohio University Press, 2014.

Allorge-Boiteau, Lucile, and Maxime Allorge. *Faune et flore de Madagascar*. Paris: Karthala/Tsipika, 2007.

Allorge-Boiteau, Lucile, and Olivier Ikor. *La fabuleuse odyssée des plantes: les botanistes voyageurs, les jardins des plantes, les herbiers*. Paris: J.-C. Lattès, 2003.

Alpers, Edward A. 'Becoming "Mozambique": Diaspora and Identity in Mauritius'. In *History, Memory and Identity*, eds. Edward A. Alpers and Vijayalakshmi Teelock. Port Louis, Mauritius: Nelson Mandela Centre for African Culture, 2001.

Andaya, Leonard Y. 'Local Trade Networks in Maluku in the 16th, 17th, and 18th Centuries'. *Cakalele: Maluku Research Journal* 2, no. 2 (1991).

The World of the Maluku: Eastern Indonesia in the Early Modern Period. Honolulu: University of Hawaii Press, 1993.

Armitage, David, and Alison Bashford, eds. *Pacific Histories: Ocean, Land, People*. New York: Palgrave Macmillan, 2014.

Armitage, David, Alison Bashford, and Sujit Sivasundaram, eds. *Oceanic Histories*. Cambridge: Cambridge University Press, 2017.

Arrow, Kenneth J. 'The Economic Implications of Learning by Doing'. *Review of Economic Studies* 29 (1962): 155–73.

Atran, Scott. *Cognitive Foundations of Natural History: Towards an Anthropology of Science*. Cambridge: Cambridge University Press, 1999.

Augusto, Geri. 'Knowledge Free and "Unfree": Epistemic Tensions in Plant Knowledge at the Cape in the 17th and 18th Centuries'. *International Journal of African Renaissance Studies – Multi-, Inter- and Transdisciplinarity* 2, no. 2 (2007): 136–82.

Banks, Kenneth J. *Chasing Empire across the Sea: Communications and the State in the French Atlantic, 1713–1763*. Montreal: McGill-Queen's University Press, 2002.

Barnwell, Patrick Joseph, and Auguste Toussaint. *A Short History of Mauritius*. London: Longmans, Green, 1949.

Barrera-Osorio, Antonio. *Experiencing Nature: The Spanish American Empire and the Early Scientific Revolution*. 1st edn. Austin: University of Texas Press, 2006.

'Experts, Nature, and the Making of Atlantic Empiricism'. In *Expertise: Practical Knowledge and the Early Modern State*, ed. Eric H. Ash, *Osiris* 25 (2010), 129–48.

Basalla, George. 'The Spread of Western Science'. *Science* 156, no. 3775 (1962): 611–22.

Batsaki, Yota, Sarah Burke Cahalan, and Anatole Tchikine, eds. *The Botany of Empire in the Long Eighteenth Century*. Washington, DC: Trustees for Harvard University, 2016.

Bauer, Ralph, and Marcy Norton. 'Introduction: Entangled Trajectories: Indigenous and European Histories'. *Colonial Latin American Review* 26, no. 1 (2017): 1–17.

Baugh, Daniel Albert. *The Global Seven Years War, 1754–1763: Britain and France in a Great Power Contest*. 1st edn. New York: Longman, 2011.

Bayly, Christopher A. *Empire & Information: Intelligence Gathering and Social Communication in India, 1780–1870*. Cambridge: Cambridge University Press, 1999.

The Birth of the Modern World, 1780–1914. Oxford: Blackwell, 2004.

Béaur, Gérard. *Histoire agraire de la France au XVIIIe siècle: Inerties et changements dans les campagnes françaises entre 1715 et 1815*. Paris: Sedes, 2000.

Beik, William. *Absolutism and Society in Seventeenth-Century France: State Power and Provincial Aristocracy in Languedoc*. Cambridge: Cambridge University Press, 1985.

Bellégo, Marine. *Enraciner l'empire. Une autre histoire du jardin botanique de Calcutta (1860–1910)*. Paris: Muséum national d'Histoire naturelle, 2021.

Béltrán, José. Scribal Practices in Natural History: The Archive of Philibert de Commerson (1727–1773). *From Florence to Goa and beyond: essays in early modern global history*, 2022 "https://hal.science/hal-04258468"⟨hal-04258468⟩

Bénot, Yves. *Les Lumières, l'esclavage, la colonisation*. Paris: Éditions la Découverte, 2005.

Berg, Maxine. 'The Genesis of "Useful Knowledge"'. *History of Science* 45, no. 2 (2007): 123–33.

'Useful Knowledge, "Industrial Enlightenment", and the Place of India'. *Journal of Global History* 8, no. 1 (2013): 117–41.

Berg, Maxine, Felicia Gottman, Hanna Hodacs, and Chris Nierstrasz, eds. *Goods from the East, 1600–1800: Trading Eurasia*. New York: Palgrave Macmillan, 2015.

Bernard-Maitre, Henri. 'Le "Petit Ministre" Henri Bertin et la correspondance littéraire de la Chine à la fin du XVIIIe siècle'. *Comptes Rendus des Séances de l'Académie des Inscriptions et Belles-Lettres* 92, no. 4 (1948): 449–51.

Bernhard, Virginia. *Slaves and Slaveholders in Bermuda, 1616–1782*. Columbia: University of Missouri Press, 1999.

A Tale of Two Colonies: What Really Happened in Virginia and Bermuda? Columbia: University of Missouri Press, 2011.

Berry, Daina Ramey. *The Price for Their Pound of Flesh: The Value of the Enslaved from Womb to Grave, in the Building of a Nation*. Boston: Beacon Press, 2017.

Bertrand, Romain. *L'Histoire à parts égales: Récits d'une rencontre Orient–Occident, XVIe–XVIIe siècle*. Paris: Seuil, 2011.

Le Long remords de la conquête. Manille–Mexico–Madrid: L'affaire Diego de Ávila (1577–1580). Paris: Seuil, 2015.

'Spirited Transactions. The Morals and Materialities of Trade Contacts between the Dutch, the British and the Malays (1596–1619)'. In *Goods from the East, 1600–1800: Trading Eurasia*, ed. Maxine Berg, 45–60. New York: Palgrave Macmillan, 2015.

Bertrand, Romain, Hélène Blais, and Emanuelle Sibeud, eds. *Cultures d'empires: Echanges et affrontements culturels en situation coloniale*. Paris: Karthala, 2015.

Bertrand, Romain, Hélène Blais, Guillaume Calafat, and Isabelle Heullant-Donat, eds. *L'Exploration du monde. Une autre histoire des grandes découvertes*. Paris: Seuil, 2019.

Bil, Geoff. 'Imperial Vernacular: Phytonymy, Philology, and Disciplinarity in the Indo-Pacific'. *British Journal for the History of Science* 51 (2018): 635–58.

Indexing the Indigenous: Plants, Peoples and Empire. Forthcoming with JHU Press.

Blais, Hélène and Rahul Markovits. 'Introduction. Le commerce des plantes, XVIe–Xxe siècle'. *Revue d'histoire moderne et contemporaine* 66 (2019): 7–23.

Bleichmar, Daniela. 'Books, Bodies, and Fields: Sixteenth-Century Transatlantic Encounters with New World Materia Medica'. In *Colonial Botany: Science, Commerce, and Politics in the Early Modern World*, eds. Londa Schiebinger and Claudia Swan, 83–99. Philadelphia: University of Pennsylvania Press, 2005.

Bödeker, Hans Erich, ed. *Wissenschaft als kulturelle Praxis: 1750–1900*. Göttingen: Vandenhoeck & Ruprecht, 1999.

Boomgaard, Peter, ed. *Empire and Science in the Making: Dutch Colonial Scholarship in Comparative Global Perspective, 1760–1830*. New York: Palgrave Macmillan, 2013.

'Introduction: From the Mundane to the Sublime: Science, Empire, and the Enlightenment, 1760s–1820s'. In *Empire and Science in the Making: Dutch Colonial Scholarship in Comparative Global Perspective, 1760–1830*, ed. Peter Boomgaard, 1–37. New York: Palgrave Macmillan, 2013.

Bose, Sugata. *A Hundred Horizons: The Indian Ocean in the Age of Global Empire*. Cambridge, MA: Harvard University Press, 2006.

Bossenga, Gail. *The Politics of Privilege: Old Regime and Revolution in Lille*. Cambridge: Cambridge University Press, 1991.

Boucheron, Patrick, ed. *Histoire mondiale de la Portugal*. Paris: Seuil, 2017.

Boulle, Pierre. *Race et esclavage dans la France de l'Ancien Règime*. Paris: Perrin, 2007.

Boulle, Pierre, and Sue Peabody. *Le droit des noirs en France au temps de l'esclavage: Textes choisis et commentés*, Paris: l'Harmattan, 2014.

Boumediene, Samir. 'L'Appropriation des remèdes Mexicains par les européens. Transferts économiques et culturels (XVIe–XVIIe siècles)'. In *Emprunts et transferts culturels dans le monde luso-hispanophone: Réalités et représentations*, eds. M. Guiraud and N. Fourtané, 249–74. Nancy: Presses Universitaires de Nancy, 2011.

La colonisation du savoir: Une histoire des plantes médicinales du Nouveau Monde (1492–1750). Vaulx-en-Velin: Editions des mondes, 2016.

Bour, Roger. 'Paul Philippe Sanguin de Jossigny (1750–1827), Artiste de Philibert Commerson. Les Dessins de Reptiles de Madagascar, de Rodrigues et des Seychelles'. *Zoosystema* 37, no. 3 (2015): 415–48.

Bourde, André. *The Influence of England on the French Agronomes, 1750–1789*. Cambridge: Cambridge University Press, 1953.

Agronomie et agronomes en France au XVIIIe siècle. Paris: SEVPEN, 1967.

Bourde de la Rogerie Henri. *Les bretons aux îles de France et de Bourbon (Maurice et la Réunion) aux XVIIe et XVIIIe siècles*. Rennes: La Découvrance, 1934.

Bourguet, Marie-Noëlle. 'Measurable Difference. Botany, Climate, and the Gardener's Thermometer in Eighteenth-Century France'. In *Colonial Botany: Science, Commerce, and Politics in the Early Modern World*, eds. Londa Schiebinger and Claudia Swan, 270–86. Philadelphia: University of Pennsylvania Press, 2005.

Le monde dans un carnet: Alexander von Humboldt en Italie (1805). Paris: Le Félin, 2017.

Bourguet, Marie-Noëlle, and Christophe Bonneuil. *De l'inventaire du monde à la mise en valeur du globe: botanique et colonisation, fin 17e siècle–début 20e siècle: dossier thématique*. Saint-Denis: Société française d'histoire d'Outre-mer, 1999.

'Présentation'. *Revue française d'histoire d'outre- mer* 86 (1999): 7–38.

Bouton, Louis. *Sur le décroissement des forêts à Maurice*. Mauritius: Impr. d'Aimé Mamarot KT, 1838.

Bravo, Michael. 'Ethnographic Navigation and the Geographical Gift'. In *Geography and Enlightenment*, ed. David N. Livingstone and Charles W. J. Withers, 199–235. Chicago: University of Chicago Press, 1999.

Bray, Francesca, Peter Coclanis, Edda Fields-Black, and Dagmar Schäfer, eds. *Rice: Global Networks and New Histories*. Cambridge: Cambridge University Press, 2015.

Breen, Benjamin. 'No Man Is an Island: Early Modern Globalization, Knowledge Networks, and George Psalmanazar's Formosa'. *Journal of Early Modern History* 17 (2013): 391–417.

Brendecke, Arndt, ed. *Praktiken der Frühen Neuzeit. Akteure-Handlungen-Artefakte*. Cologne: Böhlau, 2015.

The Empirical Empire: Spanish Colonial Rule and the Politics of Knowledge. Berlin: De Gruyter, 2016.

Brixius, Dorit. 'A Pepper Acquiring Nutmeg: Pierre Poivre, the French Spice Quest and the Role of Mediators in Southeast Asia, 1740s to 1770s'. *Journal of the Western Society for French History* 43 (2015): 68–77.

'A Hard Nut to Crack: Nutmeg Cultivation and the Application of Natural History between the Maluku Islands and Isle de France (1750s–1780s)'. *British Journal for the History of Science* 51 (2018): 585–606.

'From Ethnobotany to Emancipation: Slaves, Plant Knowledge, and Gardens on Eighteenth-Century Isle de France', *History of Science* 58 (2020): 51–75.

Brockway, Lucile. *Science and Colonial Expansion: The Role of the British Royal Botanic Gardens*. New York: Academic Press, 1979.

Brouard, N. R. *A History of Woods and Forests in Mauritius*. Port Louis, Mauritius: J. E. Félix, ISO, Govt. Printer, 1963.

Brown, Vincent. *The Reaper's Garden: Death and Power in the World of Atlantic Slavery*. Cambridge, MA: Harvard University Press, 2008.

'Social Death and Political Life in the Study of Slavery'. *American Historical Review* 114, no. 5 (2009): 1231–49.

Callon, Michel. 'Some Elements of a Sociology of Translation: Domestication of the Scallops and the Fishermen of St. Brieuc Bay'. In *Power, Action, and Belief: A New Sociology of Knowledge*, ed. John Law, 196–223. London: Routledge 1986.

Campbell, Gwyn, *The Structure of Slavery in Indian Ocean Africa and Asia*. London: Frank Cass, 2004.

'Slavery and the Trans-Indian Ocean World Slave Trade: A Historical Outline'. In *Cross Currents and Community Networks: The History of the Indian Ocean World*, eds. Himanshu Prabha Ray and Edward A. Alpers, 286–305. New Delhi: Oxford University Press, 2007.

Campbell, Gwyn, Suzanne Miers, and Joseph C. Miller. 'Women in Western Systems of Slavery: Introduction'. *Slavery & Abolition* 26, no. 2 (2005): 161–79.

'Children in European Systems of Slavery: Introduction'. *Slavery & Abolition* 27, no. 2 (2006): 163–82.

eds. *Children in Slavery through the Ages*. Athens, GA: Ohio University Press, 2009.

Cañizares-Esguerra, Jorge. *How to Write the History of the New World: Histories, Epistemologies, and Identities in the Eighteenth-Century Atlantic World*. Stanford: Stanford University Press, 2001.

'Iberian Colonial Science'. *Isis* 96, no. 1 (2005): 64–70.

Nature, Empire, and Nation: Explorations of the History of Science in the Iberian World. Stanford: Stanford University Press, 2006.

Caplan, Lionel. 'Power and Status in South Asian Slavery'. In *Asian and African Systems of Slavery*, ed. James L. Watson, 169–94. Berkeley: University of California Press, 1980.

Cardim, Pedro, Tamar Herzog, José Javier Ruiz Ibáñez, and Gaetano Sabatini, eds. *Polycentric Monarchies: How Did Early Modern Spain and Portugal Achieve and Maintain a Global Hegemony?* Brighton: Sussex Academic Press, 2012.

Carey, Daniel, and Lynn M. Festa. *Postcolonial Enlightenment, Eighteenth-Century Colonialisms and Postcolonial Theory*. Oxford: Oxford University Press, 2009.

Carney, Judith A. *Black Rice: The African Origins of Rice Cultivation in the Americas*. Cambridge, MA: Harvard University Press, 2001.

Carney, Judith A., and Richard N. Rosomoff. *In the Shadow of Slavery: Africa's Botanical Legacy in the Atlantic World.* Berkeley: University of California Press, 2009.

Carse, Ashley. 'Nature as Infrastructure: Making and managing the Panama Canal Watershed'. *Social Studies of Science* 42 (2012): 539–63.

Cartuyvels, Sabine. *'Jardin'*. In *1740, Un Abrégé du monde: Savoirs et collections autour de Dezallier d'Argenville*, ed. Anne Lafont, 130–9. Paris: Fage éditions, 2012.

Casimir, Jean. *The Haitians: A Decolonial History*, trans. Laurent Dubois. Chapel Hill: University of North Carolina Press, 2020.

Castelnau-l'Estoile, Charlotte de, and François Regourd, eds. *Connaissances et pouvoirs: Les espaces impériaux, XVIe–XVIIIe siècles, France, Espagne, Portugal.* Pessac: Presses Universitaires de Bordeaux, 2005.

Cerutti, Simona. 'Histoire pragmatique, ou de la rencontre entre histoire sociale et histoire culturelle'. *Tracés. Revue de sciences humaines* 15 (2008): 147–68.

Chakrabarti, Pratik. *Western Science in Modern India: Metropolitan Methods, Colonial Practices*. Delhi: Permanent Black, 2004.

Materials and Medicine: Trade, Conquest and Therapeutics in the Eighteenth Century. Manchester: Manchester University Press, 2010.

'Networks of Medicine: Trade and Medico-Botanical Knowledge in Eighteenth Century Coromandel Coast'. In *Science and Society in India, 1750–2000*, eds. Pratik Chakrabarti and A. Bandopadhyay, 49–82. New Delhi: Monahar Books, 2010.

Medicine and Empire: 1600–1960. Basingstoke, UK: Palgrave Macmillan, 2014.

Chambers, David Wade, and Richard Gillespie. 'Locality in the History of Science: Colonial Science, Technoscience, and Indigenous Knowledge'. In *Nature and Empire: Science and the Colonial Enterprise*, ed. Roy MacLeod, *Osiris* 15 (2000), 221–40.

Charles, Loïc, and Paul Cheney. 'The Colonial Machine Dismantled: Knowledge and Empire in the French Atlantic'. *Past & Present* 219, no. 1 (2013): 127–63.

Chateauraynaud, Francis, and Yves Cohen, eds. *Histoires pragmatiques*. Paris: Editions EHESS, 2016.

Chatterjee, Indrani, and Richard M. Eaton. *Slavery & South Asian History*. Bloomington: Indiana University Press, 2006.

Chaudenson, Robert. *Creolization of Language and Culture*. London: Routledge, 2001.

Cheke, Anthony S., and Julian Pender Hume. *Lost Land of the Dodo: An Ecological History of Mauritius, Réunion & Rodrigues*. London: T & AD Poyser, 2008.

Cheney, Paul. 'Aufklärung und die politische Ökonomie des Kolonialismus'. In *Der moderne Staat und 'le Doux commerce'– Staat, Ökonomie und internationales System im politischen Denken der Aufklärung*, ed. Olaf Asbach, 207–28. Baden-Baden: Nomos, 2014.

Chérer, Sophie. *La vraie couleur de la vanille*. Paris: L'École des loisirs, 2012.

Clark, Emily. *The Strange History of the American Quadroon: Free Women of Color in the Revolutionary Atlantic World*. Chapel Hill: University of North Carolina Press, 2013.

Coclanis, Peter, Edda Fields-Black, and Dagmar Schäfer. *Rice: Global Networks and New Histories*. Cambridge: Cambridge University Press, 2015.

Cohn, Bernard S. *Colonialism and Its Forms of Knowledge: The British in India*. Princeton, NJ: Princeton University Press, 1996.

Collins, James B. *The State in Early Modern France*. Cambridge: Cambridge University Press, 1995.

Conrad, Sebastian. *What Is Global History?* Princeton, NJ: Princeton University Press, 2016.

Cook, Alexandra. 'Linnaeus and Chinese Plants: A Test of the Linguistic Imperialism Thesis'. *Notes and Records of the Royal Society* 64 (2010): 121–38.

Cook, Harold J. 'Global Economies and Local Knowledge in the East Indies'. In *Colonial Botany: Science, Commerce, and Politics in the Early Modern World*, eds. Londa Schiebinger and Claudia Swan, 100–18. Philadelphia: University of Pennsylvania Press, 2005.

Matters of Exchange: Commerce, Medicine, and Science in the Dutch Golden Age. New Haven, CT: Yale University Press, 2007.

Cook, Malcolm. *Bernardin de Saint-Pierre: A Life of Culture*. Oxford: Modern Humanities Research Association Maney, 2006.

Cooper, Alix. *Inventing the Indigenous: Local Knowledge and Natural History in Early Modern Europe*. Cambridge: Cambridge University Press, 2007.

Cooper, Frederick. *Plantation Slavery on the East Coast of Africa*. New Haven, CT: Yale University Press, 1977.

Cordier, Henri. 'Les Marchands Hanistes de Canton'. *T'oung Pao* 3, no. 5 (1902): 281–315.

Cardoso, Alírio. 'Especiarias na Amazônia Portuguesa: Circulação Vegetal E Comércio Atlântico No Final da Monarquia Hispânica'. *Tempo* 21 (2015): 116–33.

Cornevin, Robert, and Marianne Cornevin. *La France et les Français outre-mer: de la première croisade à la fin du Second empire*. Paris: Hachette, 1993.

Costa, H. de la. 'Early French Contacts with the Philippines'. *Philippine Studies* 11, no. 3 (1963): 401–18.

Couto, Dejanirah, and Stéphane Péquignot, eds. *Les langues de la négociation: approches historiennes*. Rennes: PUR, 2017.

Craciun, Adriana, and Simon Schaffer, eds. *The Material Cultures of Enlightenment Arts and Sciences*. London: Palgrave Macmillan, 2016.

Crawford, Elisabeth, Terry Shinn, and Sverker Sörlin, eds. *Denationalizing Science: The Contexts of International Scientific Practice*. Dordrecht: Kluwer Academic, 1993.

Crestey, Nicole. 'Bernardin de Saint-Pierre, naturaliste voyageur aux Mascareignes?'. In *Bernardin de Saint-Pierre et l'océan Indien*, eds. Jean-Michel Racault, Chantale Meure, and Angélique Gigan, 353–72. Paris: Classiques Garnier, 2011.

Crosland, Maurice P. *Science under Control: The French Academy of Sciences, 1795–1914*. Cambridge: Cambridge University Press, 1992.

Cruikshank, Julie. *Do Glaciers Listen? Local Knowledge, Colonial Encounters, and Social Imagination*. Vancouver: UBC Press, 2014.

Curran, Andrew S. *The Anatomy of Blackness: Science and Slavery in an Age of Enlightenment*. Baltimore: Johns Hopkins University Press, 2011.

Curry, Helen. 'Imperilled Crops and Endangered Flowers'. In *Worlds of Natural History*, eds. Helen, Curry, Nic Jardine, Jim Secord, and Emma Sprary, 460–75. Cambridge: Cambridge University Press, 2018.

Curry, Helen, Nic Jardine, Jim Secord, and Emma Sprary, eds. *Worlds of Natural History*. Cambridge: Cambridge University Press, 2018.

Dalbine, Erwan. *Un vétérinaire sous les tropiques: François-Éloy de Beauvais, 1744–1815*. Wimereux: Sagittaire, 2008.

Damodaran, Vinita, Anna Winterbottom, and Alan Lester, eds. *The East India Company and the Natural World*. Basingstoke, UK: Palgrave Macmillan, 2015.

Daniels, Christine, and Michael V. Kennedy, eds. *Negotiated Empires: Centers and Peripheries in the Americas, 1500–1820*. New York: Routledge, 2002.

Das Gupta, Ashin. *The World of the Indian Ocean Merchant, 1500–1800: Collected Essays of Ashin Das Gupta*. New Delhi: Oxford University Press, 2001.

Daston, Lorraine. 'Type Specimens and Scientific Memory'. *Critical Inquiry* 31, no. 1 (2004): 153–82.

Daston, Lorraine, and Katharine Park. *Wonders and the Order of Nature, 1150–1750*. New York: Zone Books, 1998.

Daubigny, Eugène Théodore. *Choiseul et la France d'outre-mer après le traité de Paris avec un appendice sur les origines de la question de Terre-Neuve: étude sur la politique coloniale au XVIIIe siècle*. Paris: Hachette, 1892.

Daugeron, Bertrand. *Collections Naturalistes: Entre Sciences et Empires (1763–1804)*. Paris: Publications scientifiques du Muséum national d'histoire naturelle, 2009.

Dauser, Regina, ed. *Wissen im Netz: Botanik und Pflanzentransfer in europäischen Korrespondenznetzen des 18. Jahrhunderts*. Berlin: Akad.-Verl., 2008.

Dauser, Regina, Stefan Hächler, Michael Kempe, Franz Mauelshagen, and Martin Stuber. 'Einleitung'. In *Wissen im Netz: Botanik und Pflanzentransfer in europäischen Korrespondenznetzen des 18. Jahrhunderts*, eds. Regina Dauser, Stefan Hächler, Michael Kempe, Franz Mauelshagen, and Martin Stuber, 9–30. Berlin: Akademie Verlag, 2008.

Davids, Karel. 'The Scholarly Atlantic: Circuits of Knowledge between Britain, the Dutch Republic, and the Americas in the Eighteenth Century'. In *Dutch*

Atlantic Connections, 1680–1800: Linking Empires, Bridging Borders, eds. Gert Oostindie and Jessica V. Roitman, 224–48. Leiden: Brill, 2014.
'On Machines, Self-Organization, and the Global Traveling of Knowledge, circa 1500–1900'. *Isis* 106, no. 4 (2015): 866–74.
Dawdy, Shannon Lee. *Building the Devil's Empire: French Colonial New Orleans*. Chicago: University of Chicago Press, 2008.
Debien, Gabriel. *Les engagés pour les Antilles, 1634–1715. La Société coloniale aux XVIIe et XVIIIe siècles*. Paris: Société de l'histoire des colonies françaises Larose, 1952.
De Bruyn, Frans, and Shaun Regan, eds. *The Culture of the Seven Years' War: Empire, Identity, and the Arts in the Eighteenth-Century Atlantic World*. Toronto: University of Toronto Press, 2014.
Dechêne, Louise. *Habitants et marchands de Montréal au XVIIe siècle. Collection Civilisations et mentalités*. Paris: Plon, 1974.
Delbourgo, James. *A Most Amazing Scene of Wonders: Electricity and Enlightenment in Early America*. Cambridge, MA: Harvard University Press, 2006.
'Fugitive Colours: Shamans' Knowledge, Chemical Empire and Atlantic Revolutions'. In *The Brokered World: Go-Betweens and Global Intelligence, 1770–1820*, eds. Simon Schaffer, Lissa Roberts, Kapil Raj, and James Delbourgo. Sagamore Beach, MA: Science History Publications, 2009.
'Gardens of Life and Death'. *British Journal for the History of Science* 43, no. 1 (2010): 113–18.
'Sir Hans Sloane's Milk Chocolate and the Whole History of the Cacao'. *Social Text* 29, no. 1 (106) (2011): 71–101.
Collecting the World: Hans Sloane and the Origins of the British Museum. London: Penguin, 2017.
Delbourgo, James, and Nicholas Dew, eds. *Science and Empire in the Atlantic World*. New York: Routledge, 2008.
Deleuze, J. P. F. 'Notice sur M. De Céré', *Annales Museum d'Histoire Naturelle Paris* 16 (1810): 329–37.
Díaz-Trechuelo Spínola, María et al., eds. *La expedición de Juan de Cuéllar a Filipinas*. Barcelona: Lunwerg Editores 1997.
Dirks, Nicholas, ed. *Colonialism and Culture*. Ann Arbor: University of Michigan Press, 1992.
'Colonial Histories and Native Informants: Biography of an Archive'. In *Orientalism and the Postcolonial Predicament: Perspectives on South Asia*, eds. Carol A. Breckenridge and Peter van der Veer, 279–313. Philadelphia: University of Pennsylvania Press, 1993.
Dobie, Madeleine. *Trading Places: Colonization and Slavery in Eighteenth-Century French Culture*. Ithaca, NY: Cornell University Press, 2010.
Drayton, Richard. 'Science, Medicine and the British Empire'. In *The Oxford History of the British Empire*, ed. Robin W. Winks, 264–76. Oxford: Oxford University Press, 1999.
Nature's Government: Science, Imperial Britain, and the 'Improvement' of the World. New Haven, CT: Yale University Press, 2000.

'Synchronic Palimpsests: Work, Power and the Transcultural History of Knowledge'. In *Entangled Knowledge: Scientific Discourses and Cultural Difference*, eds. Klaus Hock and Mackenthun, 31–50. Münster: Waxmann, 2012.

Drouin, Jean-Marc. *L'Herbier des philosophes*. Paris: Seuil, 2008.

Duchet, Michèle. *Anthropologie et histoire au siècle des Lumières*. Paris: F. Maspero, 1971.

Le partage des savoirs: discours historique et discours ethnologique. Paris: La Découverte, 1985.

Dull, Jonathan R. *The French Navy and the Seven Years' War*. Lincoln: University of Nebraska Press, 2005.

Easterby-Smith, Sarah. 'Selling Beautiful Knowledge: Amateurship, Botany and the Market-Place in Late Eighteenth-Century France'. *Journal for Eighteenth-Century Studies* 36, no. 4 (2013): 531–43.

'Reputation in a Box. Objects, Communication and Trust in Late 18th-Century Botanical Networks'. *History of Science* 53, no. 2 (2015): 180–208.

'On Diplomacy and Botanical Gifts: France, Mysore and Mauritius in 1788'. In *The Botany of Empire in the Long Eighteenth Century*, eds. Yota Batsaki, Sarah Burke Cahalan, and Anatole Tchikine, 193–212. Washington, DC: Trustees for Harvard University, 2016.

Cultivating Commerce: Cultures of Botany in Britain and France, 1760–1815. Cambridge: Cambridge University Press, 2017.

'Recalcitrant Seeds: Material Culture and the Global History of Science'. *Past & Present* 242 (2019): 215–42.

'Botany as Useful Knowledge: French Global Plant Collecting at the End of the Old Regime'. In *Re-Inventing the Economic History of Industrialisation*, eds. Kristine Bruland, Anne Gerritsen, Pat Hudson, and Giorgio Riello, 276–89. Montreal: McGill-Queen's University Press, 2020.

Easterby-Smith, Sarah, and Emily Senior. 'The Cultural Production of Natural Knowledge: Contexts, Terms, Themes'. *Journal for Eighteenth-Century Studies* 36, no. 4 (2013): 471–76.

Ecott, Tim. *Vanilla: Travels in Search of the Ice Cream Orchid*. New York: Grove, 2004.

Ehrard, Jean. *Lumières et esclavage. L'Esclavage colonial et l'opinion publique en France au XVIIIE siècle*. Paris: André Versaille, 2008.

Elliot, John H. *Empires of the Atlantic World: Britain and Spain in America, 1492–1930*. New Haven, CT: Yale University Press, 2006.

Elshakry, Marwa. 'When Science Became Western: Historiographical Reflections'. *Isis* 101 (2010): 98–109.

Endersby, Jim. *Imperial Nature: Joseph Hooker and the Practices of Victorian Science*. Chicago: University of Chicago Press, 2008.

Fan, Fa-ti. *British Naturalists in Qing China: Science, Empire, and Cultural Encounter*. Cambridge, MA: Harvard University Press, 2004.

'The Global Turn in the History of Science'. *East Asian Science, Technology and Society* 6 (2012): 249–58.

Feyel, Gilles. *L'annonce et la nouvelle: La presse d'information en France sous l'Ancien Régime (1630–1788)*. Oxford: Voltaire Foundation, 2000.
Fels, Marthe de. *Pierre Poivre ou l'amour des épices*. Paris: Hachette, 1968.
Ferguson, Niall. *Civilization: The West and the Rest*. 1st US edn. New York: Penguin, 2011.
Ferrer, Ada. *Freedom's Mirror: Cuba and Haiti in the Age of Revolution*. New York: Cambridge University Press, 2014.
Filliot, Jean Michel. La Traite des esclaves vers les Mascareignes au XVIIIe siècle. 2 vols. *Tananarive: Office de la recherche scientifique, et technique d'Outre-mer*, 1970.
Findlen, Paula, ed. *Empires of Knowledge: Scientific Networks in the Early Modern World*. New York: Routledge, 2019.
Fish, Shirley. *The Manila–Acapulco Galleons: The Treasure Ships of the Pacific: With an Annotated List of the Transpacific Galleons 1565–1815*. Central Milton Keynes, UK: AuthorHouse, 2011.
Fleischer, Alette. '(Ex)changing Knowledge and Nature at the Cape of Good Hope, circa 1652–1700'. In *The Dutch Trading Companies as Knowledge Networks*, eds. Siegfried Huigen, Jan L. de Jong, and Elmer Kolfin, 243–65. Leiden: Brill, 2010.
Foucault, Michel. *Sécurité, territoire, population: Cours au Collège de France, 1977–1978*, ed. Michel Senellart. Paris: Seuil, 2004.
Foury, B. *Maudave et la colonisation de Madagascar*. Paris: Société de l'histoire des colonies françaises et Librairie Larose, 1956.
Fradera, Josep Maria. *Colonias para después de un imperio*. Barcelona: Bellaterra, 2005.
Françoise, Juliette. *L'empire de la monnaie dans les Mascareignes au XVIIIe siècle*. Editions Ithaka, 2020.
Freist, Dagmar, ed. *Diskurse – Körper – Artefakte: Historische Praxeologie in der Frühneuzeitforschung*. Bielefeld: Transcript-Verl., 2014.
'Historische Praxeologie als Mikro-Historie'. In *Praktiken der Frühen Neuzeit. Akteure-Handlungen-Artefakte*, ed. Arndt Brendeck, 62–77. Cologne: Böhlau, 2015.
Fressoz, Jean-Baptiste. 'Les politiques de la nature au début de la révolution. Sens et fonctions de l'alerte environnementale, 1789–1793'. *Annales historiques de la Révolution française* 399 (2020): 19–38.
Gabriel-Robert, Thibault. 'Bernardin de Saint-Pierre et la Physiocratie'. In *Bernardin de Saint-Pierre au tournant des Lumières*, ed. Katherine Astbury, 35–50. Leuven: Peeters, 2012.
Gainot, Bernard. *L'empire colonial français de Richelieu à Napoléon (1630–1810)*. Paris: Armand Colin, 2015.
La Révolution des Esclaves: Haïti, 1763–1803. Paris: Vendémiaire, 2017.
Galison, Peter. *Image and Logic: A Material Culture of Microphysics*. Chicago: University of Chicago Press, 1997.
Galloway, J. H. 'Agricultural Reform and the Enlightenment in Late Colonial Brazil'. *Agricultural History* 53, no. 4 (1979): 763–79.
Ganeri, Jonardon. 'Well-Ordered Science and Indian Epistemic Cultures: Toward a Polycentered History of Science'. *Isis* 104, no. 2 (2013): 348–59.

Garrod, Raphaële, and Paul J. Smith, eds. *Natural History in Early Modern France: The Poetics of an Epistemic Genre*. Leiden: Brill, 2018.
Gascoigne, John. *Encountering the Pacific in the Age of Enlightenment*. Port Melbourne: Cambridge University Press, 2014.
Geaves, Ron. 'From Pilgrimage to Tourism: Comparative Analysis of Kavadi Festivals in Tamil Diasporas'. In *Sacred Space: Interdisciplinary Perspectives within Contemporary Contexts*, eds. Steve Brie, Jenny Daggers, and David Torevell, 127–43. Cambridge: Cambridge Scholars, 2009.
Gershenhorn, Jerry. *Melville J. Herskovits and the Racial Politics of Knowledge*. Lincoln: University of Nebraska Press, 2004.
Geertz, Clifford. 'Common Sense as a Cultural System'. *Antioch Review* 33, no. 1 (1975): 5–26.
Ghobrial, John-Paul A. 'The Secret Life of Elias of Babylon and the Uses of Global Microhistory'. *Past & Present* 222 (2014): 51–93.
Girault-Fruet, Arlette. *Les voyageurs d'îles: sur la route des Indes aux XVIIe et XVIIIe siècles*. Paris: Classiques Garnier, 2010.
Givens, Bryan. 'Review of Erik Lars Myrup, Power and Corruption in the Early Modern Portuguese World'. *Bulletin for Spanish and Portuguese Historical Studies* 40, no. 1 (2015): article 10.
Gledhill, David. *The Names of Plants*. Cambridge: Cambridge University Press, 2008.
Godfroy, Marion F. *Kourou, 1763: Le dernier rêve de l'Amérique française*. Chroniques. Paris: Vendémiaire, 2011.
Golinski, Jan. *Making Natural Knowledge: Constructivism and the History of Science*. Chicago: University of Chicago Press, 1998.
Gómez, Pablo F. *The Experiential Caribbean: Creating Knowledge and Healing in the Early Modern Atlantic*. Chapel Hill: University of North Carolina Press, 2017.
Gould, Stephen Jay. *The Panda's Thumb: More Reflections in Natural History*. New York: Norton, 1980.
Grafe, Regina. *Distant Tyranny: Markets, Power, and Backwardness in Spain, 1650–1800*. Princeton, NJ: Princeton University Press, 2012.
Grafe, Regina, and Alejandra Irigoin. 'A Stakeholder Empire: The Political Economy of Spanish Imperial Rule in America'. *Economic History Review* 65, no. 2 (2012): 609–51.
Graubart, Karen B. 'Shifting Landscapes. Heterogeneous Conceptions of Land Use and Tenure in the Lima Valley'. *Colonial Latin American Review* 26, no. 1 (2017): 62–84.
Grimé, William, ed. *Ethno-Botany of the Black Americans*. Algonac: Reference Publications, 1979.
Grove, Richard H. 'Conserving Eden: The (European) East India Companies and Their Environmental Policies on St. Helena, Mauritius and in Western India, 1660 to 1854'. *Comparative Studies in Society and History* 35, no. 2 (1993): 318–51.
Green Imperialism: Colonial Expansion, Tropical Island Edens and the Origins of Environmentalism, 1600–1860. Cambridge: Cambridge University Press, 1995.

'Indigenous Knowledge and the Significance of South-West India for Portuguese and Dutch Constructions of Tropical Nature'. *Modern Asian Studies* 30, no. 1 (1996): 121–43.

'The Island and the History of Environmentalism: The Case of St Vincent'. In *Nature and Society in Historical Context*, eds. Mikuláš Teich, Roy Porter, and Bo Gustafsson, 148–62. Cambridge: Cambridge University Press, 1997.

Grove, Richard H., Mathias Lefèvre, and Grégory Quenet. *Les îles du paradis: l'invention de l'écologie aux colonies, 1660–1854*. Paris: La Découverte, 2013.

Gunn, Geoffrey C. *Historical Dictionary of East Timor*. Lanham, MD: Scarecrow Press, 2010.

Halleux, Robert. *Le savoir de la main: Savants et artisans dans l'Europe pré-industrielle*. Paris: Armand Colin, 2009.

Hansen, Lars, ed. *The Linnaeus Apostles: Global Science and Adventure*. Vol. 6. London: IK Foundation, 2007.

Harms, Robert, Bernard K. Freamon, and David W. Blight. *Indian Ocean Slavery in the Age of Abolition*. New Haven, CT: Yale University Press, 2013.

Harrison, Mark. 'Science and the British Empire'. *Isis* 96, no. 1 (2005): 56–63.

Harvey, David Allen. 'Slavery on the Balance Sheet: Pierre-Samuel Dupont de Nemours and the Physiocratic Case for Free Labor'. *Proceedings of the Western Society for French History* 42 (2014): 75–87.

Haudrère, Philippe. *La Bourdonnais: marin et aventurier*. Paris: Desjonquères, 1992.

L'empire des rois, 1500–1789. Vol. 1. Paris: Denoël, 1997.

La Compagnie Française des Indes au XVIIIe siècle. Seconde édition revue et corrigée. Paris: Indes savantes, 2005.

Les compagnies des Indes orientales: Trois siècles de rencontres entre Orientaux et Occidentaux, 1600–1858. Paris: Desjonquères, 2006.

Les Français dans l'océan Indien, XVIIe–XIXe siècle. Rennes: Presses Universitaires de Rennes, 2014.

Havard, Gilles, and Cécile Vidal. *Histoire de l'Amérique française*. Paris: Flammarion, 2014.

Hay, Peter. *A Companion to Environmental Thought*. Edinburgh: Edinburgh University Press, 2002.

Hazareesingh, K. 'The Religion and Culture of Indian Immigrants in Mauritius'. *Comparative Studies in Society and History* 8, no. 2 (1966): 241–57.

Heintzman, Kit. 'A Cabinet of the Ordinary: Domesticating Veterinary Education, 1766–1799'. *British Journal for the History of Science* 51 (2018): 239–60.

Herlitz, Lars. 'Art and Nature in Pre-Classical Economics of the Seventeenth and the Eighteenth Centuries'. In *Nature and Society in Historical Context*, eds. Mikuláš Teich, Roy Porter, and Bo Gustafsson, 163–75. Cambridge: Cambridge University Press, 1997.

Herskovits, Melville J. *The Myth of the Negro Past*. New York: Harper, 1941.

Hicks, Dan. '"Material Improvements": The Archaeology of Estate Landscapes in the British Leeward Islands, 1713–1838'. In *Estate Landscapes: Design, Improvement and Power in the Post-Medieval Landscape*, eds. F. Giles and J. Finch, 205–27. Woodbridge: Bowdell and Brewer, 2008.

Hilaire-Pérez, Liliane. 'Technology as a Public Culture in the Eighteenth Century: The Artisans' Legacy'. *History of Science* 45 (2007): 135–53.

Hodacks, Hanna, Kenneth Nyberg, and Stéphane Van Damme, eds. *Linnaeus, Natural History and the Circulation of Knowledge*. Oxford: Oxford University Studies in the Enlightenment, 201.

Hodson, Christopher. *The Acadian Diaspora: An Eighteenth-Century History*. Oxford: Oxford University Press, 2012.

Hodson, Christopher, and Brett Rushforth. 'Absolutely Atlantic: Colonialism and the Early Modern French State in Recent Historiography'. *History Compass* 8, no. 1 (2009): 101–17.

Hoquet, Thierry. *Buffon-Linné: éternels rivaux de la biologie?* Paris: Dunod, 2007.

Hörmann, Raphael, and Gesa Mackenthun, eds. *Human Bondage in the Cultural Contact Zone: Transdisciplinary Perspectives on Slavery and its Discourses*. Münster: Waxmann, 2010.

Houllemare, Marie. 'La fabrique des archives coloniales et la naissance d'une conscience impériale (France, XVIIIe siècle)'. *Revue d'histoire moderne et contemporaine* 61, no. 2 (2014): 7–31.

Humbert, H., and Jean-François Leroy. *Flore de Madagascar et des Comores: Plantes vasculaires. Publiée sous les auspices du Gouvernement Général de Madagascar et sous la Direction de H. Humbert*. Paris: Imprimérie officielle; Muséum national d'histoire naturelle, 1951.

Hünniger, Dominik. 'Sammeln, Sezieren und Systematisieren. Naturkundliche Verfahrensweisen in der Insektenkunde um 1800'. In *Akteure, Tiere, Dinge. Verfahrensweisen der Naturgeschichte*, eds. Silke Förschler and Anne Mariss, 47–60. Cologne: Böhlau Verlag, 2017.

'Nets, Labels and Boards: Materiality and Natural History Practices in Continental European Manuals on Insect Collecting 1688–1776'. In *Naturalists in the Field: Collecting, Recording and Preserving the Natural World from the Fifteenth to the Twenty-First Century*, ed. Arthur MacGregor, 686–705. Leiden: Brill 2018.

Hunt, Lynn. *Writing History in the Global Era*. New York: Norton, 2014.

Irigoin, Alejandra, and Regina Grafe. 'Bargaining for Absolutism: A Spanish Path to Nation-State and Empire Building'. *Hispanic American Historical Review* 88, no. 2 (2008): 173–209.

Jacobsohn, Antoine. 'Seed Origins: New Varieties of Fruits and Vegetables around Paris at the Turn of the Nineteenth Century'. In *Of Elephants & Roses: French Natural History, 1790–1830*, ed. Sue Ann Prince, 65–77. Philadelphia: American Philosophical Society, APS Museum, 2013.

Jardine, Nick, James A. Secord, and Emma C. Spary, eds. *Cultures of Natural History*. Cambridge: Cambridge University Press, 1996.

Jarnagin, Laura. *Portuguese and Luso-Asian Legacies in Southeast Asia, 1511–2011: Culture and Identity in the Luso-Asian World*. Pasir Panjang: Institute of Southeast Asian Studies, 2012.

Jasanoff, Sheila, ed. *States of Knowledge: The Co-Production of Science and the Social Order*. London: Routledge, 2004.

Jennings, Éric T. 'Cartels et lobbies de la vraie vanille: Marketing, genre, nostalgie et réseaux postcoloniaux'. *Revue d'histoire moderne et contemporaine* 66 (2019): 128–55.

Johnston, A. J. B. *Control and Order in French Colonial Louisbourg, 1713–1758*. East Lansing: Michigan State University Press, 2001.

Jones, Peter M. *Industrial Enlightenment: Science, Technology and Culture in Birmingham and the West Midlands, 1760–1820*. Manchester: Manchester University Press, 2008.

Agricultural Enlightenment: Knowledge, Technology, and Nature, 1750–1840. Oxford: Oxford University Press, 2016.

Kananoja, Kalle. *Healing Knowledge in Atlantic Africa: Medical Encounters, 1500–1850*. Cambridge: Cambridge University Press, 2021.

Kennedy, B. E. 'Anglo-French Rivalry in Southeast Asia 1763–93: Some Repercussions'. *Journal of Southeast Asian Studies* 4, no. 2 (1973): 199–215.

Klein, Ursula, and Emma C. Spary, eds. *Materials and Expertise in Early Modern Europe: Between Market and Laboratory*. Chicago: University of Chicago Press, 2010.

Klemun, Marianne. 'Introduction: "Moved" Natural Objects – "Spaces in Between"'. *Journal of History of Science and Technology* 5, spring (2012): 9–16.

'Live Plants on the Way: Ship, Island, Botanical Garden, Paradise and Container as Systemic Flexible Connected Spaces in between'. *Journal of History of Science and Technology* 5 (2012): 30–48.

Knörr, Jacqueline. 'Contemporary Creoleness: Or, the World in Pidginization?'. *Current Anthropology* 51 (2010): 731–59.

Koerner, Lisbet. *Linnaeus: Nature and Nation*. Cambridge, MA: Harvard University Press, 1999.

Kontler, László, Antonella Romano, Silvia Sebastiani, and Borbála Zsuzsanna Török, eds. *Negotiating Knowledge in Early Modern Empires: A Decentred View*. Basingstoke, UK: Palgrave Macmillan, 2014.

Kroupa, Šebestián. 'Ex Epistulis Philippinensibus: Georg Joseph Kamel SJ (1661–1706) and His Correspondence Network'. *Centaurus* 57 (2015): 229–59.

Kroupa, Šebestián, Stephanie Mawson, and Dorit Brixius. 'Science and Islands in the Indo-Pacific Worlds'. *British Journal for the History of Science* 51 (2018): 541–58.

Kury, Loreilai. 'Les instructions de voyages dans les expéditions dcientifiques françaises (1750–1830)'. *Revue d'histoire des sciences* 51, no. 1 (1998): 65–92.

Labrosse, Claude, and Pierre Rétat. *L'Instrument périodique: La fonction de la presse au XVIIIe siècle*. Lyon: Presses Universitaires de Lyon, 1985.

Lacour, Pierre-Yves. *'Histoire Naturelle'*. In *1740, un abrégé du Monde: Savoirs et collections autour de Dezallier d'Argenville*, ed. Anne Lafont, 112–20. Paris: Fage, 2012.

La République naturaliste: Collections d'histoire naturelle et Révolution française, 1789–1804. Paris: Muséum national d'histoire naturelle, 2014.

Lacroix, Alfred. *Notice historique sur les membres et correspondants de l'Académie des sciences ayant travaillé dans les colonies françaises des Mascareignes et de Madagascar au XVIIIe siècle et au début du XIXe: lecture faite en la séance annuelle du 17 décembre 1934*. Paris: Gauthier-Villars, 1934.

Notice historique sur les cinq de Jussieu, membres de l' Académie des sciences (1712–1853): leur rôle d'animateurs des recherches d'histoire naturelle dans les colonies françaises: leurs principaux correspondants: lecture faite en la séance annuelle du 21 décembre 1936. Paris: Gauthier-Villars, 1936.

Michel Adanson au Sénégal (1749–1753). Paris: Larose, 1938.

Lafont, Anne, ed. *1740, Un abrégé du monde: Savoirs et collections autour de Dezallier d'Argenville*. Paris: Fage, 2012.

Lagesse, Marcelle. *L'Ile de France avant La Bourdonnais*. Port-Louis: M. Coquet, 1978.

Laissus, Yves. 'Note sur les manuscrits de Pierre Poivre: (1719–1786) conservés à la Bibliothèque du Museum d'histoire naturelle'. *Proceedings of the Royal Society of Arts and Sciences of Mauritius* IV, Part II (1973): 31–56.

'Catalogue des manuscrits de Philibert Commerson 1727–1773) conservés à la Bibliothèque Centrale du Muséum National d'Histoire Naturelle (Paris)'. *Revue d'histoire des sciences* 31, no. 2 (1978): 131–62.

Lambert, David. *Mastering the Niger: James MacQueen's African Geography and the Struggle over Atlantic Slavery*. Chicago: University of Chicago Press, 2013.

Lamotte, Mélanie. *Making Race: Policy, Sex, and Social Order in the French Atlantic and Indian Oceans, 1608–1756*. Cambridge, MA: Harvard University Press, forthcoming.

Landwehr, Achim. *Policey im Alltag: Die Implementation frühneuzeitlicher Policeyordnungen in Leonberg*. Frankfurt a.M.: Vittorio Klostermann, 2000.

Larson, Pier M. 'Enslaved Malagasy and "Le Travail de la Parole" in the Pre-Revolutionary Mascarenes'. *Journal of African History* 48, no. 3 (2007): 457–79.

Ocean of Letters: Language and Creolization in an Indian Ocean Diaspora. Cambridge: Cambridge University Press, 2009.

Latour, Bruno. *Science in Action: How to Follow Scientists and Engineers through Society*. Cambridge, MA: Harvard University Press, 1987.

Launay, Adrien, ed. 'Rélation de la persécution de Cochinchine en 1750 par Mgr Lefebvre'. In *Histoire de La Mission de Cochinchine. Documents historiques*, vol. 2 (1728–71). Paris: Libraire orientale et américaine, 1924.

Lawrence, Natalie. 'Assembling the Dodo in Early Modern Natural History'. *BJHS* 48 (2015): 387–408.

Le Gouic, Olivier. 'Pierre Poivre et les épices: Une transplantation réussie'. In *Techniques et colonies (XVIe–XXe siècles)*, eds. Sylviane Llinares and Philippe Hrodej, 103–26. Paris: Publications de la Société française d'Histoire d'Outre-Mer et de l'Université de Bretagne Sud-SOLITO, 2005.

Leow, Rachel. *Taming Babel: Language in the Making of Malaysia*. Cambridge: Cambridge University Press, 2016.

Lepetit, Bernard, ed. *Les formes de l'expérience: Une autre histoire sociale*. Paris: Albin Michel, 1995.

Letouzey, Yvonne. *Le Jardin des plantes à la croisée des chemins avec André Thouin, 1747–1824*. Paris: Ed. du Muséum, 1989.

Lidwell-Durnin, John. 'Cultivating Famine: Data, Experimentation and Food Security, 1795–1848'. *British Journal for the History of Science* 53 (2020): 159–81.

Li Tana. *Nguyen Cochinchina: Southern Vietnam in the Seventeenth and Eighteenth Centuries*. Ithaca, NY: Cornell University Press, 1998.

Long, Pamela O. 'Trading Zones in Early Modern Europe'. *Isis* 106, no. 4 (2015): 840–47.

Lougnon, Albert, ed. *Correspondance du Conseil supérieur de Bourbon et de la Compagnie des Indes (1724–1750)*. Saint-Denis (Réunion): G. Daudé, 1933.

Ly-Tio-Fane, Madeleine. *Mauritius and the Spice Trade. The Odyssey of Pierre Poivre;* Port Louis: Esclapon, 1958.

'Pierre Poivre et l'expansion française dans l'Indo-Pacifique'. *Bulletin de l'Ecole Française d'Extrême-Orient* 53, no. 2 (1967): 453–512.

'Problèmes d'approvisionnement de l'Ile de France au temps de l'intendant Poivre'. *Proceedings of the Royal Society of Arts and Sciences Mauritius* 3, part 1 (1968): 101–15.

Mauritius and the Spice Trade: The Triumph of Jean Nicolas Céré and his Isle Bourbon Collaborators. The Hague: Mouton, 1970.

Pierre Sonnerat, 1748–1814: An Account of his Life and Work. Cassis: Imprimérie et Papeterie commercial, 1976.

'Contacts between Schönbrunn and the Jardin du Roi at Isle de France (Mauritius) in the 18th Century: An Episode in the Career of Nicolas Thomas Baudin'. *Mitteilungen des Österreichischen Staatsarchivs* 35 (1982): 85–109.

'Botanic Gardens: Connecting Link in Plant Transfer between the Indo-Pacific and Caribbean Regions'. *Harvard Papers in Botany* 8 (1994): 7–14.

Ly Tio Fane-Pineo, Huguette. *Île de France, 1715–1746: L'émergence de Port Louis*. Moka: Mahatma Gandhi Institute, 1993.

McClellan, James. 'The Académie Royale des Sciences, 1699–1793: A Statistical Portrait'. *Isis* 72, no. 4 (1981): 541–67.

Specialist Control: The Publications Committee of the Académie Royale des Sciences (Paris), 1700–1793. Philadelphia: American Philosophical Society, 2003.

'André Michaux and French Botanical Networks at the End of the Old Regime'. *Cast Castanea* 69, no. sp2 (2004): 69–97.

Colonialism and Science Saint Domingue in the Old Regime. Chicago: University of Chicago Press, 2010.

'Science & Empire Studies and Postcolonial Studies: A Report from the Contact Zone'. In *Entangled Knowledge: Scientific Discourses and Cultural Difference*, eds. Klaus Hock and Gesa Mackenthun, 51–74. Münster: Waxmann, 2012.

McClellan, James, and François Regourd. *The Colonial Machine: French Science and Overseas Expansion in the Old Regime*. Turnhout: Brepols, 2011.

McDonald, Christie, and Susan Rubin Suleiman, eds. *French Global: A New Approach to Literary History*. New York: Columbia University Press, 2011.

McGuire, Meredith B. *Lived Religion: Faith and Practice in Everyday Life*. Oxford: Oxford University Press, 2008.

MacKenzie, John M. *Imperialism and the Natural World*. Manchester: Manchester University Press, 1990.

MacLeod, Roy M. 'Introduction: Nature and Empire: Science and the Colonial Enterprise'. Special issue of *Osiris* 15 (2000): 1–13.

ed. *Nature and Empire: Science and the Colonial Enterprise*. Special issue of *Osiris* 15 (2000).

Madeira Santos, Catarina. 'Administrative Knowledge in a Colonial Context: Angola in the Eighteenth Century'. *British Journal for the History of Science* 43 (2010): 539–56.

Mahony, Martin. 'The "Genie of the Storm": Cyclonic Reasoning and the Spaces of Weather Observation in the Southern Indian Ocean, 1851–1925'. *British Journal for the History of Science* 51 (2018): 607–33.

Malleret, Louis. 'Pierre Poivre, L'abbé Galloys et l'introduction d'espèces botaniques et d'oiseaux de Chine à l'Ile Maurice'. *Proceedings of the Royal Society of Arts and Sciences Mauritius* 3, part 1 (1968): 117–30.

Pierre Poivre. Paris: Adrien-Maisonneuve, 1974.

Mandelblatt, Bertie. 'How Feeding Slaves Shaped the French Atlantic: Mercantilism and the Crisis of Food Provisioning in the Franco-Caribbean during the Seventeenth and Eighteenth Centuries'. In *The Political Economy of Empire in the Early Modern World*, eds. Sophus Reinert and Pernille Røge, 192–220. Basingstoke, UK: Palgrave Macmillan, 2013.

Manning, Catherine. *Fortunes à Faire: The French in Asian Trade, 1719–48*. Aldershot: Variorum, 1996.

Manning, Patrick, and Daniel Rood, eds. *Global Scientific Practice in the Age of Revolutions, 1750–1850*. Pittsburgh, PA: University of Pittsburgh Press, 2016.

Marcaida, José Ramón, and Juan Pimentel. 'Green Treasures and Paper Floras: The Business of Mutis in New Granada (1783–1808)'. *History of Science* 52, no. 3 (2014): 277–96.

Margócsy, Dániel. '"Refer to Folio and Number": Encyclopedias, the Exchange of Curiosities, and Practices of Identification before Linnaeus'. *Journal of the History of Ideas* 71, no. 1 (2010): 63–89.

Commercial Visions: Science, Trade, and Visual Culture in the Dutch Golden Age. Chicago: Chicago University Press, 2014.

Marion, Marcel. *Dictionnaire des institutions de la France aux XVIIe et XVIIIe siècles. Réimpression de l'édition originale de 1923*. Paris: Picard, 1968.

Mariss, Anne. *'A World of New Things' Praktiken der Naturgeschichte bei Johann Reinhold Forster*. Frankfurt a.M.: Campus Frankfurt, 2015.

Johann Reinhold Forster and the Making of Natural History on Cook's Second Voyage, 1772–1775. Lanham, MD: Lexington Books, 2019.

Marocci, Guiseppe. 'Too Much to Rule: States and Empires across the Early Modern World'. *Journal of Early Modern History* 20 (2016): 511–25.

Marquet, Julie. 'La médiation des dubashes. Un aspect de la politique française en Inde dans la seconde moité du XVIIIe siècle'. *La Révolution française* 8 (2015), http://lrf.revues.org/1259

Marquet, Julie, and Blake Smith, eds. 'Introduction: L'Inde et les français: pratiques et savoirs coloniaux'. *Outre-Mers. Revue historique* 388–9 (2015): 5–18.

Marsh, Kate. 'Territorial Loss and the Construction of French Colonial Identities, 1763–1962'. In *France's Lost Empires: Fragmentation, Nostalgia, and La Fracture Coloniale*, eds. Kate Marsh and Nicola Frith, 1–13. Lanham, MD: Lexington Books, 2010.

Matos, Artur Teodoro de. *Timor Português, 1515–1769: contribuição para a sua história*. Lisboa: Faculdade de Letras da Universidade de Lisboa, Instituto Histórico Infante Dom Henrique, 1974.

Matsuda, Matt K. *Pacific Worlds: A History of Seas, People, and Cultures*. Cambridge: Cambridge University Press, 2012.

Maurel, Chloé. *Manuel d'histoire globale. Comprendre le 'global turn' des sciences humaines*. Paris: Armand Colin, 2014.

Maverick, Lewis A. 'Pierre Poivre: Eighteenth Century Explorer of Southeast Asia'. *Pacific Historical Review* 10, no. 2 (1941): 165–77.

Menon, Minakshi. 'What's in a Name? William Jones, "Philological Empiricism" and Botanical Knowledge Making in Eighteenth-Century India'. *South Asian History and Culture* 13 (2022): 87–111.

Meisen, Lydia. *Die Charakterisierung der Tiere in Buffons 'Histoire Naturelle'*. Würzburg: Königshausen & Neumann, 2008.

Metcalf, Alida C. *Go-Betweens and the Colonization of Brazil, 1500–1600*. Austin: University of Texas Press, 2005.

Meyer, Jean, Jean Tarrade, Annie Rey-Goldziguer, and Jacques Thobie. *Histoire de la France coloniale. Des origines à 1914*. Paris: Armand Colin, 1991.

Miers, Suzanne. 'Slavery: A Question of Definition'. *Slavery & Abolition* 24, no. 2 (2003): 1–16.

Miller, David Philip, and Peter Hanns Reill, eds. *Visions of Empire: Voyages, Botany, and Representations of Nature*. Cambridge: Cambridge University Press, 1996.

Mintz, Sidney W., and Richard Price. *An Anthropological Approach to the Afro-American Past: A Caribbean Perspective*. Philadelphia: Institute for the Study of Human Issues, 1976.

Mokyr, Joel. *The Gifts of Athena: Historical Origins of the Knowledge Economy*. Princeton, NJ: Princeton University Press, 2002.

Enlightened Economy: An Economic History of Britain 1700–1850. New Haven, CT: Yale University Press, 2012.

Montero Sobrevilla Iris. 'Indigenous Naturalists'. In *Worlds of Natural History*, eds. Helen Anne Curry, Nicholas Jardine, James A. Secord, and Emma C. Spary, 112–30. Cambridge: Cambridge University Press.

Montessus de Ballore, Fernand-Bernard de. *Martyrologe et biographie de Commerson, médecin botaniste et naturaliste du roi, médecin de Toulon-sur-Arroux (Saône-et-Loire) au XVIIIe siècle*. Chalon-sur-Saône: impr. de L. Marceau, 1889.

Morrissey, Robert Michael. *Empire by Collaboration: Indians, Colonists, and Governments in Colonial Illinois Country*. Philadelphia: University of Pennsylvania Press, 2015.

Morton, Alan G. *History of Botanical Science: An Account of the Development of Botany from Ancient Times to the Present Day*. London: Academic Press, 1981.

Mukerji, Chandra. 'Dominion, Demonstration, and Domination: Religious Doctrine, Territorial Politics, and French Plant Collection'. In *Colonial Botany: Science, Commerce, and Politics in the Early Modern World*, eds. Londa Schiebinger and Claudia Swan, 19–33. Philadelphia: University of Pennsylvania Press, 2005.

Müller-Wille, Staffan. 'Gardens of Paradise'. *Endeavour* 25, no. 2 (2001): 49–54.

'Collection and Collation: Theory and Practice of Linnaean Botany'. *Studies in History and Philosophy of Biological and Biomedical Sciences* 38, no. 3 (2007): 541–62.

Müller-Wille, Staffan, and Isabelle Charmantier. 'Natural History and Information Overload: The Case of Linnaeus'. *Studies in History and Philosophy of Biological and Biomedical Sciences* 43, no. 1 (2012): 4–15.

Murphy, Kathleen S. 'Translating the Vernacular: Indigenous and African Knowledge in the Eighteenth-Century British Atlantic'. *Atlantic Studies* 8, no. 1 (2011): 29–48

'Collecting Slave Traders: James Petiver, Natural History, and the British Slave Trade'. *William and Mary Quarterly* 70, no. 4 (2013): 637–70.

Musselman, Elizabeth Green. 'Plant Knowledge at the Cape: A Study in African and European Collaboration'. *International Journal of African Historical Studies* 36, no. 2 (2003): 367–92.

'Indigenous Knowledge and Contact Zones: The Case of the Cold Bokkeveld Meteorite, Cape Colony, 1838'. *Itinerario* 33, no. 1 (2009): 31–44.

Myrup, Erik Lars. *Power and Corruption in the Early Modern Portuguese World*. Baton Rouge: Louisiana State University Press, 2015.

Napal, Doojendraduth, ed. *Les constitutions de l'Ile Maurice*. Port-Louis: Mauritius Printing Co., 1962.

Nappi, Carla. *The Monkey and the Inkpot: Natural History and Its Transformations in Early Modern China*. Cambridge, MA: Harvard University Press, 2010.

Nardin, Denis. 'La France et les Philippines sous l'Ancien Régime'. *Revue française d'histoire d'outre-mer* 63, no. 230 (1976): 5–43.

Newton, James E., and Ronald L. Lewis, *The Other Slaves: Mechanics, Artisans, and Craftsmen*. Boston: G. K. Hall, 1978.

Ngendahimana, Anastase. *Les idées politiques et sociales de Bernardin de Saint-Pierre*. Bern: Peter Lang, 1999.

North-Coombes, Alfred. *La découverte des Mascareignes par les Arabes et les Portugais: Rétrospective et mise au point, contribution à l'histoire de l'Océan Indien au XVIe siècle*. Port-Louis: Service Bureau, 1979.

Ogborn, Miles. *Global Lives: Britain and the World, 1550–1800*. Cambridge: Cambridge University Press, 2008.

Ogbuagu, Marc Nwosu. 'Vitamins, Phytochemicals and Toxic Elements in the Pulp and Seed of Raphia Palm Fruit (Raphia Hookeri)'. *Fruits* 63, no. 5 (2008): 297–302.

Oliver, Samuel Pasfield, and G. F. Scott Elliot. *The Life of Philibert Commerson, D.M., Naturaliste du Roi, an Old-World Story of French Travel and Science in the Days of Linnaeus*. London: Murray, 1909.

Ophir, Adi, and Steven Shapin. 'The Place of Knowledge: A Methodological Survey'. *Science in Context* 4 (1991): 3–22.

Östling, Johan, David Larsson Heidenblad, Erling Sandmo, Anna Nilsson Hammar, and Kari Nordberg, eds. *Circulation of Knowledge: Explorations in the History of Knowledge*. Lund: Lund University Publications, 2018.

'The History of Knowledge and the Circulation of Knowledge: An Introduction'. In *Circulation of Knowledge: Explorations in the History of Knowledge*, eds. Johan Östling, David Larsson Heidenblad, Erling Sandmo, Anna Nilsson Hammar, and Kari Nordberg, 9–33. Lund: Lund University Publications, 2018.

Oudin-Bastide, Caroline, and Philippe Steiner. *Calcul et morale: coûts de l'esclavage et valeur de l'émancipation, XVIIIe–XIXe siècle*. Paris: Albin Michel, 2015.

Pacini, Giulia. 'Paul et Virginie Environmental Concerns in Bernardin de Saint Pierre's "Paul et Virginie"'. *Interdisciplinary Studies in Literature and Environment* 18, no. 1 (2011): 87–103.

Parsons, Christopher M. *A Not-So-New World: Empire and Environment in French Colonial North America*. Philadelphia: University of Pennsylvania Press, 2018.

Parsons, Christopher M.; and Kathleen S. Murphy. 'Ecosystems under Sail: Specimen Transport in the Eighteenth-Century French and British Atlantics'. *Early American Studies* Fall (2012): 503–39.

Passy, Louis. *Histoire de la Société nationale d'agriculture de France, Tome premier: 1761–1793*. Paris: P. Renouard, 1912.

Paul, Louis-José. *Deux siècles d'histoire de la police à l'Île Maurice, 1768–1968*. Paris: l'Harmattan, 1997.

Peabody, Sue. *There Are No Slaves in France: The Political Culture of Race and Slavery in the Ancien Régime*. New York: Oxford University Press, 1996.

Madeleine's Children: Family, Freedom, Secrets, and Lies in France's Indian Ocean Colonies. New York: Oxford University Press, 2017.

Peabody, Sue, and Tyler Edward Stovall. *The Color of Liberty: Histories of Race in France*. Durham, NC: Duke University Press, 2003.

Peabody, Sue, and Keila Grinberg. *Free Soil in the Atlantic World*. London: Routledge, 2015.

Pearson, Michael N. *The Indian Ocean*. London: Routledge, 2003.

Pelras, Christian. *The Bugis*. Oxford: Blackwell, 1996.

Perrier de la Bathie, H. 'Les plantes introduites à Madagascar (suite)'. *Revue de botanique appliquée et d'agriculture coloniale* 11, no. 122 (1931): 833–7.

Piat, Denis. *L'île Maurice: Sur la routes des épices, 1598–1810*. Paris: les Éditions du Pacifique, 2010.

Pimentel, Juan. *Testigos del mundo: Ciencia, literatura y viajes en la ilustración*. Madrid: Marcial Pons, 2003.

Pinar García, Susana. *El Sueño de las Especias: Viaje de Exploración de Francisco Noroña por las Islas de Filipinas, Java, Mauricio y Madagascar*. Madrid: Consejo Superior de Investigaciones Científicas, Departamento de Historia de la Ciencia, 2000.

Pluchon, Pierre. 'Choiseul et Vergennes: un gâchis colonial'. In *Négoce, ports et océans, XVIe–XXe siècles: mélanges offerts à Paul Butel*, eds. Silvia Marzagalli and Hubert Bonin, 225–34. Bordeaux: Presses Universitaires de Bordeaux, 2000.

Pluchon, Pierre, and Denise Bouche. *Le premier empire colonial: Des origines à la restauration*. Vol. 1. Paris: Fayard, 1991.

Pratt, Mary Louise. *Imperial Eyes: Travel Writing and Transculturation*. London: Routledge, 1992.

Prince, Sue Ann, ed. *Of Elephants & Roses: French Natural History, 1790–1830*. Philadelphia: American Philosophical Society, APS Museum, 2013.

Pritchard, James. *In Search of Empire: The French in the Americas, 1670–1730*. Cambridge: Cambridge University Press, 2007.

Pugliano, Valentina. 'Non-Colonial Botany or, the Late Rise of Local Knowledge?'. *Studies in History and Philosophy of Biological and Biomedical Sciences* 40, no. 4 (2009): 321–8.

Pyenson, Lewis. 'An End to National Science: The Meaning and the Extension of Local Knowledge'. *History of Science* 40 (2002): 251–90.

Pyenson, Lewis, and Susan Sheets-Pyenson. *Servants of Nature: A History of Scientific Institutions, Enterprises and Sensibilities*. London: HarperCollins, 1999.

Raj, Kapil. '18th-Century Pacific Voyages of Discovery, "Big Science", and the Shaping of a European Scientific and Technological Culture'. *History and Technology* 17, no. 2 (2000): 79–98.

Relocating Modern Science: Circulation and the Construction of Knowledge in South Asia and Europe, 1650–1900. New York: Palgrave Macmillan, 2007.

'The Historical Anatomy of a Contact Zone. Calcutta in the Eighteenth Century'. *Indian Economic & Social History Review* 48, no. 1 (2011): 55–82.

'Beyond Postcolonialism … and Postpositivism: Circulation and the Global History of Science'. *Isis* 104, no. 2 (2013): 337–47.

Raj, Kapil, and Heinz Otto Sibum, eds. *Histoire des sciences et des savoirs, 2: Modernité et globalisation*. Paris: Éditions du Seuil, 2015.

Régent, Frédéric. *La France et ses esclaves: de la colonisation aux abolitions, 1620–1848*. Paris: Pluriel, 2012.

Regourd, François 'Kourou 1763. Succès d'une enquête, échec d'un projet colonial'. In *Connaissances et pouvoirs: Les espaces impériaux, XVIe–XVIIIe siècles, France, Espagne, Portugal*, eds. Charlotte de Castelnau-l'Estoile and François Regourd, 233–54. Pessac: Presses Universitaires de Bordeausx, 2005.

'Capitale savante, capitale coloniale: Sciences et savoirs coloniaux à Paris aux XVIIe et XVIIIe siècles'. *Revue d'histoire moderne et contemporaine* 55, no. 2 (2008): 121–51.

'Les lieux de savoir et d'expertise colonial à Paris au XVIIIe siècle: Institutions et enjeux savants'. In *Les mondes coloniaux à Paris au XVIIIe siècle: Circulation et enchevêtrement des savoirs*, eds. Anja Bandau, Marcel Dorigny and Rebekka von Malinckrodt, 31–48. Paris: Karthala, 2010.

Reuss, Martin, and Stephen Cutcliffe, eds. *The Illusory Boundary: Environment and Technology in History*. Charlottesville: University of Virginia Press, 2010.

Revel, Jacques. *Jeux d'échelles: La Micro-analyse à l'expérience*. Paris: Seuil, 1996.

Rice, A. L. *Voyages of Discovery: Three Centuries of Natural History Exploration*. New York: Potter, 1999.

Richard, François G. 'Hesitant Geographies of Power: The Materiality of Colonial Rule in the Siin (Senegal), 1850–1960'. *Journal of Social Archaeology* 13, no. 1 (2012): 54–79.

Ridley, Glynis. *The Discovery of Jeanne Baret: A Story of Science, the High Seas, and the First Woman to Circumnavigate the Globe*. New York: Crown, 2010.

Rigby, Nigel. 'The Politics and Pragmatics of Seaborne Plant Transportation, 1769–1805'. In *Science and Exploration in the Pacific: European Voyages to the Southern Oceans in the Eighteenth Century*, ed. Margaret Lincoln. Rochester: Boydell Press, 1998.

Roberts, Lissa. 'Situating Science in Global History: Local Exchanges and Networks of Circulation'. *Itinerario* 33, no. 1 (2009): 9–30.

'Accumulation and Management in Global Historical Perspective: An Introduction'. *History of Science* 52, no. 3 (2014): 227–46.

'"Le Centre de Toutes Choses": Constructing and Managing Centralization on the Isle de France'. *History of Science* 52, no. 3 (2014): 319–42.

'Practicing Oeconomy during the Second Half of the Long Eighteenth Century: An Introduction'. *History and Technology* 30, no. 3 (2014): 1–16.

'Producing (in) Europe and Asia, 1750–1850'. *Isis* 106, no. 4 (2015): 857–65.

Roberts, Lissa, Simon Schaffer, and Peter Dear, eds. *The Mindful Hand: Inquiry and Invention from the Late Renaissance to Early Industrialisation*. Amsterdam: Koninklijke Nederlandse Academie van Wetenschappen, 2007.

Roche, Daniel. *La France des Lumières*. Paris: Fayard, 1993.

Rodao García, Florentino. *Españoles en Siam (1540–1939): Una aportación al estudio de la presencia hispana en Asia oriental*. Madrid: Consejo superior de investigaciones científicas, 1997.

Røge, Pernille. '"La Clef de Commerce" – The Changing Role of Africa in France's Atlantic Empire ca. 1760–1797'. *History of European Ideas* 34 (2008): 431–43.

'The Question of Slavery in Physiocratic Political Economy'. In *L'economia come linguaggio della politica nell'Europa del settecento*, ed. Manuela Albertone, 149–69. Milan: Feltrinelli, 2009.

'A Natural Order of Empire: The Physiocratic Vision of Colonial France after the Seven Years' War'. In *The Political Economy of Empire in the Early Modern World*, eds. Sophus A. Reinert and Pernille Røge, 32–52. Basingstoke, UK: Palgrave Macmillan, 2013.

Romaniello, Matthew P. 'True Rhubarb? Trading Eurasian Botanical and Medical Knowledge in the Eighteenth Century'. *Journal of Global History* 1 (2016): 3–23.

Romano, An tonella. *Impressions de Chine: L'Europe et l'englobement du monde, XVIe–XVIIe siècle*. Paris: Fayard, 2016.

'The History Manifesto, History of Science, and Big Narratives: Some Pending Questions'. *Isis* 107, no. 2 (2016): 338–40.

Romano, Antonella, and Stéphane Van Damme. 'Science and World Cities: Thinking Urban Knowledge and Science at Large (16th–18th Century)'. *Itinerario* 33, no. 1 (2009): 79–95.

Rönnbäck, Klas. 'Enlightenment, Scientific Exploration and Abolitionism'. *Slavery and Abolition* 34 (2013): 425–45.

Rony, Abdul Kohar, and Iêda Siqueira Wiarda. *The Portuguese in Southeast Asia: Malacca, Moluccas, East Timor*. Hamburg: Abera, 1997.

Roos, Robert. 'The Dutch as Globalisers in the Western Basin of the Indian Ocean?'. In *Globalisation and the South-West Indian Ocean*, eds. Sandra Evers and Vinesh Y. Hookoomsing, 7–16. Réduit: University of Mauritius, 2000.

Rouillard, Guy. *Le Jardin des Pamplemousses: 1729–1979 histoire et botanique*. Les Pailles: Henry, 1983.

Rouillard, Guy, and Joseph Guého. *Les plantes et leur histoire à l'Ile Maurice*. Mauritius: MSM, 1999.

Ruggiu, François-Joseph. 'India and the Reshaping of the French Colonial Policy (1759–1789)'. *Itinerario* 35, no. 2 (2011): 25–43.

Ruppel, Sophie. 'Das Grünende Reich der Gewächse: Vom vielfältigen Nutzen der Pflanzen im bürgerlichen Diskurs (1700–1830)'. In *'Die Natur ist überall bey uns': Mensch und Natur in der Frühen Neuzeit*, eds. Sophie Ruppel and Aline Steinbrecher, 109–24. Zurich: Chronos, 2009.

Botanophilie: Mensch und Pflanze in der aufklärerisch-bürgerlichen Gesellschaft um 1800. Cologne: Böhlau, 2019.

Rushforth, Brett. *Bonds of Alliance: Indigenous and Atlantic Slaveries in New France*. Chapel Hill: University of North Carolina Press, 2012.

Safier, Neil. 'Fruitless Botany. Joseph de Jussieu's South American Odyssey'. In *Science and Empire in the Atlantic World*, eds. James Delbourgo and Nicholas Dew, 203–24. New York: Routledge, 2008.

Measuring the New World: Enlightenment Science and South America. Chicago: University of Chicago Press, 2008.

'Spies, Dyes and Leaves: Agro-Intermediaries, Luso-Brazilian Couriers, and the World They Sowed'. In *The Brokered World: Go-Betweens and Global intelligence, 1770–1820*, eds. Simon Schaffer, Lissa Roberts, Kapil Raj, and James Delbourgo, 239–65. Sagamore Beach, MA: Science History Publications, 2009.

'Global Knowledge on the Move: Itineraries, Amerindian Narratives, and Deep Histories of Science'. *Isis* 101, no. 1 (2010): 133–45.

'Masked Observers and Mask Collectors: Entangled Visions from the Eighteenth-Century Amazon'. *Colonial Latin American Review* 26 (2017): 104–30.

Said, Edward W. *Culture and Imperialism*. London: Chatto & Windus, 1993.

Saldaña, Juan José, ed. *Science in Latin America: A History*. 1st edn. Austin: University of Texas Press, 2006.

Sargent, Matthew. 'Global Trade and Local Knowledge: Gathering Natural Knowledge in Seventeenth-Century Indonesia'. In *Intercultural Exchange in Southeast Asia: History and Society in the Early Modern World*, 144–60, eds. Tara Alberts and David Irving. London: I. B. Tauris, 2013.

Saussol, Alain, and Joseph Zitomersky, eds. *Colonies, territoires, sociétés: L'enjeu français*. Paris: L'Harmattan, 1996.

Savage, Victor R. 'Southeast Asia's Indigenous Knowledge: The Conquest of the Mental Terra Incognitae'. In *Asia, Europe, and the Emergence of Modern Science Knowledge Crossing Boundaries*, ed. Arun Bala, 253–70. New York: Palgrave Macmillan, 2012.

Schaffer, Simon. 'Enlightenment Brought down to Earth'. *History of Science* 41 (2003): 257–68.

'Instruments and Cargo in the China Trade'. *History of Science* 44 (2006): 217–45.

Schaffer, Simon, Lissa Roberts, Kapil Raj, and James Delbourgo, eds. *The Brokered World: Go-Betweens and Global Intelligence, 1770–1820*. Sagamore Beach, MA: Science History Publications, 2009.

Schiebinger, Londa. *Nature's Body: Gender in the Making of Modern Science*. Boston: Beacon Press, 1993.

Plants and Empire: Colonial Bioprospecting in the Atlantic World. Cambridge, MA: Harvard University Press, 2004.
ed. 'Focus: Colonial Science'. *Isis* 96, 2005.
'Prospecting for Drugs: European Naturalists in the West Indies'. In *The Postcolonial Science and Technology Studies Reader*, ed. Sandra Harding, 110–26. Durham, NC: Duke University Press, 2011.
Secret Cures of Slaves: People, Plants, and Medicine in the Eighteenth-Century Atlantic World. Stanford: Stanford University Press, 2017.
Schiebinger, Londa, and Claudia Swan. 'Introduction'. In *Colonial Botany: Science, Commerce, and Politics in the Early Modern World*, eds. Londa Schiebinger and Claudia Swan, 1–16. Philadelphia: University of Pennsylvania Press, 2005.
Schreiber, Roy E. 'Colonial Botany and Tropical Agriculture'. *Itinerario* 29, no. 3 (2005): 114–18.
Schumann, Matt, and Karl W. Schweizer. *The Seven Years War: A Transatlantic History*. London: Routledge, 2009.
Scott Parrish, Susan. 'Diasporic African Sources of Enlightenment Knowledge'. In *Science and Empire in the Atlantic World*, eds. James Delbourgo and Nicholas Dew, 281–310. New York: Routledge, 2008.
Sebastiani, Silvia, and Jean-Frédéric Schaub. 'Between Genealogy and Physicality: A Historiographical Perspective on Race in the Ancien Régime'. *Graduate Faculty Philosophy Journal* 35, nos. 1–2 (2014): 23–51.
Secord, James A. *Victorian Sensation: The Extraordinary Publication, Reception, and Secret Authorship of Vestiges of the Natural History of Creation*. Chicago: University of Chicago Press, 2000.
'Knowledge in Transit'. *Isis* 95, no. 4 (2004): 654–72.
Seeber, Edward D. *Anti-Slavery Opinion in France during the Second Half of the Eighteenth Century*. Baltimore: Johns Hopkins Studies, 1937.
Segalen, Jean-Jacques. *Plantes et fruits tropicaux des Îles de La Réunion et de Maurice*. Sainte-Marie: Jade Editions, 2013.
Selvon, Sydney. *A Comprehensive History of Mauritius: From the Beginning to 2001*. 2nd edn. Port Louis: MDS, 2005.
Servan-Schreiber, Catherine, ed. *Indianité et créolité à l'île Maurice*. Paris: Editions de l'EHESS, 2014.
Seth, Suman. 'Putting Knowledge in Its Place: Science, Colonialism, and the Postcolonial'. *Postcolonial Studies* 12, no. 4 (2009): 373–88.
Shaw, Edward P. *Problems and Policies of Malesherbes as Directeur de La Libraire in France (1750–1763): The Schools of the Imperial Age*. Albany, NY: SUNY Press, 1966.
Simon, Josep, and Néstor Herrán. *Beyond Borders: Fresh Perspectives in History of Science*. Cambridge: Cambridge Scholars, 2008.
Sivasundaram, Sujit. *Nature and the Godly Empire: Science and Evangelical Mission in the Pacific, 1795–1850*. Cambridge: Cambridge University Press, 2005.
'Introduction: Global Histories of Science'. *Isis* 101 (2010): 95–7.
'Sciences and the Global: On Methods, Questions, and Theory'. *Isis* 101 (2010): 146–58.
Islanded: Britain, Sri Lanka, and the Bounds of an Indian Ocean Colony. Chicago: University of Chicago Press, 2013.

'Science'. In *Pacific Histories: Ocean, Land, People*, eds. David Armitage and Alison Bashford, 237–60. Basingstoke, UK: Palgrave Macmillan, 2014.

'Oils of Empire'. In *Worlds of Natural History*, eds. Helen Curry, Nic Jardine, Jim Secord, and Emma Sprary, 379–98. Cambridge: Cambridge University Press, 2018.

Waves across the South: A New History of Revolution and Empire. London: William Collins, 2020.

Skuncke, Marie-Christine. *Carl Peter Thunberg, Botanist and Physician: Career-Building across the Oceans in the Eighteenth Century*. Uppsala: Swedish Collegium for Advanced Study, 2014.

Smith, Pamela H. *The Body of the Artisan: Art and Experience in the Scientific Revolution*. Chicago: University of Chicago Press, 2004.

ed. *Entangled Itineraries: Materials, Practices, and Knowledges across Eurasia*. Pittsburgh: University of Pittsburgh Press, 2019.

Smith, Pamela. *From lived experience to the written word. Reconstructing practical knowledge in the early modern world*. Chicago: The University of Chicago Press, 2022.

Smith, Pamela H., and Paula Findlen, eds. *Merchants and Marvels: Commerce, Science and Art in Early Modern Europe*. New York: Routledge, 2001.

Smith, Pamela H., and Benjamin Schmidt, eds. *Making Knowledge in Early Modern Europe – Practice: Practices, Objects, and Texts, 1400–1800*. Chicago: University of Chicago Press, 2008.

Smith, Vanessa. 'Give Us Our Daily Breadfruit: Bread Substitution in the Pacific in the Eighteenth Century'. *Studies in Eighteenth Century Culture* 35 (2006): 53–75.

Intimate Strangers: Friendship, Exchange and Pacific Encounters. Cambridge: Cambridge University Press, 2010.

Sorenson, Richard. 'The Ship as a Scientific Instrument in the Eighteenth Century'. *Osiris* 11 (1996): 221–36.

Spary, Emma C. *Utopia's Garden: French Natural History from Old Regime to Revolution*. Chicago: University of Chicago press, 2000.

'"Peaches which the Patriarchs Lacked": Natural History, Natural Resources, and the Natural Economy in France'. *History of Political Economy* 35 (2003).

'Of Nutmegs and Botanists: The Colonial Cultivation of Botanical Identity'. In *Colonial Botany: Science, Commerce, and Politics in the Early Modern World*, eds. L. Schiebinger and Claudia Swan, 187–203. Philadelphia: University of Pennsylvania Press, 2005.

'Botanical Networks Revisited'. In *Wissen im Netz: Botanik und Pflanzentransfer in europäischen Korrespondenznetzen des 18. Jahrhunderts*, ed. Regina Dauser, 1–18. Berlin: Akad.-Verl., 2008.

Eating the Enlightenment: Food and the Sciences in Paris. Chicago: University of Chicago Press, 2012.

Feeding France: New Sciences of Food, 1760–1815. Cambridge: Cambridge University Press, 2014.

Stanziani, Alessandro. 'Free Labor – Unfree Labor: An Uncertain Boundary?' *Kritika* 9, no. 1 (2008): 27–52.

Bondage: Labor and Rights in Eurasia from the Sixteenth to the Early Twentieth Centuries. New York: Berghahn, 2014.

Starr, Douglas. 'The Making of Scientific Knowledge in an Age of Slavery: Henry Smeathman, Sierra Leone and Natural History'. *Journal of Colonialism and Colonial History* 9, no. 3 (2008).

Steiner, Benjamin. *Colberts Afrika: Eine Wissens- Und Begegnungsgeschichte in Afrika Im Zeitalter Ludwigs XIV*. Munich: De Gruyter Oldenbourg, *2014.*

Stern, Philip J. *The Company-State: Corporate Sovereignty and the Early Modern Foundations of the British Empire in India*. Oxford: Oxford University Press, 2011.

Stockland, Etienne. '"La Guerre aux Insectes": Pest Control and Agricultural Reform in the French Enlightenment'. *Annals of Science* 70, no. 4 (2013): 435–60.

'Policing the Oeconomy of Nature: The Oiseau Martin as an Instrument of Oeconomic Management in the Eighteenth-Century French Maritime World'. *History and Technology* 30, no. 3 (2014): 1–25.

Stoler, Ann Laura. *Along the Archival Grain: Epistemic Anxieties and Colonial Common Sense*. Princeton, NJ: Princeton University Press, 2008.

Stroup, Alice. *A Company of Scientists: Botany, Patronage, and Community at the Seventeenth-Century Parisian Royal Academy of Sciences*. Berkeley: University of California Press, 1990.

Stuchtey, Benedikt, ed. *Science across the European Empires, 1800–1950*. Oxford: Oxford University Press, 2005.

Subrahmanyam, Sanjay, ed. *Maritime India*. Oxford: Oxford University Press, 2004.

From the Tagus to the Ganges: Explorations in Connected History. Oxford: Oxford University Press, 2005.

Mughals and Franks: Explorations in Connected History. Oxford: Oxford University Press, 2005.

Sweet, James. *Domingos Álvares, African Healing, and the Intellectual History of the Atlantic World*. Chapel Hill: University of North Carolina Press, 2011.

Taiz, Lincoln, and Lee Taiz. *The Discovery & Denial of Sex in Plants*. Oxford: Oxford University Press, 2017.

Terrall, Mary. 'Biography as Cultural History of Science'. *Isis* 97(2006): 306–13.

Catching Nature in the Act: Réaumur and the Practice of Natural History in the Eighteenth Century. Chicago: University of Chicago Press, 2014.

'Experimental Natural Knowledge'. In *Worlds of Natural History*, eds. Helen, Curry, Nic Jardine, Jim Secord, and Emma Sprar, 170–85. Cambridge: Cambridge University Press, 2018.

Tharoor, Shashi. *Inglorious Empire: What the British Did to India*. London: Hurst, 2017.

Thiébaut, Rafael. 'An Informal French–Dutch Alliance: Trade and Diplomacy between the Cape Colony and the Mascarenes, 1719–1769'. *Journal of Indian Ocean World Studies* 1 (2017): 128–47.

Thomas, Adrian P. 'The Establishment of Calcutta Botanic Garden: Plant Transfer, Science and the East India Company, 1786–1806'. *Journal of the Royal Asiatic Society* 16, no. 2 (2006): 165–77.

Thomas, Nicholas. *Entangled Objects: Exchange, Material Culture and Colonialism in the Pacific*. Cambridge, MA: Harvard University Press, 1991.

Touchet, Julien. *Botanique & colonisation en Guyane française, 1720–1848: Le jardin des Danaïdes*. Petit-Bourg: Ibis rouge, 2004.

Toussaint, Auguste. *Port-Louis, deux siècles d'histoire 1735–1935*. Port-Louis: Impr. de la Typographie moderne, 1936.

'Les débuts de l'imprimérie aux îles Mascareignes'. *Revue d'histoire des colonies* 35, no. 122 (1948): 1–26

Early Printing in the Mascarene Islands, 1767–1810. London: University of London Press, 1951.

L'administration française de l'Ile Maurice et ses archives, 1721–1810. Port Louis: Imp. commerciale, 1965.

Histoire des îles mascareignes. Paris: Berger-Levrault, 1972.

Tricoire, Damien. *The Colonial Dream: Imperial Knowledge and the French–Malagasy Encounters in the Age of Enlightenment*. Berlin: De Gruyter, 2023.

Trivellato, Francesca. 'Is There a Future for Italian Microhistory in the Age of Global History?'. *California Italian Studies* 2 (2011): np.

Turner, Sasha. *Contested Bodies: Pregnancy, Childrearing, and Slavery in Jamaica*. Philadelphia: University of Pennsylvania Press, 2017.

Van Damme, Stéphane. *Paris, capitale philosophique: de la Fronde à la Révolution*. Paris: O. Jacob, 2005.

'Capitalizing Manuscripts, Confronting Empires: Anquetil-Duperron and the Economy of Oriental Knowledge in the Context of the Seven Years' War'. In *Negotiating Knowledge in Early Modern Empires. A Decentred View*, eds. László Kontler, Antonella Romano, Silvia Sebastiani, and Borbála Zsuzsanna Török, 109–28. Basingstoke, UK: Palgrave Macmillan, 2014.

Vaughan, Megan. *Creating the Creole Island: Slavery in Eighteenth-Century Mauritius*. Durham, NC: Duke University Press, 2005.

Vergès, Françoise. 'Creolization and the Maison des civilisations et de l'unité réunionnaise'. *Journal of Visual Culture* 5 (2006): 29–51.

Vidal, Cécile. *Caribbean New Orleans: Empire, Race, and the Making of a Slave Society*. Chapel Hill: University of North Carolina Press, 2019.

Voicua, Cristina-Georgiana. 'Caribbean Cultural Creolization'. *Procedia – Social and Behavioral Sciences* 149 (2014): 997–1002.

Vos, Paula De. 'The Science of Spices: Empiricism and Economic Botany in the Early Spanish Empire'. *Journal of World History* 17, no. 4 (2006): 399–427.

Wanquet, Claude. 'Joseph-François Charpentier de Cossigny et le projet d'une Colonisation "Eclairée" de Madagascar à la fin du XVIIIe siècle'. In *Regards sur Madagascar et la Révolution française*, ed. Guy Jacob, 71–85. Madagascar: CNAPMAD, 1990.

Warde, Paul. *The Invention of Sustainability: Nature and Destiny, c.1500–1870*. Cambridge: Cambridge University Press, 2018.

Weulersse, Georges. *Le mouvement physiocratique en France*. 2 vols. Paris: Félix Alcan, 1910.

White, Richard. *The Middle Ground: Indians, Empires, and Republics in the Great Lakes Region, 1650–1815*. New York: Cambridge University Press, 1991.

White, Sophie. *Voices of the Enslaved. Love, Labor, and Longing in French Louisiana*. Chapel Hill: University of North Carolina Press, 2019.

Wickremesekera, Channa. *Kandy at War: Indigenous Military Resistance to European Expansion in Sri Lanka 1594–1818*. New Delhi: Manohar, 2004.

Widjojo, Muridan Satrio. *The Revolt of Prince Nuku: Cross-Cultural Alliance-Making in Maluku, c.1780–1810*. Leiden: Brill, 2009.

Williams, Roger L. *Botanophilia in Eighteenth-Century France: The Spirit of the Enlightenment*. Dordrecht: Kluwer Academic, 2001.

Wilson, Jon E. *India Conquered: Britain's Raj and the Chaos of Empire*. London: Simon & Schuster, 2016.

Winterbottom, Anna. *Hybrid Knowledge in the Early East India Company World*, Basingstoke, UK: Palgrave Macmillan, 2016.

Winterbottom, Anna, and Facil Tesfaye. *Histories of Medicine and Healing in the Indian Ocean World*. 2 vols. Basingstoke, UK: Palgrave Macmillan, 2016.

Withers, Charles. *Placing the Enlightenment: Thinking Geographically about the Age of Reason*. Chicago: University of Chicago Press, 2007.

Wu, Huiyi. 'Entre curiosité et utilité: Les traductions d'"herbiers chinois" dans les *Lettres édifiantes et curieuses* et dans la *Description de l'Empire de la Chine et de la Tartarie chinoise*'. In *Les savoirs-mondes. Mobilités et circulation des savoirs depuis le Moyen Âge*, eds. Pilar González Bernaldo and Liliane Hilaire-Pérez, 183–96. Rennes: Presses Universitaires de Rennes, 2015.

Yoo, Genie. 'Wars and Wonders: The Inter-Island Information Networks of Georg Everhard Rumphius'. *British Journal for the History of Science* 51 (2018): 559–84.

Yuste López, Carmen. *Emporios transpacíficos: comerciantes mexicanos en Manila, 1710–1815*. Mexico: Universidad Nacional Autónoma de México, 2007.

Zancarini-Fournel, Michelle. *Les luttes et les rêves: une histoire populaire de la France de 1685 à nos jours*. Paris: Zones, 2016.

Zanco, Jean-Philippe, ed. *Dictionnaire des ministres de la marine: 1689–1958*. Paris: Éd. SPM, 2011.

Zumbroich, Thomas J. 'The Introduction of Nutmeg (Myristica Fragrans Houtt.) and Cinnamon (Cinnamomum Verum J.Presl) to America'. *Acta Bot. Venez*. 28, no. 1 (2005): 155–60.

Zupanov, Ines G., and Ângela Barreto Xavier. 'Quest for Performance in the Tropics: Portuguese Bioprospecting in Asia (16th–18th Centuries)'. *Journal of the Economic and Social History of the Orient* 57 (2014): 511–48.

Unpublished PhD Theses and Manuscripts

Beltrán, José. 'Nature in Draft: Images and Overseas Natural History in the Work of Charles Plumier (1646–1704)'. PhD thesis, European University Institute, 2017.

Boomgaard, Peter. 'A Hub of Plant Exchange: Batavia (Java), the Dutch East India Company, and the Networks of European Botanists, 1620s to 1850s', by permission of the author.

De Cambiaire, Elisabeth. 'Enlightened Alliance. Nature, Botany and France's Expansion to the East Indies: The Colonization of the Mascarenes (1665–1775)'. PhD thesis, UNSW, 2016, http://handle.unsw.edu.au/1959.4/5619.

Dumoulin-Genest, Marie-Pierre. 'L'introduction et l'acclimatation des plantes chinoises en France au XVIIIe siècle'. PhD thesis, EHESS, 1994.

Fleischer, Alette. 'Rooted in Fertile Soil: Seventeenth-Century Dutch Gardens and the Hybrid History of Material and Knowledge Production'. PhD thesis, University of Twente, 2010.

Kroupa, Šebestián. 'Georg Joseph Kamel (1661–1706): A Jesuit Pharmacist in Manila at the Borderlines of Erudition and Empiricism', by permission of the author.

'Georg Joseph Kamel (1661–1706): A Jesuit Pharmacist at the Frontiers of Colonial Empires'. PhD thesis, University of Cambridge, 2019.

Menon, Minakshi. 'Making Useful Knowledge: British Naturalists in Colonial India, 1784–1820'. PhD thesis, University of California, 2013.

Otremba, Eric. 'Enlightened Institutions: Science, Plantations, and Slavery in the English Atlantic, 1626–1700'. PhD thesis, University of Minnesota, 2012.

Parsons, Christopher M. 'Plants and People: French and Indigenous Botanical Knowledges in Colonial North America, 1600–1760'. PhD thesis, University of Toronto, 2011.

Regourd, François. 'Sciences et colonisation sous l'Ancien Régime: Le cas de La Guyane et des Antilles françaises, XVIIe–XVIIIe siècles'. PhD thesis, Michel de Montaigne Bordeaux 3, 2000.

Spary, Emma C. 'Eating Beyond Reason', by permission of the author.

'Fruits of Paradise? Exotics, Environments and Enlightenments in Eighteenth-Century France', by permission of the author.

Index

Printed in the United States
by Baker & Taylor Publisher Services